The Development of Bioethics in the United States

Philosophy and Medicine

VOLUME 115

For further volumes:
http://www.springer.com/series/6414

Jeremy R. Garrett • Fabrice Jotterand
D. Christopher Ralston
Editors

The Development of Bioethics in the United States

Editors
Jeremy R. Garrett
Children's Mercy Bioethics Center
Children's Mercy Hospital
Kansas City, MO, USA

Fabrice Jotterand
Institute for Biomedical Ethics
University of Basel
Basel, Switzerland

D. Christopher Ralston
Department of Philosophy
Rice University
Houston, TX, USA

ISSN 0376-7418
ISBN 978-94-007-4010-5 ISBN 978-94-007-4011-2 (eBook)
DOI 10.1007/978-94-007-4011-2
Springer Dordrecht Heidelberg New York London

Library of Congress Control Number: 2012955942

Printed on acid-free paper

Springer is part of Springer Science+Business Media (www.springer.com)

Contents

Contributors

George J. Annas Department of Health Law, Bioethics and Human Rights, School of Public Health, Boston University, 715 Albany Street, Boston, MA 02118, USA

Howard Brody Institute for the Medical Humanities, University of Texas Medical Branch at Galveston, Galveston, TX 77555-1311, USA

Eric J. Cassell Department of Public Health (Emeritus), Weill Medical College, Cornell University Medical Center, P.O. Box 96, Shawnee on Delaware, PA 19356, USA

H. Tristram Engelhardt, Jr. Department of Philosophy, Rice University, 6100 S. Main Street, MS-14, Houston, TX 77005-1892, USA

Baylor College of Medicine (Emeritus), Houston, TX, USA

Edmund L. Erde School of Medicine, University of Medicine and Dentistry of New Jersey (Retired), 409 Loral Dr., Cherry Hill, NJ 08003, USA

Jeremy R. Garrett Children's Mercy Bioethics Center, Children's Mercy Hospital, 2401 Gilham Road, Kansas City, MO 64108, USA

Department of Pediatrics, Department of Philosophy, University of Missouri-Kansas City, Kansas City, MO, USA

John Collins Harvey Center for Clinical Bioethics, Georgetown University, Washington, DC 20057-1409, USA

Albert R. Jonsen Program in Medicine and Human Values, California Pacific Medical Center, 2395 Sacramento Street, San Francisco, CA 94115, USA

Fabrice Jotterand Institute for Biomedical Ethics, University of Basel, Missionstrasse 24, 4055 Basel, Switzerland

Department of Philosophy and Humanities, University of Texas Arlington, 305 Carlisle Hall, Box 19527, Arlington, TX 76019, USA

Loretta M. Kopelman Department of Bioethics and Interdisciplinary Studies (Emeritus), Brody School of Medicine, East Carolina University, East Fifth Street, Greenville, NC 27858-4353, USA

Kennedy Institute of Ethics, Georgetown University, Washington, DC, USA

Laurence B. McCullough Center for Medical Ethics and Health Policy, Baylor College of Medicine, One Baylor Plaza, MS 420, Houston, TX 77030-3411, USA

Edmund D. Pellegrino Center for Clinical Bioethics, Georgetown University, Box 571409, 4000 Reservoir Road NW, Suite 238, Washington, DC 20057-1409, USA

D. Christopher Ralston Department of Philosophy, Rice University, 6100 S. Main Street, MS-14, Houston, 77005-1892, USA

Warren T. Reich Department of Theology, Georgetown University, 120 New North, Box 571135, Washington, DC 20057, USA

Carson Strong Department of Medicine, College of Medicine, University of Tennessee Health Science Center, 956 Court Avenue, Suite G212, Memphis, TN 38163, USA

Robert M. Veatch Kennedy Institute of Ethics, Georgetown University, 423 Healy Building, Washington, DC 20057, USA

Richard M. Zaner Center for Biomedical Ethics and Society (Emeritus), Vanderbilt University Medical Center, 2525 West End Ave., Suite 400, Nashville, TN 37203, USA

Chapter 1
The Development of Bioethics in the United States: An Introduction

Jeremy R. Garrett, Fabrice Jotterand, and D. Christopher Ralston

In the last four decades bioethics has experienced tremendous development, which can be attributed to various factors. First, advances in the biomedical sciences and biotechnology increasingly raise ethical, legal, and social issues that concern society at large. The complexity of the current health care system and the development of powerful biotechnologies necessitate the critical and interdisciplinary analysis that bioethics can provide. Second, the Joint Commission on Accreditation of Healthcare Organizations, which accredits health care organizations in the United States, requires hospitals to have a mechanism that offers clinical ethics consultations (Lo 2005, 111), and funding agencies expect the inclusion of bioethics components in research protocols. Finally, there is a need for bioethics education for health care professionals (physicians, nurses, chaplains, social workers, administrators, etc.) seeking advancement, IRB/Hospital Ethics Committee members working on health policies/ethics programs, and students seeking to include bioethics in their primary training or to pursue a career in bioethics.

J.R. Garrett, Ph.D. (✉)
Children's Mercy Bioethics Center, Children's Mercy Hospital,
2401 Gilham Road, Kansas City, MO 64108, USA

Department of Pediatrics, Department of Philosophy,
University of Missouri-Kansas City, Kansas City, MO, USA
e-mail: jgarrett@cmh.edu

F. Jotterand, Ph.D., M.A.
Institute for Biomedical Ethics, University of Basel,
Missionstrasse 24, 4055 Basel, Switzerland

Department of Philosophy and Humanities, University of Texas Arlington,
305 Carlisle Hall, Box 19527, Arlington, TX 76019, USA
e-mail: fabrice.jotterand@unibas.ch

D.C. Ralston, Ph.D., M.A.
Department of Philosophy, Rice University,
6100 S. Main Street, MS-14, Houston, TX 77005-1892, USA
e-mail: ralston@alumni.rice.edu

J.R. Garrett et al. (eds.), *The Development of Bioethics in the United States*,
Philosophy and Medicine 115, DOI 10.1007/978-94-007-4011-2_1,

While recognizing that the rise of bioethics has been quite remarkable, the field is still a work in progress—particularly in relation to its nature, scope, and methodologies. To assess its current state and to understand the reasons for its emergence in American culture, we brought together a collection of essays by seminal figures in the fields of medical ethics and bioethics. We invited these scholars to reflect on the field of bioethics around the following five framing questions: (1) *Are there precise moments, events, socio-political conditions, legal cases, and/or works of scholarship to which we can trace the emergence of bioethics as a field of inquiry in the United States?* (2) *What is the relationship between the historico-causal factors that gave birth to bioethics and the factors that sustain and encourage its continued development today?* (3) *Is it possible and/or useful to view the history of bioethics in discrete periods with well-defined boundaries?* (4) *If so, are there discernible forces that reveal why transitions occurred when they did? What are the key concepts that ultimately frame the field and how have they evolved and developed over time?* and (5) *Is the field of bioethics in a period of transformation into biopolitics*?[1]

In light of the responses the authors gave to these questions, we structured the volume into four main parts. The first part, entitled "The Birth of Bioethics: Historical Analysis," provides an historical examination of the emergence of bioethics and an account of the leading institutions that took part in the early phases of its development. The second part, "The Nature of Bioethics: Cultural and Philosophical Analysis," examines two issues: (1) the particular social-cultural context and the intellectual climate that has shaped the ongoing development of bioethics, and (2) the nature of the field as a discipline in its own right. The third part, "The Practice of Bioethics: Professional Dimensions," addresses issues pertaining to the professionalization of bioethics and the tensions between medical ethics and bioethics. The final part, "The Future of Bioethics: Looking Ahead," offers reflections on the future of bioethics and calls for a return to more robust philosophical inquiry as new biotechnologies increasingly stretch our moral imagination and our ability to define what it means to be human.

Before we turn to a more detailed outline of the volume, we would like to include a brief word on its development. The individual essays were written over a period of approximately 6 years. Consequently, the references cited by some authors may be more recent than others. Nevertheless, we believe the long gestational period for this volume has produced a work rich in reflection, both theoretical and personal, on the historical development of the field of bioethics, encompassing within its ambit a consideration of where the field has been, where it is now, and where it may very well go in the future.

1.1 The Birth of Bioethics: Historical Analysis

The birth of bioethics, everyone seems to agree, took place in the tumultuous period of the late 1960s and early 1970s. The four essays in Part I provide a unique historical perspective of this occurrence and this time period. Each essay presents a rich and lively picture of the field's early days by weaving together first-person remembrance and third-person reflection. Together, they describe the particular social-cultural context and the intellectual climate that gave birth to bioethics, while introducing the reader to the influential people and institutions that shaped its early development. Especially notable in this regard is the way in which these essays tell the story of bioethics' birth through the histories of three important institutions: (1) the Hastings Center, (2) the Institute for the Medical Humanities at the University of Texas Medical Branch (UTMB), and (3) the Kennedy Institute of Ethics at Georgetown University.

In the volume's first paper, "The Beginnings of Bioethics," Eric J. Cassell offers a highly personal window into the early years of bioethics. In particular, he interweaves a historical account of the evolution of the interdisciplinary approach to bioethics with his personal experiences with research working groups organized by the newly-created Hastings Center. Cassell, an established internist, became involved with the newly developing field of bioethics in 1970, after being invited to comment on a chapter of Paul Ramsey's *The Patient as Person* at an early conference sponsored by the Center. His recounting of events at this conference illustrates just how exciting and intellectually invigorating bioethics must have been for scholars in this formative period. It also illustrates just how revolutionary and alien bioethics must have seemed to ordinary clinicians at the same time. Cassell's anecdotes remind us that practices that now are regular and deeply embedded aspects of daily clinical practice, such as obtaining informed consent, are actually recent developments in medical ethical thinking.

In addition to outlining his personal development as a bioethicist, Cassell sheds light on the changing societal expectations that coincided with, and in part encouraged, the emergence of bioethics as a field. Some of the changes that he emphasizes include (1) the general reaction against the status of science and medicine as unchallengeable, paternalistic, and authoritarian, (2) an acknowledgment of the impact of physicians' personal values on the practice of medicine, and (3) the shift to thinking of medicine as a commodity. As obvious as some of these trends may seem in hindsight, Cassell notes that they were not especially well-defined or obvious at the time. These lessons are instructive for bioethicists today. While it is tempting to believe that we can achieve sufficient detachment from our own cultural, social, and political context and see the field for what it really is, Cassell's reflections encourage a healthy degree of caution and skepticism.

In "Teaching at the University of Texas Medical Branch, 1971–1974: Humanities, Ethics, or Both?" Howard Brody outlines the history of the creation of the Institute for the Medical Humanities at the University of Texas Medical Branch (UTMB). The genesis of the Institute goes back to the 1940s and 1950s where the first reflections on medical humanities and medical history appeared in the minutes of the UTMB archives. As a result, Dr. Chauncey Leake, vice-president and dean of UTMB between 1942 and 1955, hired philosopher Dr. Patrick Romanell to join the faculty of UTMB to teach philosophy and ethics. However, the real push for the creation of a unit focusing on medical humanities took place in the 1970s. P.L. Hendricks, a medical student, wrote a letter advocating for the creation of a department of medical humanities. The letter turned out to offer many insightful suggestions concerning American medicine of the day and paved the way for the establishment of the Institute itself in 1973, a year after H. Tristram Engelhardt, Jr. joined the History of Medicine Division. That period coincided also with the attempt of President Blocker to recruit a prominent scholar in medical humanities, Edmund D. Pellegrino. Pellegrino, however, declined the offer due to his position as Vice President and Director of the Health Sciences Center, State University of New York at Stony Brook. Ultimately, the Institute hired William B. Bean, a Professor of Medicine at the University of Iowa Medical School, as director, who retired in 1980 and was replaced in 1982 by Ronald A. Carson, who served in that capacity until 2005. The history of UTMB is relevant beyond the institution itself or the UT system. It sheds light on the struggle to determine the role of medical humanities and medical ethics in medical education in the United States. This struggle is far from over, and many other medical schools around the country could draw inspiration from the UTMB experiment in developing their own programs in medical humanities and bioethics.

In "André Hellegers, the Kennedy Institute, and the Development of Bioethics: The American-European Connection" John Collins Harvey (2013) narrates the creation of the Kennedy Institute under the visionary leadership of André Hellegers. Hellegers was an obstetrician on faculty at Georgetown University who had interests beyond medicine, particularly theology and philosophy. He established a relationship with the Shriver family, who shared his desire to create a bioethics institute (a desire that came to fruition in 1971). To this end, they recruited an ecumenical group of philosophers and theologians from various religious traditions to serve as scholars in the Institute. The Kennedys also established fellowships which turned out to be crucial for the expansion of bioethics in Europe. Rev. Dr. Francesc Abel, S.J., a Jesuit priest-obstetrician from Barcelona, spent three and a half years at the Kennedy Institute before going back to his native Spain. Upon his return he established the Borja Institut de Bioètica with the support of the Father General of the Jesuit Order, Padre Pedro Arrupe and two of his Jesuit priest colleagues.

In his analysis, Collins Harvey makes the important observation that the Kennedy Institute relocated the traditional field of medical ethics within the boundaries of a new discipline—bioethics—which in turn helped to provide academic legitimacy. However, as Laurence B. McCullough has argued elsewhere (2012), this shift may have inadvertently deprofessionalized medical ethics and undermined the fiduciary nature of medical practice.

The final contribution in Part I comes from H. Tristram Engelhardt, Jr., who is well-known for his original and often iconoclastic analyses of the social, cultural, and intellectual milieu in which bioethics emerged. In his provocatively titled contribution, "Bioethics as a Liberal Roman Catholic Heresy: Critical Reflections on the Founding of Bioethics," Engelhardt considers the subject anew, reaffirming many of his past conclusions while seeking to fill in dimensions that he now views as insufficiently appreciated hitherto. His perspective continues to be informed by the conviction that bioethics emerged to fill the moral vacuum in health care policy and practice created by (1) a deflation of medicine as a guild with the consequent marginalization of professional medical ethics and of physician medical ethics experts, (2) the secularization of American society with the subsequent cultural discounting of medical moral theology and medical moral theologians, and (3) the emergence of a general hermeneutic of suspicion directed towards traditional cultural authorities. However, his present essay follows Collins Harvey in emphasizing the cardinal contributions made to the emergence of bioethics by Roman Catholic moral theological and moral philosophical assumptions as well as by the cultural consequences of the Second Vatican Council, which Engelhardt is now convinced have been previously underappreciated.

According to Engelhardt, the moral theological and moral philosophical assumptions of Roman Catholicism made it plausible for the Roman Catholic founders of bioethics to hold that humans share a common morality, and that moral philosophers are capable of disclosing its character on analogy with the contemporary Roman Catholic understanding of natural law. Also, in the wake of the theological and spiritual disorientation precipitated by the Council, many Roman Catholic clerics emerged as dissidents who regarded bioethics as having the capacity to reorient moral reflection and redirect society. More broadly, many of the founders of bioethics who had substantive intellectual roots in theology and the ministry concurred with this Roman Catholic view that there was one common human morality and that there would be one common bioethics. For Engelhardt, then, bioethics emerged as an intellectual offspring spawned from a very particular tradition of moral-theological assumptions, as well as commitments to late-Enlightenment aspirations for moral philosophy that engendered bioethics, encouraged in great measure by the displaced post-religious passions of the formerly religious.

1.2 The Nature of Bioethics: Cultural and Philosophical Analysis

The essays in Part II of the book, while also historical in many ways, seek to conceptualize the nature of bioethics using the tools of cultural and philosophical analysis. Hence, these four essays focus less on describing the historical factors that gave rise to and influenced bioethics and more on analyzing the nature of the field and what it has become. In so doing, they offer accounts of the identity of bioethics as it overlaps with, and remains autonomous from, the wider culture and other fields and activities, such as medicine, science, theology, philosophy, and politics.

In his contribution entitled "A Corrective for Bioethical Malaise: Revisiting the Cultural Influences that Shaped the Identity of Bioethics," Warren T. Reich addresses three main issues. First, he provides a sobering and very personal account of the current state of bioethics. Reich contends that the early days of intellectual excitement in bioethics are over. Bioethics, he says, has become boring. Already in 1983 Reich had become increasingly discontent with the field due to its reductionist approach (Reich 2013, 80). In his view, bioethical investigations have adopted a minimalist approach that does not integrate the cultural, historical, philosophical, religious, theological, social, and legal insights of disciplines in the humanities and social sciences. To his own astonishment, he remarks that it would have been ridiculous to predict that bioethics could become "boring" considering the ethical challenges raised by biotechnological innovations. To address this "malaise" of bioethics, Reich suggests that we ought to re-examine the field and promote a renewal and broadening of the scope and methodologies of current bioethics. Bioethics ought to be "a systematic study of the moral dimensions… of the life sciences and health care" and not limited to medical ethics (Reich 2013, 83).

Second, Reich addresses how the socio-cultural forces of the 1960s shaped the field of bioethics. Bioethics, he contends, was not just a phase in the development of applied moral philosophy to respond to advancements in biomedical sciences. Bioethics emerged within a context in which society questioned traditional sources of moral authority and the role of social institutions, including marriage, the church, and the state. The field arose with its own unique intellectual, moral, and social characteristics in response to the particular social context of the 1960s. In concluding his reflections, Reich offers a framework based on John O'Malley's taxonomy of concepts of culture (the prophetic culture, the academic/professional culture, and the humanistic culture) to assess how contemporary culture shapes today's field of bioethics. Ultimately, Reich asks readers to reflect on what we can learn from the early stages of bioethics in order to recover from the current "malaise" of bioethics.

In "American Biopolitics," George J. Annas provides an account of the nature of American bioethics which he depicts as intrinsically political, i.e., biopolitics. Influenced by the political philosophy of John Stuart Mill, Annas remarks, American bioethics seeks to limit the power of the state and enhance individual liberties through a range of measures the government takes to promote public health. But without an understanding of America itself, Annas contends, it is difficult to understand American biopolitics or bioethics. First, Annas shows the influence of Mill on the American political landscape, which grounds its identity on three principles quintessential to a functional democracy: equality, liberty, and self-protection. Based on these premises, American biopolitics reflects the core concept that "the government has no business interfering with an individual's liberty to decide how to conduct his or her life (or healthcare) unless an individual's actions are likely to be harmful to others…" (Annas 2013, 104). What Annas calls the Mill doctrine finds its applications in various public health care contexts (e.g., vaccination, sterilization, abortion, obesity, and health care reform) to protect negative rights, i.e., the noninterference by the state with individual rights, or to uphold positive rights, i.e., the right to adequate health care provided by the government. Fundamental liberty interests define American culture as reflected in American healthcare: individualistic, technologically-driven, death-denying, and wasteful (Annas 2013, 111). American bioethics qua biopolitics is a work in progress that continuously evolves. Current trends focus more on economic issues than public health concerns.

Whereas Annas focuses on the interaction between bioethics and political philosophy, Carson Strong analyzes the relationship between philosophy and medicine. In "Medicine and Philosophy: The Coming Together of an Odd Couple," Strong gives an account of how the interaction between the two disciplines resulted in the birth of the field of bioethics as we know it today. Historically, there has always been interaction between philosophy and medicine, in various forms and to varying degrees. In the 1960s and 1970s the medical profession sought input from philosophers and humanists in addressing various perplexing issues in medicine, such as (1) challenges in the patient-physician relationship, (2) the rise of patient autonomy, (3) professional misconduct in biomedical research involving human subjects, (4) advancements in biomedical technology, and (5) dissatisfaction with the model of medical education. But the marriage between philosophy and medicine did not come without problems. Physicians and philosophers have a very "different cognitive style" (Strong 2013, 126), the former geared toward the tangible and the latter toward the abstract/conceptual. In the early stages, Strong writes, "many in the field of medicine did not understand what is involved in medical ethics. It was regarded as focusing on a search for the 'right answers,' and when there is no clear right answer, a natural conclusion is that ethics is a waste of time" (Strong 2013, 127).

The initial tension died down once a better synergy developed between medicine and philosophy, and physicians and philosophers learned to work together. Physicians learned to appreciate better the nature of philosophical inquiry into medical moral issues, and philosophers began making contributions in the clinical setting. This interaction gave birth to the field of bioethics, which, in turn, influenced medicine—and not only in the clinical setting. Specialty medical professional organizations created their own bioethics committees that produced position statements on ethics issues. To oversee medical research the federal government requested recommendations from the National Commission for the Protection of Human Subjects of Biomedical and Behavioral Research, appointed by the Secretary of the Department of Health, Education and Welfare. The Commission produced what is known today as the Belmont Report, which was influenced by the contributions of philosophers such as H. Tristram Engelhardt, Jr., Tom L. Beauchamp, and Stephen Toulmin. The creation of Hospital Ethics Committees also required the hiring of clinical ethicists who provided ethics consultation.

Medicine also changed moral philosophy in many ways. Philosophical investigations became relevant to real-world experience (applied philosophy); medicine raised philosophical questions worth exploring; and, finally, the practice of medicine challenged normative ethics and forced moral philosophers to rethink how to apply theoretical frameworks to clinical cases. This marriage of medicine and philosophy gave birth to bioethics.

In the final contribution to Part II, entitled "The Growth of Bioethics as a Second-Order Discipline," Loretta M. Kopelman offers a historically-grounded analysis of the nature of bioethics as a field of inquiry and practice. She begins with a detailed history of the evolution of early bioethics organizations, including the Society for Health and Human Values, the American Association of Bioethics, and the Society for Bioethics Consultation, and traces this history into the period in which these organizations merged as the American Society for Bioethics and Humanities (for which Kopelman served as the founding president). Her rich recounting offers both first-person and third-person observations of these organizations' development, membership, problems, and mergers.

With this historically-informed background in place, Kopelman proceeds analytically to consider four basic approaches to characterizing the nature of bioethics.

Here she summarizes, and offers critical analyses of, the following views of bioethics: (1) bioethics as pure public discourse, with no experts, core texts, common methods, or common standards; (2) bioethics as unique to a single existing discipline (e.g., philosophy, medicine, etc.), with bioethics expertise limited to those who have undertaken extensive study in that field; and (3) bioethics as evolving into a new, independent discipline with its own core texts, methods, area of expertise, and common standards for scholarship and consultation. After carefully analyzing and rejecting each of these approaches, Kopelman turns to summarizing and defending her own view that (4) bioethics is an interdisciplinary and second-order discipline with no unique area of expertise. In support of this view, she points out that bioethicists tend to expand their subject matter (unlike most new fields, which splinter off from, and narrow the subject matter of, existing disciplines), think of bioethics problems as complex and solutions as interdisciplinary, welcome various perspectives from different professions, and allow those various professions to set their own standards of competency.

1.3 The Practice of Bioethics: Professional Dimensions

The influence of bioethics has not been limited to purely intellectual and academic settings. Indeed, one of the more remarkable features of the roughly 40 year lifespan of bioethics is the way in which it has made deep inroads into the *practice* of medicine and the life sciences. In many ways, bioethics has transformed these practices, both at the systemic/structural level (e.g., the widespread introduction of ethical regulations, ethics committees, and institutional review boards, etc.) and at the everyday/personal level (e.g., the physician-patient relationship). One notable consequence of bioethics' rapid ascendancy is the tension it has created with traditional views of medical ethics and medical professionalism. The essays in Part III of the book focus on precisely this conflict. They offer perspective on the historical foundations and development of the conflict, and they advance alternative prescriptions regarding the direction of biomedical practice.

In "The Development of Bioethics: Bringing Physician Ethics into the Moral Consensus," Robert M. Veatch analyzes the prehistory and early development of bioethics as a practice and scholarly discipline. According to Veatch, bioethics emerged in the 6-year period surrounding 1970 and resulted from a unique confluence of scientific and cultural developments related to medicine and the biological sciences. However, Veatch argues that understanding this period properly requires setting it within a larger historical context. Thus, he retraces some of the major historical developments in "physician ethics" and "medical ethics" and explains how they gradually gave way to "bioethics." Crucial to his analysis here is the role that humanists have played in informing the ethical thinking of both physicians and the wider public. Where communication between physicians and humanists has been close (here Veatch cites the Scottish Enlightenment specifically), the connection between medical ethics and the wider ethics of society has also been tightly fitted. However, when cultural and scientific changes "[drive] physicians apart from the leaders in ethics" (as with the roughly 150 year period following the Scottish Enlightenment), the prevailing medical ethic not only separates from the wider cultural ethic, but also can come into direct tension with it (Veatch 2013, 166). For Veatch, bioethics emerged when physicians and humanists were brought back together by the collapse of an "isolated professionally-articulated medical ethics" that was out of touch with the dominant moral thinking of the nineteenth and early twentieth centuries and unable to withstand the strong scientific and cultural trends of the 1960s (Veatch 2013, 168). This collapse was precipitated by the emergence of chronic disease as the dominant concern of medicine, which left patients in a unique position to question the decisions of physicians, as well as by the general challenges to authority that were prevalent in the various protests and countercultural movements of the time period. Bioethics emerged to fill the resulting void, offering a cross-disciplinary dialogue better suited to address the new moral issues and ethical domains opened up by rapid advances in medicine and the life sciences.

In "Professionalism vs. Medical Ethics in the Current Era: A Battle of Giants?" Edmund L. Erde focuses on the question, "Which norms should govern the practice and practitioner of medicine—and why those?" (Erde 2013, 180). Erde critically examines the partially overlapping, yet competing, answers given to this question by two rival paradigms: (1) "secular academic medical ethics," and (2) medical professionalism.

The first paradigm, according to Erde's historical and philosophical analysis, emerged in the 1960s as a result of vast changes in society and in medicine. This paradigm drew on "the methods of humanities and the social sciences to clarify and resolve conceptual and normative dilemmas" and produced a manageable set of secular norms intended to appeal to a diverse, pluralistic society (Erde 2013, 182). Gradually, the principle of autonomy gained preeminence within this paradigm and brought the negative rights of patients and research participants to the forefront of medical practice and research. For those committed to more traditional norms about the role of doctors, this was not a welcome development, as many of these norms either were directly challenged by secular academic medical ethics (e.g., the de facto model of medical paternalism in which physicians determined and pursued what was in the "best interests" of patients) or were wholly dismissed or ignored in terms of their ethical significance (e.g., maintaining a certain public image for medicine).

The second paradigm, professionalism, began to emerge as a rival to secular academic medical ethics in the 1980s and 1990s. According to Erde, this paradigm is "two-sided," in that it functions both as the "essential ideology of medicine as a tradition and an institution" and also as a "character trait—a virtue—of individual practitioners" (Erde 2013, 185). Professionalism seeks to define what it means to be a doctor, to have "the disposition to sustain life and maximize health" and "to improve the skills one must have to achieve these goals" (Erde 2013, 185). This ethos encompasses all facets of being a "true" professional, many of which run counter to or lie beyond the interest of secular academic medical ethics.

After clarifying the nature of professionalism by examining how it would resolve or ameliorate a set of ten problems that worry medicine's leaders, Erde makes a multifaceted case for favoring the norms of secular academic medical ethics. His case emphasizes professionalism's tendency toward medical paternalism, its encouragement of a "peace in the house" of medicine culture that resists accountability to those outside of medicine, and its insistence on fiduciary standards that are overly demanding and perhaps unworkable in our current era of medicine.

In "The Role of an Ideology of Anti-Paternalism in the Development of American Bioethics," Laurence B. McCullough advances a provocative thesis that might potentially shed light on the aforementioned tension between traditional medical ethics and bioethics as it has developed in the American context. McCullough's central contention is that contemporary American bioethics was both shaped in its founding and continues to be captivated by an "ideology of anti-paternalism" (McCullough 2013, 207), an ideology marked by the claims that "medical paternalism was well documented empirically and [that it] was the accepted norm in the Western history of medical ethics" (McCullough 2013, 218). Both of these claims, McCullough argues, were fundamentally mistaken.

Medical paternalism, understood as involving two individually necessary and jointly sufficient conditions—namely, (1) an overriding of patient autonomy, (2) justified on the grounds of benefit to the patient (i.e., beneficence)—became what McCullough terms the "*bête noire*" of bioethics, "which required slaying and which bioethics understands itself indeed to have slain" (McCullough 2013, 209). The "central front" in this "anti-paternalistic" struggle developed in the context of medical decision making. Here, two lines of evidence were adduced to support the claim that physicians had, from ancient times onward, engaged systematically in medical paternalism, overriding patient autonomy by "not telling the truth" to them about their conditions and prognoses and, thereby, not involving them in decisions about their own medical care. First, an empirical study published by Donald Oken (1961) was cited repeatedly in the early bioethics literature as evidence for the claim that physicians routinely and systematically engaged in medical paternalism. Oken studied what physicians told cancer patients about their "diagnosis, prognosis, and treatment planning" (McCullough 2013, 211). Classic sources in the bioethics literature used Oken's findings to support their claims of pervasive medical paternalism. According to McCullough, however, a "close reading" of Oken's findings actually reveals just the opposite—namely, that neither of the two necessary conditions for medical paternalism were satisfied by the findings of Oken's study. Consequently, McCullough argues, the reliance upon Oken's study in the early bioethics literature was mistaken and led to erroneous conclusions about the pervasiveness of medical paternalism. A second major line of evidence that was taken to support the claim of pervasive medical paternalism was allegedly paternalistic elements from major texts in the history of medical ethics—specifically, the writings of John Gregory (1772 [1998]) and Thomas Percival (1803 [1975]). In response to these claims, McCullough similarly argues that the two necessary and sufficient components of medical paternalism *cannot* be found in the writings of either Gregory or Percival.

In the final analysis, McCullough believes he has developed a "correct" reading of the Oken, Gregory, and Percival texts. On his reading, "[p]hysicians were not systematic paternalists before bioethics heroically exposed and corrected medical paternalism" (McCullough 2013, 219). Instead, "[t]he origin of this view in bioethics, and the heroic image of American bioethicists as reformers, even revolutionaries, was empirically and historically flawed" (McCullough 2013, 219). It is, McCullough concludes, "long past time to abandon" this ideology of anti-paternalism "as the defining trope of bioethics" (McCullough 2013, 219). Such an abandonment would, in turn, "free the field for the important work of empirically and historically well-informed conceptual investigation of the nature and limits of physicians' intellectual and moral authority and therefore professional integrity, especially in the context of, and as shaped by, organizational culture" (McCullough 2013, 219).

1.4 The Future of Bioethics: Looking Ahead

Appropriately enough, the book concludes by looking ahead to the future of bioethics. However, in so doing, the three essays in Part IV must necessarily look back as well. Making credible predictions about where the field is headed requires an accurate and insightful sense of where it has been. And offering compelling prescriptions for where the field ought to go demands a sense of what has been good or bad about its existence thus far. What unites the final three papers in the volume is precisely this attempt to draw on the history of bioethics in order to envision and influence its future.

The first essay in Part IV, Richard M. Zaner's "Themes and Schemes in the Development of Biomedical Ethics," follows the shift in focus of medical moral issues from the importance of the physician-patient relationship that was so central to medicine in the 1960s, to the issues of personal identity brought on by developments in genetic research. This shift in focus reflects a larger trend wherein the primacy of end-of-life issues has begun to give way to the primacy of life-before-birth issues.

According to Zaner, the rapidity and extensiveness of this transition have stretched our medical and moral imaginations "beyond traditional limits" (Zaner 2013, 224). Medically speaking, the new paradigm of molecular medicine "bodes radical changes in the way health, disease, treatment, and the like will come to be understood" (Zaner 2013, 224). For example, "health" becomes less a matter of fending off sickness and death and more about creating and maintaining healthy genes (Zaner 2013, 225). Morally speaking, the implications of genomic medicine are equally radical and far-reaching. It forces us to reconsider, and perhaps fundamentally reconceive, the nature of personal identity and individual uniqueness, including the extent to which these have genetic bases. Here we also confront anew the specter of eugenics. As Zaner reminds us, genetic research might easily reach beyond the accepted goal of therapeutic applications and turn its eye toward creating a utopian future in which fundamental improvements are made to the human species, thereby calling into further question the very nature of who we are as persons. If Zaner's analysis is accurate, then not only will genetics-based ethical issues loom large in the future of bioethics, but they may ultimately redefine the field altogether.

The chapter by Edmund D. Pellegrino, "Medical Ethics and Moral Philosophy in an Era of Bioethics," addresses four specific issues. First, it examines the socio-cultural context that gave birth to bioethics, which is characterized by the rejection of traditional moral authority and the rapid development of biomedical sciences and biotechnology. According to Pellegrino, this particular milieu recast traditional medical ethics outside its philosophical foundation and paved the way for the emergence of bioethics. The second issue relates to the decline of medical ethics as the source for the professional ethics of physicians. Bioethics reconfigured medical ethics within the particular socio-cultural and scientific context of the 1960s. Pellegrino deplores this shift because it redefines the patient-physician relationship in term of social mores instead of the traditional foundations of medical ethics. Third, Pellegrino looks at the meaning of the word "ethics" in the terms "medical *ethics,*" and "bio*ethics,*" each presupposing different moral visions. Medical ethics, he contends, presupposes rigorous classical philosophical ethics whereas bioethics, in its latest iteration (i.e., "progressivist bioethics"), combines the values of liberalism and pragmatism to advance its socio-political agenda. Pellegrino sees the latter development of bioethics as problematic because it conflates social mores and political ideology with ethics. In his view "ethical discourse must go beyond activism or political ideology," whether in its progressivist or conservative conceptualization. The fourth and final issue Pellegrino addresses is the plea for a "more rigorous adherence to classical philosophical ethics" to ground ethical reflections in concepts such as the good, the right, and the just, rather than in particular ideologies. To this end, he makes a call for a reconsideration of the potential role of moral philosophy in bioethical and medical ethics debates.

In the final contribution to the volume, "Prolegomena to Any Future Bioethics," Albert R. Jonsen calls for bioethics education and scholarship to return to examining the "commonplaces" of moral philosophy. These commonplaces, or sets of "definitions and arguments common to a certain proposal or problem," were a regular feature of ethics scholarship and pedagogy for centuries, but have receded considerably since the birth of bioethics (Jonsen 2013, 256). While Jonsen believes bioethicists would benefit from reconsidering a wide-ranging set of commonplaces, he focuses this paper on three in particular: (1) finality, (2) free will, and (3) relativism.

To demonstrate how these three commonplaces can contribute to our understanding of pressing bioethical issues, Jonsen applies each commonplace to a particular issue for which it seems especially illuminating. First, he considers how the debate over stem-cell research might benefit from exploring the old commonplace of "finality." Here he explains why the idea of "purpose inscribed in structure" is likely inevitable in debates regarding the moral status of embryos, as well as the valuable role it can play in opening up new vantage points from which to assess practices that alter the "destined" processes (Jonsen 2013, 258–259). Next, he analyzes the ways in which a reconsideration of "free will" might shed light on issues in modern neuroethics. In particular, he focuses on how the emerging data of the neurosciences forces ethicists to fundamentally reconsider what central concepts like "choice, freedom, and responsibility might mean in light of them" (Jonsen 2013, 259–261). Finally, he investigates the potential usefulness of "relativism" in navigating the murky waters of cultural bioethics. Here he suggests that reexploring the general philosophical arguments for and against ethical relativism might contribute to "a more penetrating understanding of the philosophical problem of moral objectivity" and, hence, the diversity of values among and within cultures (Jonsen 2013, 261–262).

1.5 Concluding Remarks and Acknowledgments

Any account of a phenomenon as complex and far-reaching as the birth and development of bioethics is bound to have some limitations. We recognize that not every perspective is included in this volume. Likewise, we acknowledge that there are many worthy issues and questions that are not addressed here. However, despite these limitations, this collection of essays, written by founding figures in the field, offers a rich and unique perspective of where bioethics came from, where and what it has been, and where it might go and what it might be in the future. We sincerely hope, and optimistically believe, that this volume will be of benefit for those looking to better understand, in the words of one contributor, this "great wonder of recent cultural history" (Engelhardt 2013, 55).

In bringing this book together, we have acquired many debts. First and foremost, we'd like to thank H. Tristram Engelhardt, Jr., Senior Editor of the *Philosophy and Medicine* book series, for his enthusiastic support of this project, which is one part of a larger collection of perspectives on the emergence and development of bioethics being gathered from around the globe. Though the project has been challenging, and far more so than any of us could have imagined at the outset, Tris' confidence and encouragement have been invaluable throughout the process. We are each grateful to him for his mentorship and friendship.

Additionally, we'd like to thank each of our esteemed contributors for their insightful and high-quality essays. As leading senior figures in the field of bioethics, they are each very busy and faced with more requests and invitations than they could possibly take on (though many of them appear to be trying, nonetheless!). Several contributors dealt with serious medical or familial issues while working to deliver their manuscripts. We cannot say enough how grateful we are that they saw in this project something worthy of their time and energy.

We'd also like to share our appreciation for the assistance of the team at Springer Academic Publishers who helped us along the way. In particular, special thanks go to Marion van Wagennaar, Yolanda Voogd, Maja deKeijzer, Nicoline Ris, and Chris Wilby, each of whom have been involved with the project at various points, and who have exhibited extraordinary patience in what has turned out be a very long-lived project. Additionally, thanks goes to Lisa Rasmussen for her assistance with this project as well.

Finally, we'd like to thank our families for their support and encouragement during the production of this volume. Working on the book often competed with time spent with them, and we are grateful to them for making this work more bearable and enjoyable.

Note

1. *Biopolitics* is defined as a "gradual transformation in bioethics... characterized by an increasing politicization of bioethical issues, that is, one's 'bio-*ethical* views' will reflect one's *political* assumptions concerning the nature, goals and values that should guide the biomedical sciences" (Bishop and Jotterand 2006, 205).

References

Annas, G.J. 2013. American biopolitics. In *The development of bioethics in the United States*, ed. J.R. Garrett, F. Jotterand, and D.C. Ralston, 101–115. Dordrecht: Springer.

Bishop, J.P., and F. Jotterand. 2006. Bioethics as biopolitics. *The Journal of Medicine and Philosophy* 31: 205–212.

Engelhardt Jr., H.T. 2013. Bioethics as a liberal Catholic heresy: Critical reflections on the founding of bioethics. In *The development of bioethics in the United States*, ed. J.R. Garrett, F. Jotterand, and D.C. Ralston, 55–76. Dordrecht: Springer.

Erde, E.L. 2013. Professionalism vs. medical ethics in the current era: A battle of giants? In *The development of bioethics in the United States*, ed. J.R. Garrett, F. Jotterand, and D.C. Ralston, 179–206. Dordrecht: Springer.

Gregory, J. 1772 [1998]. *Lectures on the duties and qualifications of a physician*. Edinburgh: W. Strahan and T. Cadell. Reprinted in *John Gregory's writings on medical ethics and philosophy of medicine*, ed. L.B. McCullough, 161–248. Dordrecht: Kluwer Academic Publishers.

Harvey, J.C. 2013. André Hellegers, the Kennedy Institute, and the development of bioethics: The American-European connection. In *The development of bioethics in the United States*, ed. J.R. Garrett, F. Jotterand, and D.C. Ralston, 37–54. Dordrecht: Springer.

Jonsen, A.R. 2013. Prolegomena to any future bioethics. In *The development of bioethics in the United States*, ed. J.R. Garrett, F. Jotterand, and D.C. Ralston, 255–263. Dordrecht: Springer.

Lo, B. 2005. *Resolving ethical dilemmas. A guide for clinicians*. Philadelphia: Lippincott Williams and Wilkins.

McCullough, L.B. 2013. The role of an ideology of anti-paternalism in the development of American bioethics. In *The development of bioethics in the United States*, ed. J.R. Garrett, F. Jotterand, and D.C. Ralston, 207–220. Dordrecht: Springer.

McCullough, L.B. 2012. Bioethics and professional medical ethics: Mapping and managing an uneasy relationship. In *Bioethics critically reconsidered: Having second thoughts*, ed. H.T. Engelhardt Jr., 71–84. Dordrecht: Springer.

Oken, D. 1961. What to tell cancer patients. A study of medical attitudes. *Journal of the American Medical Association* 175: 1120–1128.

Percival, T. 1803 [1975]. *Medical ethics: A code of institutes and precepts, adapted to the professional conduct of physicians and surgeons*. London: J. Johnson & R. Bickerstaff. C.R. Burns (ed.). 1975. Huntington: Robert E. Krieger Publishing Company.

Reich, W.T. 2013. A corrective for bioethical malaise: Revisiting the cultural influences that shaped the identity of bioethics. In *The development of bioethics in the United States*, ed. J.R. Garrett, F. Jotterand, and D.C. Ralston, 77–100. Dordrecht: Springer.

Strong, C. 2013. Medicine and philosophy: The coming together of an odd couple. In *The development of bioethics in the United States*, ed. J.R. Garrett, F. Jotterand, and D.C. Ralston, 117–136. Dordrecht: Springer.

Veatch, R.M. 2013. The development of bioethics: Bringing physician ethics into the moral consensus. In *The development of bioethics in the United States*, ed. J.R. Garrett, F. Jotterand, and D.C. Ralston, 163–177. Dordrecht: Springer.

Zaner, R.M. 2013. Themes and schemes in the development of bioethics in the United States. In *The development of bioethics in the United States*, ed. J.R. Garrett, F. Jotterand, and D.C. Ralston, 223–239. Dordrecht: Springer.

Part I
The Birth of Bioethics: Historical Analysis

Chapter 2
The Beginnings of Bioethics

Eric J. Cassell

In December 1970, Daniel Callahan invited me to present a paper on the care of the dying at the Institute of Society, Ethics, and the Life Sciences (now The Hastings Center). He asked that I base my paper on a chapter called, "On (Only) Caring for the Dying," in Paul Ramsey's book, *The Patient as Person* (Ramsey 1970). I did not know Daniel Callahan, had never heard of the Institute, and had not read Ramsey's book nor heard of his name. Callahan said that Leon Kass (whom I also did not know) had read my essay on the care of the dying in *Commentary* (then a liberal Jewish intellectual journal) and believed that the Institute's Task Force on Dying needed the perspective of a practicing physician. Although I was wary, Callahan sounded interesting and so did the Institute. I had never given direct thought to the problem of caring for the dying, although, as an internist, I had considerable experience with patients dying. In addition, both of my parents had recently died in their age just a month apart.

My essay in *Commentary* was not primarily about death, but about the changing disease pattern in the United States; and, more broadly, about how culture influences patterns of disease and how disease changes culture. The essay opened with the sentence, written before HIV-AIDS, "For most of us in the Western world, premature death is no longer imminent, the death of children rare, and the death of young adults so improbable that it must be removed from the realistic possibilities of young life" (Cassell 1969). The essay suggested that this change had a profound effect on young people, so that lives lived without the realistic danger of early death might not have the urgency of young lives lived, say, in the time of rampant infectious diseases. What the essay called "Priority One" thinking in medicine gave highest place to warding off death, and was the basis for the excitement of shows about medicine

E.J. Cassell, M.D., M.A.C.P. (✉)
Department of Public Health (Emeritus), Weill Medical College, Cornell University
Medical Center, P.O. Box 96, Shawnee on Delaware, PA 19356, USA
e-mail: eric@ericcassell.com

J.R. Garrett et al. (eds.), *The Development of Bioethics in the United States*, Philosophy and Medicine 115, DOI 10.1007/978-94-007-4011-2_2,

on TV like *ER*, and much of the public interest in medicine, and (especially) in young physicians. Its preeminent place in concerns about health had not changed much, I said, despite the change in time and cause of death. "Priority Two" thinking, concerned with defending against disability, did not receive as much attention, I wrote, despite its increasing importance. Alas, 40 years later, these mixed up priorities remain true in both medicine and the public, despite the overwhelming preponderance of chronic disabling diseases and the aging of the population. In addition, disability receives far less attention in the bioethics literature than often arcane issues about death. I wrote from this perspective despite being a high technology (for the time) Bellevue-trained internist because my Fellowship after residency was in a department of public health under the direction of Walsh McDermott. He understood the cultural and social dimensions of medicine, and drummed these ideas into his faculty until we never forgot them and never left them out of our ideas and teaching of medicine.

Thus, I approached my first meeting of (what became) the Hastings Center already seeing medicine on a much larger canvas than most physicians. Nonetheless, I had never before read anything like Paul Ramsey's book. Both its tone and language were foreign and especially its concern with the moral aspects of the problems presented by dying patients. In brief, Ramsey was raising moral issues that he believed cast doubt on the unquestioning adherence to the technological imperative. That in itself was interesting because in 1970 medicine (and the public) were still riding the wave of enthusiasm and optimism generated by advances in medical science and the achievements of physicians. I was born with what I think of as an inborn error of cognition that makes persons philosophically inclined from childhood – which I think you are or are not, and choice is not a part of it. So despite being so different than the language of science, medicine, and public health that was my daily fare, what Ramsey said was very interesting. Such questions were afoot in the United States at the time and I was aware of, but not captured by, them.

It intrigued me that Ramsey did not always mean what I meant by the word "dying." So, as with other problems, I started asking my patients what they thought the word "dying" meant. I put a large microphone on my desk between the patients and me and recorded their answers to the question, "What do you mean by the word 'dying?'" (I did not ask their permission to make the recordings, it had never occurred to me to do so.) When physicians (including myself) said that a patient was "dying," it meant that the patient had a disease, or was in a stage of a disease that indicated that the patient was surely becoming dead. It did not mean that death was necessarily imminent, but rather that death was inevitable. A physician could look at a chest X-ray and see the large shadow of what appeared to be a cancerous mass and say, "he's dead," even though at that time the patient might be free of symptoms. The meaning of the word "dying" to patients was quite different. Generally, patients meant that the patient was in a state of mind from which death usually ensued, but not necessarily. For example, a man said of his father, "He started dying when he stopped listening to the Brooklyn Dodger's baseball games, and that was years before he died." Similarly, a patient could become dead without ever having been dying. As an instance, a man said of his friend who died of cancer, "He wasn't dying until the moment he died. Before that he was alive as you and me." My paper discussed the impact of these meanings on the care of the dying by physicians and by patients (Cassell 1971, 1972).

The proceedings did not go smoothly the Friday evening of our presentations. At first, things were fine. Paul Ramsey gave his presentation laying out the points he made in his chapter. He was, to say the least, an imposing speaker. I had never heard anyone actually harrumph before, nor speak in the stentorian tones Paul Ramsey used. He was strikingly effective. When it was my turn to speak, I started to play the audio tapes that illustrated the points I was making. Henry Beecher stopped me, asking whether I had obtained signed consents for those recordings and whether the patients know to what purpose they would be put. I had not and they did not. Beecher would not permit me to continue using the recordings even though I argued the point. This was my unforgettable introduction to signed informed consent. (In addition to Paul Ramsey and Henry Beecher, others present included Dan and Sydney Callahan, Will Gaylin, Leon Kass, Renee Fox, Bill May, and Robert Veatch.) After the presentations, there was an excellent discussion period – the best I had ever heard. The evening ended with further conversation about the subject around drinks. On Saturday morning, I apologized to Beecher for not realizing how important the process of consent was and admitted to having had a significant learning experience. I do not recall the content of the Saturday and Sunday morning session. I left the meeting drunk on the conversation. The level of conversation and discussion, and the quality of the thinking in evidence during the weekend, was intoxicating (and intimidating).

I continued to be a part of the Institute's activities and was an eager participant in each meeting. The quality of thought evident on that first occasion continued meeting after meeting and year after year. Other participants included H. Tristram Engelhardt, Jr., philosopher and physician; Richard McCormack, Catholic theologian; James Gustafson, Protestant theologian; Hans Jonas, philosopher; and Robert Morison, physician educator. Daniel Callahan was a taskmaster and our work led to many publications. Philosophers, theologians, physicians, biologists, and social scientists were all part of the discussions, with guests invited to provide expertise or experience the core group lacked. The interdisciplinary nature of the group worked well; people respected the knowledge and seriousness of the others and, as a result, we all learned from each other. The staff – Arthur Caplan, Robert Veatch, Peter Steinfels, Peggy Steinfels, and others – kept the operation moving and created an effective and productive organization. For those early years, the quality continued at the high level that had been established initially. Generally, we did not focus on classical or even modern theories of ethics. There were scholars whose knowledge of ethics, as it was still taught in departments of philosophy, was considerable. Utilitarians, deontologists, consequentialists, virtue ethicists, and so on talked from their traditions. These philosophies were painfully foreign to me, but that largely subsided within years. Others tried to construct theories of ethics that would work in bioethics, and I believe it is safe to say that none of these theories have had a lasting impact. When John Rawls' *A Theory of Justice* was published in 1971, it created quite a stir. His ideas were debated at a number of meetings, but, again, it was difficult to fit them within the framework of medicine and science, and eventually interest died down. We were mostly pragmatists. The new problems that medicine and science presented to ethics were exciting and seemed to fall outside the scope of both classical and contemporary moral theories. For example, on one occasion, Daniel Callahan invited Derek Parfit and other analytical philosophers to discuss some of the kinds of cases that were commonly encountered at the Hastings Center. Parfit presented a case whose biological basis was invented and impossible. When this was pointed out, Parfit responded that the specific case did not matter; it was the argument that counted. Even in the beginnings of bioethics, most of the scholars involved had left that kind of artificial argument behind. It was this kind of introduction of the *real* into our discussions that prompted Steven Toulmin to write the paper, "How Medicine Saved the Life of Ethics" (Toulmin 1982).

What were we actually doing with all the meetings and the publications? We were effectively challenging the idea that science and medicine (particularly medicine in my group, the Task Force on Death) were value-free. We were insisting that the persons involved with sickness, whether patients or physicians, had an impact because they were persons with individual values. Today, that may seem a small task because the belief that medicine or any of the sciences are value-free no longer has currency. At the time, however, the widespread belief in a value-free science was part of the foundation of the authority of science and medicine. When doctors spoke in those days, they were believed to speak with the authority of science. The United States had emerged from World War II with a belief that science and medicine were the source of answers that would free us from human fallibility. The idea that something larger than a moment, an idea, a conceptualization, or even a machine (once it became part of the lived lives of individuals) could be free of human values was a conceit of positivism. From this perspective, human values are irrational, and the source, therefore, of irrational human behavior. Logical positivism and its child, analytic philosophy, lived in support of this concept. They promised an opening to a rational life lived on the basis of science, and a medicine similarly unencumbered with conflicts based in human frailty. Showing the impossibility of such a world was not, of course, our conscious intent. However, on whatever subject we focused our varied disciplines, we exposed the presence of human values in every human interaction. It seemed to me that we were primarily after basic ideas, rather than mere commentary.

The effect of our discussions was a wider and wider view of medicine (in our case) and the function of physicians. It is difficult to remember the specific details of issues we tackled then because the old programs and schedules are archived and unavailable. In my case, the ideas eventuated in articles that in retrospect seem almost primitive, but represent the arc of my thought. In 1973, I presented a paper at one of meetings held at Dartmouth called, "Making and Escaping Moral Decisions" (Cassell 1973). The point of the essay was that medicine was a moral profession (or moral-technical profession) because it had fundamentally to do with the welfare of persons. It is difficult to believe, but that was perhaps the first published expression of the idea. I remember telling my wife over the phone that I was about to give the paper and she warned me to be careful, if I stepped in the swamp of the moral (because it was so imprecise and hard to define) I might never get out again. In a way, I guess, I never have gotten away from it. In the early years, I wrote paper after paper and so did my colleagues. These efforts were, in retrospect, largely phenomenological. Each new discussion exposed problems in medicine that could be examined by looking closely at what actually happened to patients and how doctors dealt with the many problems that the care of the sick raised. These phenomenological insights led us to challenge theories and ideas in medicine that previously seemed certain and settled. I once wrote a paper called, "The Newly Dead Patient," the basis of which was that the way that physicians, hospitals, and patients behaved in relation to the *newly* dead was as though there was something actually different about being dead for a day or so than about being dead for a longer period. This opened a discussion about attitudes of both physicians and non-physicians toward death and the dead that was extremely productive. Despite this, it is probably a good thing that the paper has been lost – in those days I wrote longhand – because I doubt that its insights would have survived. The choice of topic (others also wrote on the subject), however, shows the eclecticism of the Hastings Center group. Writing for this group was not an easy matter. The people around the table were friendly, kind, and usually nice, but they were a tough audience who did not let lapses of thinking or reasoning survive their fierce questioning. One member once gave a paper that was nice, but not rigorous, and he was not given an inch in the painfully frank discussion. Summing up, he said, "I can see the comments on my paper, but I can also see the grade: F." The brilliance of Daniel Callahan and Will Gaylin was knowing that interdisciplinary effort was not only possible, but also that it was necessary to solve the problems posed by the advances of medicine and the life sciences.

The effect on me personally was profound. I was so aware of how generally poor my thinking was in relation to others in the group. My father was correct when he said a few years earlier that I was the most illiterate educated person he knew. I knew medicine, and I was able to utilize my training, knowledge, and experience in my contributions to our discussions. Otherwise, however, I was ignorant. In the discussion of my paper, "Being and Becoming Dead," Robert Morison said that he personally did not believe in the Cartesian duality. I did not know what that was! The group discussions were, however, an education, and with those discussions and the reading that they necessitated, I gradually became more knowledgeable. But it wasn't enough. In the late 1970s, The Hastings Center started a project called *The Basis of Ethics in the Life Sciences*. The project lasted for 4 years and followed the best structure I have known. In the fall of each year, the speakers would present a paper on the assigned subject of that year's conference. A commentator would present also. In the spring of the following year, the speaker would present the same paper, but with revisions in response to the remarks of the commentator and the ideas brought up in the general discussion. The subsequent discussion would be richer, and the papers, gathered into four volumes, are worth reading still. Among the members of the group were Alasdair MacIntyre and Stephen Toulmin. The quality of their thought made my deficiencies even more evident to me. In 1978, largely at the urging of Rabbi Jack Bemporad, I started reading philosophy systematically. Rabbi Bemporad insisted that I start with Alfred North Whitehead because he believed that I could never accomplish the changes in the underlying theory of medicine I was after without changing the metaphysics on which my thought depended. I read philosophy every morning for 20 or 30 min without fail. It took me 5 years to read and reread Whitehead's corpus (although not the *Principia Mathematica).* Then I read the logics of F.H. Bradley, Bernard Bosanquet, and John Dewey. This daunting task (especially Dewey) unquestionably changed the quality of my thinking and my writing. I went on, starting in 1987, to reading Georg Hegel as I had read Whitehead – thoroughly. I believe that working in bioethics had a similar effect on a number of other physicians who entered ethics – it forced us to educate ourselves and to become better thinkers. All of this furthered my own project, which was to place the person, rather than the disease, in the center of medicine.

In the early years, there were not only meetings, but also a wonderful group of scholars available to call on for help. They previously had been working on similar projects on their own, but without the support of the group that the Hastings Center provided. There were virtually no philosophers of medicine to talk with at the time, so I went to California to spend several days with Otto Guttentag, a physician and philosopher at University of California Medical School in San Francisco, who supported the idea of the importance of theory in medicine.

The attack on medical paternalism and sterile ideas of autonomy – what Daniel Callahan called "the desert of autonomy and paternalism" – that characterized the early days of bioethics (and too often still does) was really an attack on the authority of medicine and science to dictate to quotidian concerns. The best aspect of this attack was that it ultimately changed medicine from a profession that did things *to* sick persons to one that works *with* the sick in helping them to get well. The more common picture of medicine that gradually emerged, however, treated doctors as if they dispensed a commodity – as though a visit to a doctor was a commodity. That idea fit in with the changes in the delivery of medical care that were widespread in the United States. Managed care and other prepaid healthcare plans placed their emphasis on *cost* and *product*. Older ideas of medicine, as a *relational* profession where the relationship between patient and physician was crucial, lost currency. The newer idea, promoted in bioethics, gradually displaced older, more basic ethical precepts such as benevolence. I have argued elsewhere that the net effect of all these changes has not been a better medicine (Cassell 2000). One of the problems evident from the beginnings of bioethics has been the primacy of individualism in the absence of an understanding of what sickness does to individuals. I have also argued that bioethics has not made progress in its initial basic ideas because it has failed to pursue the ethical nature of relationships and the problem of responsibility (Cassell 2007).

When we do things in our personal and professional lives, we are often unaware of the cultural, social, and political context in which these things take place. That was true of my experience in the formative days of bioethics. I suppose I was aware that there was considerable flux in American society in those early years, but that the two were related probably did not occur to me. In Albert Jonsen's excellent book, *The Birth of Bioethics*, he fails to connect the events that led to bioethics' emergence with what was occurring in American society (Jonsen 1998). Things were happening both in science and in society, and the old order that had existed before WWII and come under liberalizing fire in the 1960s was giving way. In 1960, the development of birth control pills had a profound effect on the place of women in society. In 1961, President John F. Kennedy established the President's Commission on the Status of Women (Eleanor Roosevelt was the chair). Betty Friedan's seminal book, *The Feminine Mystique,* was published in 1963. Coronary care units came into being in the 1960s. President Kennedy was assassinated in November 1963. Christiaan N. Bernard performed the first heart transplant in 1967. (An earlier attempt at the University of Missouri using a chimpanzee heart had occurred in 1964, but the heart only beat for 1 h.) Martin Luther King was assassinated in 1968. In 1972, the ERA (Equal Rights Amendment) was passed by Congress (though it failed to pass in the states). The Supreme Court decision in *Roe v. Wade* was handed down in 1973, legalizing early term abortion. The Computerized Tomographic Scan (CT) was invented in 1974. It has often been said that one of the effects of all these changes was to diminish the status of authority – the authority of government and the authority of medicine and doctors. The birth of bioethics fits naturally into this time of flux, as it challenged medicine's authority and heightened the status of individuals and individualism.

The emergence of bioethics changed us all: American society, medicine and doctors, and me personally.

References

Cassell, E.J. 1969. Death and the physician. *Commentary* 47: 73–79.

Cassell, E.J. 1971. *On caring for the dying – working paper*. Hastings on Hudson: The Hastings Center, Institute of Society, Ethics, and the Life Sciences. Series on caring for the dying.

Cassell, E.J. 1972. Being and becoming dead. *Social Research* 39: 528–542.

Cassell, E.J. 1973. Making and escaping moral decisions. *The Hastings Center Studies* 1: 53–62.

Cassell, E.J. 2000. The principles of the Belmont report revisited: How have respect for persons, beneficence, and justice been applied to clinical medicine? *The Hastings Center Report* 30: 12–21.

Cassell, E.J. 2007. Unanswered questions: Bioethics and human relationships. *The Hastings Center Report* 37: 20–23.

Jonsen, A.R. 1998. *The birth of bioethics*. New York: Oxford University Press.

Ramsey, P. 1970. *The patient as person: Explorations in medical ethics*. New Haven: Yale University Press.

Toulmin, S. 1982. How medicine saved the life of ethics. *Perspectives in Biology and Medicine* 25: 736–750.

Chapter 3
Teaching at the University of Texas Medical Branch, 1971–1974: Humanities, Ethics, or Both?

Howard Brody

3.1 Introduction

The papers of the Institute for the Medical Humanities at the University of Texas Medical Branch (UTMB) in Galveston contain the following undated, handwritten letter (probably written between 1971 and 1973) to Charles LeMaistre, M.D., chancellor of the University of Texas (UT):

> Doctor Le Maistre:
>
> During the day I was at the T[exas] M[edical] A[ssociation] meeting, I made a couple of comments to Dr. Blocker [Truman Blocker, President of UTMB] about the proposed ethics program at UTMB and he asked me if I would repeat them to you. Since I missed you after the dinner I hope you don't mind if I convey them in this note. The summation of my thoughts was to tell Dr. Blocker that as a student I welcomed and also felt I needed more exposure to ethics and humanities in medicine. I and most of my classmates come to medical schools with various degrees of idealism, mostly on the high side. I realize this idealism must be tempered with realism but the increasing complexity of medicine coupled with the average American[']s drug attitude (I feel largely due to TV) turns this idealism not to moderation but tends more to skepticism and cynicism. I admit the high pace of medical training doesn't help these feelings either. I look forward to a department whose main purpose is help us [*sic.*] settle and look at problems such as—who pulls the plug, do you give a patient an Rx because she expects it and will leave if you don't, do you give placebo, do you alter genetics, how to work out conflicts within yourself about feeling you have for and against patients.
>
> I am not saying these questions are not addressed now but I feel it is important to have a department whose sole job is to help us "understand" medicine. I would encourage the support of the university system to this project.

H. Brody, M.D., Ph.D. (✉)
Institute for the Medical Humanities, University of Texas Medical Branch at Galveston, Galveston, TX 77555-1311, USA
e-mail: habrody@utmb.edu

J.R. Garrett et al. (eds.), *The Development of Bioethics in the United States*, Philosophy and Medicine 115, DOI 10.1007/978-94-007-4011-2_3,

Thank you
P.L. Hendricks
Medical Student, UTMB
(Hendricks to LeMaistre, IMH 1971–1974, NR 39 Box 10 File 78, Blocker Papers, UTMB archives)

What events at UTMB prompted the student to write such a cogent summary of topics in American medicine of that day that required ethical scrutiny? How did UTMB respond to his call for action?

In 1973, the unit this student was advocating, the Institute for the Medical Humanities, was established at UTMB. Soon after, a formal ethics course was initiated, covering most of the topics the student recommended. Ironically, however, the people leading this effort at UTMB seldom included the word "ethics" in any of their deliberations. The reasons for this may reflect the unique history of UTMB, but may also shed some light on the history of medical humanities and medical ethics in American medical education.

3.2 Early Teaching of Ethics at UTMB

UTMB, originally the Medical Department of UT, opened its doors in 1891 as one of the first medical schools west of the Mississippi (Burns 2003). Flexner reported in 1910 that it was the only one of four schools then in Texas that was of satisfactory teaching quality, though he worried that research would languish due to the separation from the parent university at Austin (Flexner 1910, pp. 309–311).

Scattered efforts were made during its first half-century to include ethics and related subjects in the curriculum. For example, the UTMB School of Medicine catalogue for 1926–1927 included a course, "History of Medicine," taught by emeritus professor of medicine Marvin L. Graves. The course included a section, "Medical Ethics," based primarily on the Hippocratic Oath and the principles of medical ethics of the American Medical Association (Catalog 1926–1927, UTMB archives). Graves resigned as lecturer for this course in 1928 ("Items in the Barker Center, President's Records, VF10/D: Medical School Files 1946–1950," UTMB archives).

3.3 Chauncey Leake

Chauncey Leake, a distinguished research pharmacologist, was brought in as vice-president and dean of UTMB's medical school in 1942, when the school's accreditation was in jeopardy and faculty morale was low. Leake stayed until 1955, setting the school on a solid footing and becoming popular with both the faculty and the Galveston community (Burns 2003, pp. 54–61).

As editor and author of the introduction to a reprinting of Percival's *Medical Ethics* in 1927, Leake became one of the few American scholars to write in an analytic way about medical ethics in the years between 1900 and the appearance of Joseph

Fletcher's *Morals and Medicine* in 1954 (Leake 1927). Leake also published extensively on the history of anesthesiology, including a volume on the subject in verse (Leake 1947).

In 1948, Leake proposed to UT President Dr. Theophilus S. Painter that a Department of Medical Jurisprudence and Sociology be established at UTMB, to be headed by Dr. Hubert W. Smith, who had previously taught Legal Medicine at the University of Illinois ("Items in the Barker Center, President's Records, VF10/D: Medical School Files 1946–1950," UTMB archives). Such a Department apparently was less Leake's own vision than an accommodation to the desires of Painter and Smith. Though nothing came of the Smith proposal, the UTMB faculty approved in March, 1950, a Curriculum Committee recommendation for "a course or courses in Medical History, Ethics and Economics, part of which will be given in the Senior year" (Minutes, UTMB Faculty, 3/7/50, UTMB archives).

While the faculty had approved the creation of these courses within the existing departmental structure, Leake had other ideas. He told his children in a letter dated May 15, 1952,

> [y]esterday I took a quick trip to Austin.... We are setting up a new department of legal and cultural medicine, to cover the history of medicine, medical economics, medical jurisprudence, medical ethics and general medical philosophy. We're hoping to get Patrick Romanell to join our staff, but he got a Fulbright fellowship and will be away for a year (Leake to Leake Children, 5/15/52, Leake Papers, National Library of Medicine, copy in UTMB archives).

Leake explained to UT Chancellor James P. Hart,

> We are fortunate in having available for appointment as Associate Professor of the History and Philosophy of Medicine, Dr. Patrick Romanell, at present the Professor of Philosophy at Wells College, Aurora, New York. Dr. Romanell has participated with me in professional studies involving ethics, and has been a welcome visitor at the Medical Branch. Our staff and students are warmly stimulated by him, and we are confident that he would aid materially in promoting the general cultural coordination of our effort, particularly in the field of Medical Ethics, and the general philosophy of scientific activity with relation to health. Dr. Romanell received his Ph.D. in Philosophy from Columbia University... (Leake to James P. Hart, 4/9/52 ["Budget, 1952–1954, 1954–1955," VFA/3, Chancellor's Records, Barker Center], copy in UTMB archives).

The "professional studies" Leake refers to is a book he and Romanell co-authored in 1950, sub-titled "A scientist and a philosopher argue about ethics" (Leake and Romanell 1950). Leake thus showed himself a couple of decades ahead of his time by hiring a philosopher to join the faculty of a medical school to teach philosophy and ethics.

Initially the UT Regents approved the new Department. Apparently for budgetary reasons, Chancellor Hart objected, prompting Leake to propose instead the addition of the needed faculty positions to UTMB's existing Department of Preventive Medicine and Public Health, including an annual salary of $6000 for Romanell as Associate Professor of Medical and Public Health, Philosophy and Ethics ("Budget Items, 1951–1952," VFA/3, Chancellor's Records, Barker Center, copy in UTMB archives). In July, 1954, the Curriculum Committee approved changing the title of the "History of Medicine" course in the second year to one semester of "History of

Medicine" and a second of "Philosophy of Medicine" (Box 6, Folder 88, Duncan Papers, UTMB archives). The following year, Leake recommended that Romanell be promoted to full professor, quoting the chair of Preventive Medicine: "It is our opinion that he has done outstanding work, that he is making a significant contribution to our teaching effort..." (President's [Chancellor] Office Records, VF E/3, F: "Med [B&D] Teaching and Research," Barker Center, copy in UTMB archives).

In 1955, Leake left UTMB, frustrated by the loss of autonomy caused by the creation of a more centralized governance system for UT (Burns 2003, pp. 60–61). Later events showed his prescience in wishing to create a secure departmental home for efforts in medical ethics and history. Dean John B. Truslow proposed in 1961 that "two of the eight tenure positions currently in the Department (of Preventive Medicine) be declared as representing disciplines indefensible in the present design of the mission of the Medical Branch...." One targeted faculty was Romanell, whom Truslow described in terms quite different from the glowing praise previously employed by Leake: "His lectures in Medical History are regarded as exceptionally monotonous...." Truslow offered as proof positive of Romanell's unsuitability, "his response to the telephone call reporting the Red Building on fire, but everything was being done to save his books and papers...: 'I'll be finished with the lawn in just a few minutes and will get over there right after my hot bath. You go ahead and take care of them'" (John B. Truslow to Chancellor Harry H. Ransom, 7/20/1961, "Teaching and Research, 9/1/1960–8/31/1962, Part I," VF L/6, Chancellor's Records, Barker Center, copy in UTMB archives).

3.4 Truman Blocker

During Leake's deanship, Truman Blocker, an innovative plastic surgeon with extensive wartime experience, was rising through UTMB's faculty ranks. Under Leake, he served as dean of the medical faculty (1953–1955), and later became President of UTMB in 1964. Blocker had been influenced by Leake in understanding the importance of the history of medicine; in particular, he developed a love of rare books and made the expansion of the historical books collection of UTMB's Moody Medical Library a top personal priority. Under his leadership, the institutionalization of history of medicine and related subjects at UTMB resumed in the late 1960s.

Blocker's first move was to recruit Chester R. Burns as an assistant professor in history of medicine. Burns, an M.D. who had completed an internship and then proceeded to Johns Hopkins to complete a Ph.D. in history of medicine, was notable as the first American-born graduate of the distinguished Hopkins history of medicine graduate program. Blocker's surviving correspondence indicates that at this time he envisioned Burns's position as relating specifically to history of medicine (History of Medicine, 1962–1969, MS 39, Series 2, Box 8, File 61, Blocker Papers, UTMB archives).

The annual report of the History of Medicine Division for 1970–1971 lists the topics of an elective seminar series taught by Burns, "Major Issues in American

Medical Ethics." One of the topics was "How Long Do You Want Your Life Prolonged?" Burns also noted that he addressed the Galveston County Medical Society on "The Recent Surge of Ethical Concerns in Medicine: Plethora or Renaissance?" (IMH Records, History of Medicine Division, RG9, Box 1, Folder 2, UTMB archives). These topics may reflect Burns's growing scholarly interest in the history of American medical ethics, popular demand within the community, or both.

The 1971–1972 annual report of the Division begins with a striking statement by Burns: "After 2 years of teaching experiences at UTMB, I was convinced that (1) medical history was not 'the' social science or humanities in health professional education any longer—if it ever had been, and (2) that episodic or systematic lecturing to a multi-faceted audience in a 'hip-hip-Hippocrates' way was the least effective teaching method" (IMH Records, History of Medicine Division, RG 9, Box 1, File 3, UTMB archives). Burns, one might surmise, had emerged from his graduate training at Hopkins with a very expansive view of what a medical historian could accomplish as a solo faculty member, and a very constrained view of what were appropriate teaching methods for medical and health professions students. His experience in the trenches at UTMB caused him now to question both assumptions.

Burns's response to his soul-searching was to establish three advisory committees, one of which, the Medical Humanities Committee, met during the entire academic year of 1971–1972 and laid the groundwork for the Institute for the Medical Humanities.

3.5 UTMB and National Trends

To a large extent, Chauncey Leake's earlier efforts in medical ethics and philosophy of medicine were idiosyncratic activities, mostly divorced from any national developments in American medical education. If Veatch is correct that the middle years of the twentieth century represented a "disrupted dialogue" between medicine and philosophical ethics, then Leake was the exception to that rule (Veatch 2005). What of the later UTMB efforts in 1969–1972?

Fox argues that the national medical humanities movement emerged in the late 1960s as a critique of the existing trends in medical education. The critique found its institutional home in the Committee on Medical Education and Theology, formed by some staff of the national Methodist and Presbyterian churches in 1966, which evolved in 1969 into the Society for Health and Human Values (which in turn, in 1998, became a part of the American Society for Bioethics and Humanities) (Fox 1985).

What was wrong with American medical education, according to this critique, was depersonalization, the centrality of molecular biology, and the teaching of mechanistic medicine—what many later would summarize under the term reductionism. The group that launched this critique consisted largely of chaplains attached to medical centers and a few sympathetic physicians. Perhaps imbued with the spirit of the 1960s, the critics spent relatively little time analyzing how or why medical education had gotten that way, and little time in mapping out a detailed program to fix it. Rather the group set about as quickly as possible to identify demonstration projects that seemed capable of countering the undesirable trends and of being widely emulated.

The Society for Health and Human Values, at its inception, sought grant support from the National Endowment for the Humanities. The new organization proposed to identify the values that were currently absent or underrepresented in medical education, and to remedy the deficiencies identified with what Fox wryly called "a dose of the humanities" (Fox 1985, p. 334). Fox noted the chutzpah required to assert its capability of undertaking this task when the Society at the time numbered among its membership only a handful of academic physicians, hardly any social scientists, and no one whose primary responsibility was teaching or scholarly work in the traditional humanities disciplines.

Perhaps surprisingly, the Society encountered considerable success in receiving funding for its demonstration project strategy, thereby giving its leadership the academic equivalent of patronage to be dispensed. So long as all agreed as to what was wrong with medical education, reductionism could remain the common enemy, and no one need question closely just what was meant by "humanities" and what the various critics of medical education actually had in common. As Fox noted, "the only forces that could tear this Society apart were clarity and poverty" (Fox 1985, p. 337).

Just as Fox attributed many of the assumptions of the founders of the Society for Health and Human Values to the environment of the 1960s, his own analysis, presented to one Society section in 1984, reflected the times in which it was formulated. After a period of camaraderie and common purpose as a small but growing organization in the 1970s, the now-enlarged and more diverse Society entered into a series of disputes about purpose and priorities around 1981–1982 that led to internal splintering into sections and interest groups. By the time Fox spoke, some pined for the good old days when everyone knew who the enemy was and no one asked overly searching questions.

3.6 Planning the Institute at UTMB, 1971–1972

The Society for Health and Human Values, and its closely affiliated Institute on Human Values in Medicine (Fox's so-called patronage arm), formed the national reference group as Burns's Committee on Medical Humanities began its work in October, 1971. The Committee elected to divide its work into three phases. First they would review the content required to resolve "the dilemma of creating an educated person and a technically trained one … attitudes that make it possible for health professionals to understand and to deal with the human environment in a more effective way" (Minutes of the Medical Humanities Committee, 10/13/71, IMH Records, RG9, Box 2, Folder 18, UTMB archives). Phase II would analyze the current activities at UTMB around the humanities disciplines, while Phase III would address practical implementation of the Committee's recommendations.

The Committee divided Phase I of its discussion into five disciplinary headings—religion; philosophy; history; law; and language, literature and art. In the discussion of philosophy, the group was told, "[i]n medicine, philosophical activity is most prominently reflected in the area called medical ethics.... Although medical ethics is central to the consideration of philosophy and medicine, other areas are equally important" (Minutes of Medical Humanities Committee, 11/1/71, IMH Records, RG9, Box 2, Folder 18, UTMB archives). As illustrations of philosophically informed discussions of medical ethics, the group heard segments of two audiotaped lectures, on abortion and genetic intervention, respectively. (The first was by Dr. H. Tristram Engelhardt, Jr., soon to join Burns as a UTMB faculty member. The second was by "Dr. Fletcher," presumably Joseph Fletcher, who had lectured at UTMB in 1962 [Burns 2003]). The segments "demonstrated the philosopher's primary goal as one of reflection, critical analysis, and probing into complex questions that are viewed in terms of certain principles which are discussed within a technical language framework. They also demonstrated the difficulty of leaping from biological data to a framework of ethical theory, the somewhat personal nature of philosophical stances, and the vigorous disagreement and intellectual conflict that can be generated in a philosophical discussion" (Minutes of Medical Humanities Committee, 11/1/71, IMH Records, RG9, Box 2, Folder 18, UTMB archives).

The Committee leadership prepared an 8-page summary of Phase I of the discussion. All five disciplinary topics were reviewed, along with some observations about their interrelationships. The summary contained one mention of "ethics": "[Health professionals] should be cognizant of the philosophies undergirding their activities and they should be capable of considering the ethical issues of medicine in a relatively sophisticated manner" (Medical Humanities Committee, Summary of Phase I, n.d., IMH Records, RG9, Box 2, Folder 18, UMTB archives).

In Phase II, the Committee began to query members of various constituencies on the state of current teaching at UTMB. The categories medical students were asked to address, to identify unmet needs in their curriculum, included "religion" and "philosophy" but not "ethics," consistent with the categories employed in Phase I. In questions that apparently reflected the Committee's own image of what sort of physician would be produced by a more "humanities"-rich environment, patient representatives were asked whether it mattered to them if their physicians were religious and/or liked music; the patients answered "no" (Minutes, Medical Humanities Committee, 1/26/1972, 2/16/1972, IMH Records, RG9, Box 2, File 18, UTMB archives).

On March 1, 1972, the Committee demonstrated its linkages with the national medical humanities scene by hosting Al Vastyan, founding chairman of the Department of Humanities at the Hershey Medical Center of Pennsylvania State University. On being asked, "[w]hy are people now interested in adding the humanities to medical education?" Vastyan offered a very "60s" answer:

> I think that our era has been one of movements towards participatory democracy, of racial justice, towards radical criticism of existing institutions—all in order to counter the technological momentum. I think that we are beginning to realize that technology itself does not build its own sense of values. How can we harness these dramatic achievements for human ends? (Minutes, Medical Humanities Committee, 3/1/1972, IMH Records, RG9, Box 2, Folder 18, UTMB archives).

Vastyan candidly admitted some teaching failures at Hershey and reinforced Burns's conclusion that small group teaching was generally superior to lectures. Once again, in a 10-page summary of Vastyan's discussion with the Committee, the word "ethics" hardly appears.

Overall, one finds few *explicit* mentions of medical ethics in the minutes of the Committee. For example, ethics is mentioned less often than the suggestion that all UTMB students and faculty should be required to learn Spanish (which emerged as the first final recommendation from the Committee). Also, when mentioned explicitly, ethics is always categorized under "philosophy" and never under "religion."

"Ethics" appears twice in the final recommendations of the Committee—once as an admissions requirement, and once as the major example of what would be included in a curriculum in "Philosophy and Medicine." The Committee ended up recommending seven curricular areas—Religion and Medicine, History and Medicine, Philosophy and Medicine, Law and Medicine, Literature and Medicine, Languages and Medicine, and Arts and Medicine (Final Recommendations of the Medical Humanities Committee [Spring 1972], IMH Records, RG9, Box 2, Folder 18, UTMB archives).

3.7 Ethics Arrives at the Institute

In September, 1972, H. Tristram Engelhardt, Jr., Ph.D., M.D., joined the History of Medicine Division as Assistant Professor of the Philosophy of Medicine (History of Medicine Division Final Report of Activities, 1972–1973, IMH Records, History of Medicine Division, RG9, Box 1, File 4, UTMB archives). His position then became a part of the new Institute upon its creation in the summer of 1973. His arrival launched a year of frenetic activity by the Burns-Engelhardt team (assisted by an adjunct instructor in Religion in Medicine, Rev. Gammon Jarrell) as they inaugurated a series of new courses across the medical, nursing, and allied health schools of UTMB, often in the time-consuming but favored small-group-discussion format. In January, 1974, a course on medical ethics for the freshman medical students was begun on a pilot basis, and was approved as a required course shortly thereafter by the Curriculum Committee (History of Medicine Division Final Report of Activities, 1973–1974, IMH Records, History of Medicine Division, RG9, Box 1, File 5, UTMB archives).

3.8 Leadership of IMH

While the young faculty of the new Institute were accomplishing a great deal, President Blocker felt the need for more senior leadership to bring stature and perhaps more outside funding. In 1972–1973 he attempted to recruit Dr. Edmund D. Pellegrino, then Vice President and Director of the Health Sciences Center, State University of New

York at Stony Brook. This effort shows that UTMB remained in synchrony with the field of medical humanities as it was emerging nationally, as Pellegrino was a powerful force and recognized leader within that field. Pellegrino wrote back: "You are distinctly in the lead nationally; you have the best opportunity I know of to succeed" (ED Pellegrino to TG Blocker, 8/6/73, Blocker Papers, "IMH-1971–1974," NR 39, Box 10, File 78, UTMB archives). It was practically a foregone conclusion, however, that Pellegrino was not going to abandon a career track leading to a university presidency for what amounted to a demotion as director of an institute.

For his next choice as Institute director, Blocker turned in a quite different direction. One of Burns's first acts as faculty at UTMB had been to organize, along with Houston allergist and medical historian Dr. John P. McGovern, a symposium to mark the 50th anniversary of the death of Sir William Osler, held in Galveston April 21–22, 1970. (Chauncey Leake attended and wrote a poem for the occasion; McGovern and Burns 1973). One of the speakers was Dr. William B. Bean, Sir William Osler Professor of Medicine at the University of Iowa medical school. While the members of the American Osler Society considered themselves carriers of the torch of humanistic medicine, there was at that time virtually no overlap between that group and the Society for Health and Human Values; and the young humanities scholars who were beginning to fill the ranks of the latter Society tended to regard the former group as hobbyists rather than as fellow scholars. Bean, nonetheless, popularly known as "the sage of Iowa City," had an enviable reputation as a medical writer and editor, student of medical history, and (especially attractive to Blocker) rare book collector.

Bean, at the time mulling over Blocker's offer to come to Galveston as Institute director, admitted in December, 1973, "I would need to study the field to see what scholars to seek out as early recruits" (WB Bean to TG Blocker, 12/3/73, Blocker Papers, "IMH-1971–1974," NR 39, Box 10, File 78, UTMB archives). When Bean finally accepted Blocker's offer in January, 1974, he proposed that "[w]ithin the Institute itself, there need to be sections, or divisions, dealing with (1) history and medical history, (2), <u>philosophy</u>, (3) <u>morals and ethics</u>, (4) <u>religion</u>, and probably the law, though I look upon this as a later development" (WB Bean to TG Blocker, 1/15/74, Blocker Papers, "IMH-1971–1974," NR 39, Box 10, File 78, UTMB archives). He thus disagreed with the UTMB Committee in seeing "morals and ethics" as needing separate and equal billing alongside philosophy. He offered the following as his overall vision:

> First of all, my view is that an Institute of the Humanities has as its primary function the involvement of the whole medical community, which means premedical students, undergraduate medical students, interns and residents, fellows, and the junior and senior staff in all the clinical and basic science departments. In short, we need to inspire and then utilize the participating support and enthusiasm for humanism in medicine at every level. It is obvious that an Institute will do no good if merely superimposed from without. It needs to be a vivifying force which is exemplified, not by lip service, but by dedicated support which has to be a real and moving force thriving in the Medical Branch faculty in their day-by-day and, indeed, hour-by-hour experience. Frankly, I am not sure how one gets this done, but we will find out (WB Bean to TG Blocker, 1/15/74, Blocker Papers, "IMH-1971–1974," NR 39, Box 10, File 78, UTMB archives).

By this date, it was standard practice among the Society for Health and Human Values members to distinguish carefully between the terms "medical humanities" and "humanism in medicine," and the UTMB Committee on Medical Humanities also noted the importance of this distinction in their discussions (Medical Humanities Committee, 1971–1972, Summary of Phase I, n.d., IMH Records, RG9, Box 2, Folder 18, UTMB archives). Thus, by describing the major function of the Institute as "humanism in medicine," Bean indicated once again his lack of contact with the field as it was then evolving.

By November of 1974, Bean was feeling even less confident. He wrote privately to Blocker, "I need to talk to you. In fact we probably need to have two or three informal as well as formal sessions.... This is a field in which I have had no experience and have small competence" (WB Bean to TG Blocker, Blocker Papers, "IMH-1971–1974," NR 39, Box 10, File 78, UTMB archives). Throughout this period Bean's correspondence with Blocker suggests much more enthusiasm and interest in adding to the rare book collection at the UTMB library than in the workings of the Institute. Bean retired as Director in 1980 (Burns 2003, pp. 234–236).

With the next permanent Director of the Institute, UTMB rejoined the national medical humanities movement. Ronald A. Carson, a Ph.D. graduate in Divinity from the University of Glasgow and leader of the nascent medical humanities program in Gainesville, Florida, was recruited to UTMB in 1982 and served until 2005. By the time of Dr. Carson's arrival, there was no longer any question that medical ethics formed one of the key topics and areas of research to occupy the Institute (Carson Interview, May/June 1986, Centennial Project, NR 73, Box 24, Folder 12, UTMB archives).

3.9 Conclusion

Today the American Society for Bioethics and Humanities is sometimes referred to half-humorously as a "big B, little h" organization. In most medical schools, bioethics dominates the humanities, if the latter is represented at all. Bioethics seems to get the lion's share of the curriculum time, faculty positions, funding, and media attention.

Given that context, it is interesting to be reminded that the implementation of a bioethics curriculum at UTMB in 1969–1974 was very much a "little b, big H" operation. Ethics was viewed as a subcategory of philosophy, which was in turn a subcategory of the humanities. While the medical student P.L. Hendricks clearly identified ethics as a critical missing component of the school's curriculum, the faculty who planned the new curriculum hardly made mention of it. Either they assumed that ethics would come along for the ride, or else they simply did not think much about it.

With the emergence more recently of graduate programs specifically called "bioethics" (as opposed to graduate study in philosophical or religious ethics with a bioethics concentration), and of fellowships in bioethics, some have expressed concern that an interdisciplinary aspect of the field that was present in its early days will be lost. In actuality, the disciplinary or interdisciplinary nature of bioethics has been a contested issue throughout its recent history. Daniel Fox's political history of the field in 1984 stressed the lack of input from the psychological and social sciences (Fox 1985). This weakness of bioethics has been stressed more recently by Renee Fox and Judith Swazey (2008).

I have suggested elsewhere that at least three basic models exist for 'medical humanities'—as a list of disciplines; as a kind of great books, cultural enrichment program suggested by the life and legacy of William Osler; and as a historical continuum from the *studia humanitatis* movement of the Renaissance (Brody 2009a). Each model has both strengths and limitations. A peculiarity of the history of bioethics at UTMB is that two models appeared to dominate early conceptions of 'medical humanities' at that school. The Medical Humanities Committee clearly adopted a list-of-disciplines model, and the early influence of the American Osler Society through the 1970 Galveston conference and the later selection of William Bean as Director was also evident. It took a number of years, and probably the arrival of Ronald Carson as Director, for the Institute at UTMB to turn in the direction of the *studia humanitatis* model. Yet that model, which argues for an interdisciplinary humanities geared to the needs and problems of the real world, may provide the most appropriate home for the full development and flourishing of the field of bioethics (Brody 2009b, pp. 21–48).

Historians commonly date the emergence of the modern era of bioethics in the late 1960s and early 1970s (Rothman 2003). I have tried here to show how different some things looked at one U.S. medical school during those years, compared to contemporary views of bioethics and its role in the medical curriculum. This perspective may ultimately assist us in better understanding the development of both bioethics and medical humanities.

References

Brody, H. 2009a. Defining the medical humanities: Three conceptions and three narratives. *Journal of Medical Humanities*. E-published 19 Nov 2009. doi:10.1007/s10912-009-9094-4. Available at http://www.springerlink.com/content/1041-3545/?k=Howard+Brody

Brody, H. 2009b. *The future of bioethics*. New York: Oxford University Press.

Burns, C.R. 2003. *Saving lives, training caregivers, making discoveries: A centennial history of the University of Texas Medical Branch at Galveston*. Austin: Texas State Historical Association.

Flexner, A. 1910. *Medical education in the United States and Canada*. New York: Carnegie Foundation.

Fox, D.M. 1985. Who we are: The political origins of the medical humanities. *Theoretical Medicine* 6: 327–342.

Fox, R.C., and J.P. Swazey. 2008. *Observing bioethics*. New York: Oxford University Press.

Leake, C.D. (ed.). 1927. *Percival's medical ethics*. Baltimore: Williams and Wilkins.

Leake, C.D. 1947. *Letheon: The cadenced story of anesthesia*. Austin: University of Texas Press.

Leake, C.D., and P. Romanell. 1950. *Can we agree? A scientist and a philosopher argue about ethics*. Austin: University of Texas Press.
McGovern, J.P., and C.R. Burns (eds.). 1973. *Humanism in medicine*. Springfield: Charles C. Thomas.
Rothman, D.J. 2003. *Strangers at the bedside: A history of how law and bioethics transformed medical decision making*. New York: Aldine de Gruyter.
Veatch, R.M. 2005. *Disrupted dialogue: Medical ethics and the collapse of physician-humanist communication, 1770–1980*. New York: Oxford University Press.

Chapter 4
André Hellegers, the Kennedy Institute, and the Development of Bioethics: The American–European Connection

John Collins Harvey

4.1 The Socio-Cultural Forces that Shaped the Field

At the beginning of the 1970s, bioethics came into existence as if from nowhere. This rapid emergence of the field was powered by dramatic cultural changes. Although initially made in America, the genesis of bioethics was tied to intellectual and cultural developments with deep roots in Western Europe. Often such accounts of intellectual developments are presented as *Ideengeschichten* in which the force of particular crucial ideas is considered to be decisive. However, it is not just ideas, but also persons and their choices that shape history. This essay takes the latter position and is directed primarily to the significance of the persons and their decisions for the emergence of bioethics. This account of bioethics thus takes a particular historiographical standpoint with regard to the history of bioethics, namely, that in the shaping of history it is persons who are masters of ideas, not ideas of persons. Importantly, this essay explores the role of André Hellegers in the creation of the field of bioethics.

Second, this exploration of the beginnings of bioethics shows the central place played by Roman Catholic institutions in the genesis of bioethics. Bioethics is the way it is because of the Center for Bioethics of the Kennedy Institute of Ethics at Georgetown University, the Borja Institute of Ethics (Barcelona), the St. John of God Hospital (Barcelona), and the International Study Group in Bioethics of the International Federation of Catholic Universities, as well as the European Association of Centers of Bioethics (primarily an association of centers within the Roman Catholic tradition). Given the emphasis on natural law in this tradition and its faith

J.C. Harvey, Ph.D. (✉)
Center for Clinical Bioethics, Georgetown University, Washington, DC 20057-1409, USA
e-mail: harveyjc@georgetown.edu

J.R. Garrett et al. (eds.), *The Development of Bioethics in the United States*, Philosophy and Medicine 115, DOI 10.1007/978-94-007-4011-2_4,

in moral-philosophical reflection, one would fully expect that a field such as bioethics would be possible. André Hellegers and Francesc Abel, his student and collaborator, chose a position in the progressive spectrum of this tradition, and through their choices effected the genesis and influenced the character of both American and European bioethics.

4.2 The Crucial Power of a Name

With the goal of supporting a field he termed bioethics, André Hellegers founded an institute at Georgetown University in 1971 that bore the name "The Kennedy Institute for the Study of Human Reproduction and Ethics." Along with a group of scholars and sponsors he gathered together, Hellegers coined the term 'bioethics,' which came to be part of the title of the Institute's Center for Bioethics. His goal was to describe better their project of understanding the relationship of philosophical and theological ethics to the sciences of genetics, human biology, health, disease, and population growth.[1] They had no knowledge at the time that Fritz Jahr had first used the term in 1927 (Jahr 1927), nor that Van Rensselaer Potter, professor of chemistry and oncology at the University of Wisconsin School of Medicine and Director of the McArdle Laboratory for the study of cancer at the School of Medicine, had re-coined the same term 'bio-ethics' in 1949. Potter fashioned the term to identify his studies of the effects of humankind's actions on the long-term survival of both humanity and the character of the environment, the results of which he published in *Global Bioethics: Building on the Leopold Legacy* (Potter 1949). Aldo Leopold was among a group of scientists who for many years were concerned with the continuing detrimental changes in our planet's environment caused by irresponsible human activity. Based on their observations, these scientists predicted that adverse changes would occur. Early in 1949, Leopold had published *A Sand County Almanac: With other essays on conservation from Round River* (Leopold 1949). Potter found other scientists who were gravely concerned about the prospects for the future of humankind in the face of a gradual deterioration of the planet's environment, due to what this group held were irresponsible and destructive human activities consequent on a personal indifference to the consequences of such personal actions. Besides Leopold, this group included Teilhardde Chardin (1971a, b), Paul Ehrlich (1968), Margaret Mead (1957), and Rachel Carson (1962),[2] among others.

Potter again used the term 'bioethics' in 1970 in an article entitled "Bioethics, the Science of Survival," published in the journal *Perspectives in Biology and Medicine* (Potter 1970). He employed the term again in 1971 in the title of his book, *Bioethics: Bridge to the Future* (1971). Here Potter had developed a concern with the connections between the development of certain cancers in human beings and exposure of these individuals to environments contaminated by specific chemicals as well as by radioactivity. Potter maintained the term 'bioethics' in order to focus on the relationship between the destructive activities of humans and the subsequent effects of such actions on the environment. Such behavior, he and his associates predicted, would lead gradually to deleterious environmental changes of such severity that it would result in an increase in the death rate of the human population. Over time, they were able to demonstrate some of these changes and reflected on the effects that these changes would have on the future of humankind. For Potter, "bioethics" was a "science of survival" that could help humankind prevent ecological disaster. It highlighted the effects of human indifference regarding the effects of irresponsible and destructive activities, which, he held, would cause adverse changes in the world's flora and fauna, as well as have an adverse impact on the chemical composition of its air, water, and land masses, such that the world would finally be changed to a point at which it would no longer be able to support its human population. Potter's approach consisted in raising awareness about the importance of environmental preservation and the interrelatedness of all living beings.

This was not the definition of "bioethics" developed by Hellegers and his associates. They engaged the term bioethics to identify the study of the benefits and harms of human action on the physical and emotional health and well-being of humans and animals. They proposed bioethics in order to link moral reflection and the practice of medicine (Lolas 1999). As Reich has shown, the two groups gave two different specific meanings to the term "bioethics" (Reich 1995, 1999). The bioethics that came into existence at the Kennedy Institute at Georgetown University and then spread internationally was focused on what should count as a proper ethos for medicine and the biomedical sciences.

4.3 The Precursors of Bioethics: Daniel Callahan, William Gaylin, and the Hastings Center

About 2 years before Hellegers founded the Kennedy Institute, Daniel Callahan and his friend, the psychiatrist William Gaylin, founded the Hastings Center. In 1965 Callahan earned his doctorate in philosophy at Harvard University and on graduation accepted an invitation to join the philosophical faculty of Yale University. His academic project was the study of procured abortion. This study required a consideration of abortion from many different perspectives—ethical, physical, legal, theological, medical, and social. As Callahan quickly discovered, universities impose demands on their faculty that can distract from research goals. Faculty membership exists within a hierarchical structure that requires certain behaviors and services dictated by faculty rank and department chairpersons. To study a subject such as abortion from all relevant perspectives in a university setting involves not only intellectual labor but also intense physical labor and time. Chores such as arranging appointments to meet specific scholars; finding the location of faculty offices in specific schools, departments, and divisions; planning daily activities while allowing enough time to keep appointments; recording or otherwise enabling preservation of data from conversations—all these can detract greatly from meeting the required faculty obligations of departmental teaching. Committee service and the various obligations consequent on being a professor impose the significant burdens of administrative tasks on scholars with limited time and energy. The young scholar Callahan found obtaining and interpreting the data needed for his study of abortion to be quite burdensome, given the other obligations of his academic setting.

Callahan concluded that if the individuals with the knowledge requisite for such a study were in close proximity one to another, there would be more time for interdisciplinary conversations permitting an easy exchange of information, fruitful discussions, and the development of academic friendships. In turn, this would promote productive scholarship. The organizational structures of colleges and universities simply do not permit the housing of individuals in different academic disciplines in close proximity to one another. Such institutions are generally not amenable to allowing time for extensive interdisciplinary scholarly conversations. He was being brought to the conclusion that he could never create in a college or university setting the sort of institute he now envisioned and that it would be much easier to accomplish the scholarly work he wished to undertake in an academic structure quite different from that possible in a college or university setting. He conceived the idea of an "ethics center," which could have as members individuals from different scholarly fields who could work together on a single ethical project or dilemma, bringing to bear input from many disciplines.

Callahan as a young faculty member in Yale's philosophy department had great promise and tremendous opportunities. Yet, he made a crucial decision for himself and for the yet-to-be-created field of bioethics: he gave up his faculty appointment at Yale and all its promise for an outstanding academic career (which other philosophers believed were ahead of him), and founded, along with his longtime friend Willard Gaylin, the *Institute of Society, Ethics and the Life Sciences* in September 1970 at Hastings, New York, a small community on the Hudson River. This institute, better known as the Hastings Center or "Hastings-on-the-Hudson," had no relationship with any college or university. It was a freestanding institution established to study in a seriously intense manner various ethical problems that arose from the novel medical methodologies, technologies, and pharmaceuticals which were developed from 1950 onwards and which were revolutionizing medical care. In addition, it is important to recall that physicians were using all these advances in a new social milieu marked by a salient individualism and concern for individuals and self-determination, so that individual autonomy came to govern the appreciation of the choices and rights of patients, the actions and rights of professionals, and the decisions of both groups regarding the initiation, use, and discontinuation of treatments, as well as the assessment of its costs and benefits. In addition to abortion, the members of the Hasting Center began to consider other issues, about which few philosophers and theologians had written. Callahan and Gaylin insisted that all these issues were best described under the rubric of "medical ethical" issues.

Most of the members of the Hastings Center were theologians or philosophers who formerly had been faculty members of various universities. These members of the Institute had adopted Callahan's philosophy about "centers." Their decision to join Callahan and Gaylin led to a novum: Callahan's and Gaylin's Hastings Center was the first organization to devote itself solely to the study of ethics in relation to medical and nursing practices, research, and service. Despite not being recognized as an academic institution by those educational organizations which evaluated universities and colleges for "official" accreditation and certification, this freestanding and independent scholarly organization helped create an academic focus on "medical ethics" as a legitimate field. The Kennedy Institute of Ethics, building upon the efforts of the Hastings Center, went one step further. It relocated medical ethics, a traditional field of moral concern, within the novel rubric of bioethics. In the process, it helped forge a new area of legitimate academic enterprise.

4.4 André Hellegers and the Kennedy Center at Georgetown University: The Legitimization of "Bioethics" as an Academic Enterprise

André Hellegers was born in 1926 in the Netherlands, as was Potter (although in the year 1919). Hellegers was 13 years of age when World War II began. His family had lived for many years in the Flemish part of Belgium. When the Nazis invaded the Netherlands, the Hellegers family was able to escape safely by boat to England, leaving Belgium just about four hours ahead of the invading German armies. After the family arrived in England, Hellegers' father enrolled André and his brothers in the prestigious Catholic public school, Stonyhurst College (under the direction of the Fathers of the English Jesuit province) to complete their secondary education. Hellegers studied both philosophy and theology at Stoneyhurst for all 5 years that he was there. He graduated in 1944. He then matriculated in the University of Edinburgh Medical School and earned his Bachelor's degree in Medicine in 1952. During his first 2 years at Edinburgh, while pursuing the collegiate pre-medical part of the medical curriculum, he had the opportunity to continue his study of philosophy and theology. After graduation from Edinburgh, he enrolled in the Sorbonne in Paris and studied aviation medicine for the academic year 1952–1953, earning a master's degree from the University of Paris. His investigative work at the Sorbonne addressed the transfer of oxygen across the lung membranes.

Hellegers then made a crucial decision. His choice not only had a far-reaching impact on his career, but also shaped what was to become the field of bioethics. He decided to complete his education in the United States. In July of 1953, he began an internship in obstetrics at the Johns Hopkins Hospital in Baltimore, Maryland. His brilliant record as an intern earned him a residency position in obstetrics, which permitted him 2 more years of in-hospital specialized training in obstetrics. He was appointed Chief Resident in obstetrics at the Johns Hopkins Hospital for the academic year 1955–1956. During his 3-year period of residency, he arranged to have some free time to work in the department of physiology at Yale Medical School. He had earned a Kennedy scholarship to support[3] himself and his family while working on his research project—investigating the placental transfer of oxygen in pregnant sheep. After he had completed his residency training, Nicholson Eastman, professor of obstetrics in the Johns Hopkins University and Director of the Department of Obstetrics in the Johns Hopkins Hospital, invited him to become a member of the distinguished full-time faculty of the Hopkins' obstetrical department. While teaching and practicing obstetrics, he continued his research regarding the placental transfer of oxygen with particular emphasis on what the placental tissue originating from the fetus contributed to the transfer process. At the Hopkins, he was a popular teacher and a well-recognized research investigator. He rose rapidly in rank and was a full professor of obstetrics by the time he left Hopkins in 1968 to join the obstetrical faculty at Georgetown University in Washington, DC. Its newly appointed chairman Paul Bruns also came from the Hopkins. He had been Hellegers' chief resident in the academic year 1953–1954 (Hellegers' internship year).

This decision offered Hellegers an interdisciplinary context within which he could bring together the diverse dimensions of his education. From early in his life, Hellegers had had an interest in theology and philosophy. He had first studied these disciplines while attending Stoneyhurst during his secondary education, and had continued to study them while at the University of Edinburgh during his first 2 years of premedical studies. His lifelong interest in these two disciplines led him, when he was a faculty member at the Hopkins, to establish with other Catholic physician faculty members a Catholic "think tank," which was called Carroll House in honor of the first Archbishop of Baltimore, John Carroll. Hellegers first obtained the ecclesiastical approval of Cardinal Lawrence Sheehan, then the Roman Catholic Archbishop of Baltimore. He also obtained financial backing for the project from the Henry Knott Family Foundation, a Baltimore philanthropic institution supporting Catholic intellectual activities exclusively in Baltimore. Carroll House was located in the neighborhood of the Johns Hopkins School of Medicine and the Johns Hopkins Hospital. Philosophers and students from the faculty of the Johns Hopkins University, Catholic theologians from St. Mary's Seminary, members of Johns Hopkins Hospital medical staff and staff of the School of Nursing, together with interested medical and nursing students, met jointly to consider philosophical and Catholic theological aspects related to their medical and nursing practices and research. A fuller account of the founding and operation of Carroll House has been published elsewhere (Harvey 2004).

Between the years 1964 and 1966, Hellegers took a leave of absence from his duties in the Department of Obstetrics at Hopkins to serve as the Deputy Secretary General of the Papal Commission on Population and Birth Control (McClory 1997). He played a very prominent role in the work of the Commission, becoming the confidant of cardinals, bishops, and theologians in Roman pontifical universities. His appreciation of theology and philosophy matured through his study of the many then newly promulgated decrees of the Second Vatican Council, offering him access to the "newest" Roman Catholic theology. His acquaintance with various Vatican authorities and faculty members of the pontifical universities was very significant for Hellegers when, later, he organized the Kennedy Institute of Bioethics. He saw himself building on recent Roman Catholic dogmatic developments in order to confront the novel challenges of the time. His decision to accept this involvement proved crucial. With the knowledge and connections he derived from these experiences, he was able to recruit a first-rate staff of Catholic and non-Catholic philosophers and theologians for the Kennedy Institute.

As all this was occurring, Eunice Kennedy Shriver was President of the Kennedy Foundation. She served in this capacity well before and during the period that Hellegers was at the Hopkins. She continued in this role for some two decades, leading the Foundation with great skill and much vigor to a fruitful involvement with Georgetown University. She and her husband, the Honorable Sergeant Shriver, lived in Potomac, Maryland, a suburb of Washington, DC. They responded positively to Hellegers' wish to establish an institute similar to Hastings-on-the-Hudson—but one devoted to "bioethics" as Hellegers defined the discipline rather than to "medical ethics" (as had been the focus of Callahan and Gaylin at the Hastings Center). The Shrivers and the other members of the Kennedy Foundation (mostly members of the Kennedy family) were committed to establishing an academic enterprise to honor Mrs. Shriver's parents, the Honorable Joseph Patrick Kennedy and Rose Fitzgerald Kennedy. They had known the work of Hellegers, had financially supported the Papal Birth Control Commission, and had come to know him well in his capacity as its Deputy Secretary. When he decided to join the faculty of the Georgetown University as a member in the department of obstetrics, the stage was set for Hellegers and the Shrivers to establish their mutually long-desired "Bioethics" Institute. Their interest in Roman Catholic concerns with bioethics was transformed to an interest articulated in general secular terms, a transition that not only appeared unproblematic, but positively to be expected, given their appreciation of natural law as well as the bond Roman Catholicism offered between faith and reason.

Georgetown University was an ideal location for this collaborative project between Hellegers, the Kennedy Foundation, and the larger Kennedy family. The eminent philosopher, Robert Henle, S.J., president of Georgetown University at the time, endorsed Hellegers' proposals and directed the Shrivers' financial backing towards the realization of such an institution. The Shrivers had many conversations with Hellegers after he joined the faculty at Georgetown Medical School, during which they planned how this institute could successfully be devoted to this new academic discipline. The Shrivers, prominent Roman Catholics in the United States, knew many of the members of the hierarchy in both the United States and Europe. Having served on the Papal Commission for Reproduction and Birth Control, Hellegers also knew very well many of the Vatican officials and professors of theology in the pontifical universities in Rome. He had become a leading intellectual leader of the liberal Roman Catholic commitment to social justice and cultural transformation that gave accent to human rights and human dignity.

Working together, Hellegers and the Shrivers were able to assemble a group of prominent Catholic theologians to serve on a council for the Institute. They recruited an outstanding ecumenical group of theologians and philosophers as scholars. The trustees of the Kennedy Foundation promised funding for three professorships at the Institute. Mr. Shriver became President of the Advisory Council. They recruited five individuals as members of the council, including the prestigious Professor of Moral Theology at the Pontifical Lateran University, Fr. Bernard Haring, a priest of the Redemptorist Order; a professor of religion at Princeton University, Paul Ramsey[4]; a professor of genetics at Stanford University and Nobel Laureate, Joshua Lederberg; and Nobel Laureate Jacques Monod, at that time Director of the Pasteur

Institute of Paris. The academic staff of the Institute chosen by Hellegers included a professor of obstetrics at Georgetown University, his old friend and mentor, Paul Bruns, and one representative from the School of Medicine, the School of Law, and the School of Foreign Service.

The three professorships of ethics, the original staffing of the Institute, compassed a Catholic scholar-ethicist, a Protestant scholar-ethicist, and a scholar-ethicist of the Jewish tradition. They set in motion a novel intellectual venture. Hellegers named these professorships for Mrs. Shriver's parents and oldest sister. He invited Fr. Richard McCormick, S.J., the prominent Catholic moral theologian, to accept the Roman Catholic professorship. A young doctoral graduate of the Yale Divinity School who was valedictorian of his class, LeRoy Walters, was invited for the Protestant position, and Leon Kass, professor of philosophy at the University of Chicago, for the Jewish one. In addition, other scholars were recruited to the staff, including Roman Catholic moral theologians such as Frs. Charles Curran, Warren Reich, and John Connery, and moral philosophers of various Protestant denominations, including Gene Outka, Roy Branson, James Childress, and Frederich Carney, as well as Isaac Franck of the Jewish tradition. To this group one must also add Tom Beauchamp and also H. Tristram Engelhardt, Jr., who from 1977 to 1982 was the Rosemary Kennedy Professor of the Philosophy of Medicine. Hellegers realized that the medical-ethical dilemmas of the future would certainly involve questions of resource distribution and social justice, underscoring for him the need to recruit scholars in health care economics and demography. However, the work involving population studies never came to be fully appreciated by the other ethicists of the Institute. After Hellegers' premature death in 1979, demography receded into the background. By the time of his death, clinical bioethics had become a legitimate field of study, and bioethics had achieved intellectual respectability.

4.5 The Movement Expands: Francesc Abel and the European Connection

Both the Hastings Institute and the Kennedy Institute established fellowships enabling scholars to spend 6 months to 2 years at either one of these institutions so that they could immerse themselves in the field of bioethics. This enabled these scholars to return to their own academic institutions to introduce the new academic discipline—bioethics—within their own academic centers. It was a period of intense proselytizing. Among the first scholars at the Kennedy Institute to join vigorously in this movement was a young Jesuit priest-obstetrician from Barcelona, Spain, the Rev. Dr. Francesc Abel, S.J. The decision to support Fr. Abel's work and his decision to join in the development of this field led to an important nexus of collaboration that shaped the field of bioethics and bound the Kennedy Institute with scholars in Europe, especially Roman Catholic scholars. Once again, it was persons and their decisions who shaped the character of the emerging field.

Abel was a very patriotic Catalonian who did not consider himself Spanish. Having been born in Barcelona 7 years before the Spanish Civil War, he was not old enough to experience the political situation in Spain that led to the civil war. He nevertheless did live through some of the most profound horrors of the period, because Franco was determined among other things to crush the Catalonian goal of independence. These dreadful experiences had a profound effect on this very brilliant boy. For 12 years, Abel attended a public secondary school in Barcelona. As a public institution, the Franco regimen determined the school's operating policies. There was a climate of tolerance in the school, which permitted intellectual freedom for the students. The courses offered in this school were very rigorous. Faculty demanded much of the students. Among the subjects Abel studied were Latin and Greek, philosophy and theology, history, biology, chemistry, Spanish literature, English and French. He graduated with highest honors in 1951. He was a very idealistic person and chose medicine as a career, intending to do social medicine. He enrolled in the University of Barcelona's School of Medicine in 1951, and earned his bachelor's degree in medical science in 1957. In his fourth year of medicine, he reported that his life "changed." He became more aware of poverty in Catalonia and began to be engaged in volunteer work in the poor community of Barcelona, assisting many members to obtain medical care for themselves or their families. He began to think about a possible religious vocation. By this time, he was in graduate education in the hospital attached to his medical school.

Abel had a required rotation in obstetrics and his experiences during this rotation had the effect of bringing him to awareness of the existential nihilism of many of the women he saw as patients, all Catholics. These women experienced deep conflicts regarding the Church's position on contraception. Out of all of this, Abel decided to become an obstetrician. He hoped that by practicing in this specialty he could combine his earlier plans to do social medicine with a focus on his new goal of assisting poor pregnant women in learning natural family planning. During this time, he became acquainted with Dr. Luis Campos, one of the most prominent obstetricians in Barcelona. Campos came from a poor family in Barcelona, but by dint of hard work he was able to put himself through medical school, obtain a superb internship and residency training in obstetrics, and develop a private practice in Barcelona which was the envy of all the obstetricians in the city. Campos was Obstetrician-in-Chief at the Hospital *San Juan de Deu* (The Hospital Saint John of God), which was the primary obstetrical and pediatric hospital in Barcelona. He was also professor of obstetrics at the Medical School of the University of Barcelona. Abel had clerked for Campos and, before graduating *summa cum laude* from medical school, he had arranged to do an internship and residency training in obstetrics with Campos. During these years, Abel continued to ponder a religious vocation. He concluded that in following his "feelings" he could combine his work as a priest-physician in obstetrics and his desire to do "social medicine" among the poor of Barcelona.

Abel made the decision to join the Jesuit Order and study for the priesthood, another choice that framed the emerging field of bioethics. He made his first vows in August, 1960. His superiors arranged his studies so that he would still have time to do some obstetrical practice for poor women in Barcelona. Then in 1965, the Father Provincial of the Catalonian province arranged for Abel to study philosophy, theology, and other religious fields as a preparation for his ordination. During the course of his novitiate, he did some of his studies in Rome at the Pontifical Gregorian University. During this period, he also organized "Medicus Mundi" (Medicine for the World), a medical missionary activity carried out in the Spanish Sahara, Mauritania, and Algeria. He recruited young Spanish, German, and French physicians to work gratis for a month or two in these areas among some of the poorest tribes of these regions. This program was very successful and continued for well over 10 years. Abel was ordained to the Roman Catholic priesthood in 1967 at the Jesuit Community in San Cugat del Valles, Barcelona Province, Catalonia.

Again, a fateful decision was made that led from Europe to the United States and, in particular, to connections with the Kennedy Institute. Abel obtained a postgraduate fellowship in obstetrics at St. Vincent's Hospital in New York City. This period extended from the beginning of 1970 to the end of 1972. In the first week of October, 1971, the Kennedy Institute opened at Georgetown. The *New York Times* reported this event a few days later. The news item included the background of the Institute, a report on its mission, and a list of the members of the Board of Trustees along with its staff of scholars. It also provided a description of the impressive opening ceremonies. Abel read the story, immediately called Hellegers, and with the permission of his mentors at St. Vincent's Hospital applied for one of the visiting scholar positions at Georgetown. After reviewing his application, Hellegers quickly invited Abel to Washington for a visit and interview. Hellegers wanted Abel to see the Institute and meet the senior scholars. Hellegers asked each senior scholar to evaluate Abel. After half an hour's conversation with Abel, Hellegers concluded that Abel was just the type of scholar that the Institute should accept. He wrote in his evaluation that he was of the view that Abel had the "right type of mind" and came with the "right type of attitude" and "the right type of background," which would make him a brilliant scholar who would enrich the life of the Institute. At the conclusion of his interviews, Hellegers offered him one of the three funded visiting scholarships on the spot.

At about the same time, Hellegers invited Dr. Robert Cefalo, a Lt. Commander in the Naval Medical Corps Reserve then serving on active duty on the Obstetrical Service of the National Naval Medical Center in Bethesda, Maryland, to spend his free time assisting Hellegers with his physiological research on the placental transfer of oxygen. Hellegers was conducting this research in a laboratory at the National Institutes of Health, located just across the street from the National Naval Medical Center in Bethesda, Maryland. Hellegers kept his goats at the farm of the NIH in Gaithersburg, Maryland. At that time, the department of obstetrics at Georgetown University had no laboratory space in the Georgetown Hospital or in the Medical School building proper. In fact, the Medical School then did not have any facilities to house any experimental animals bigger than rats. Cefalo, an obstetrician on leave from the faculty of the University of North Carolina, had been serving at the National Naval Medical Center in order to complete his 2 years of required military service as payback for the Navy medical scholarship he held as an Ensign in the Naval

Medical Corps Reserve. This had allowed him to complete his medical school education and residency training in obstetrics at the University of North Carolina. Cefalo completed his service obligations for the Navy about 6 months after Abel joined the Kennedy Institute. At this time, Cefalo, at the request of his chief at the University of North Carolina, joined both the Kennedy Institute as a postgraduate student to study bioethics, and the department of physiology as a postdoctoral trainee in that department in Georgetown Medical School in order to work toward a Ph.D. degree in physiology. His chief at the University of North Carolina promised Cefalo that, if he accomplished these tasks, his faculty position would be held open for his return and upon same, Cefalo would be promoted to Vice-Chief of the department. Cefalo fulfilled his promise to his Chief. Hellegers suggested to Cefalo that he invite Abel to become engaged in some experimental laboratory work with him. Cefalo introduced Abel to his research in the physiological laboratory as he (Cefalo) continued to work now full time with Hellegers on their research project investigating the transport of oxygen across the placenta in pregnant sheep.

Abel spent three-and-a-half years at the Kennedy Institute, during which he studied bioethics and became very familiar with various philosophical approaches to bioethics under such rubrics as principalism, utilitarianism, pragmatism, deontology, and virtue ethics, among others. Abel also studied demography under the distinguished demographer Conrad Tauber, who after retiring as the chief demographer of the United States Census Bureau accepted Helleger's invitation to join the staff of the Kennedy Institute and establish its Division of Demography and Population Studies. Hellegers argued that such studies were essential to fleshing out the field of bioethics by achieving a more complete understanding of the philosophical and theological issues underlying medical practice and the conduct of the biomedical sciences. Murray Feschbach and Murray Gendell, well-recognized academic demographers, joined Tauber and formed the faculty of the Division of Demography at the Kennedy Institute. Within this intellectual environment, Abel completed his master's degree in demography under the direction of Tauber and Feschbach. Georgetown University awarded him this degree in demography in June of 1975. Cefalo completed his physiological studies concerning the transfer of oxygen across those placental membranes specifically arising from the fetus. Georgetown University awarded Cefalo his Ph.D. degree in physiology in September, 1975. Following this, Cefalo returned to the University of North Carolina where he spent his entire subsequent academic career. Cefalo and Abel remained fast friends, collaborating in Europe as well as in the United States, in fashioning the emerging field of bioethics. The result was a further strengthening of an early bridge between scholars in the United States and in Europe. Cefalo died in 2007.

Abel left the Kennedy Institute and returned to Barcelona in early September, 1975. His work with André Hellegers helped him to introduce the new academic discipline "bioethics" to Europe. He quickly began to plan for a bioethics institute at the Jesuit House in San Cugat del Valles, which he modeled after the Center for Bioethics at Georgetown University. In this, he had the support of the Father General of the Jesuit Order, Padre Pedro Arrupe, who was well acquainted with Abel's work at the Kennedy Institute. The new Father Provincial of the Jesuit's Province of

Catalonia also supported Abel enthusiastically. Abel began working with two of his Jesuit brother-priests, Fr. Manuel Cuyas, S.J., an applied moral theologian, and Fr. Jordi Escude, S.J., a fundamental moral theologian, to establish the Borja Institut de Bioethica, at the Provincial House of the Catalonian Province of the Society of Jesus at San Cugat. Fr. Arrupe approved the name of Abel's new institute, which honored the third Father General of the Jesuit Order, Francisco Borja, who was also a Catalonian from Barcelona. By the year 1975, there were four bioethics institutes in the world. Three were in North America; one was in Europe. Those in North America were the Institute of Society, Ethics and the Life Sciences, at Hastings-on-the-Hudson, New York; the Center for Bioethics of the Kennedy Institute of Ethics at Georgetown University, Washington, DC; and the Institute of Ethics, newly established late in 1975 at the Montreal General Hospital in Canada. The noted Canadian theologian, David Roy, with the help of Maurice de Wachter, a Belgian theologian then working in Canada (once again with a connection to Europe), was the founder of the latter. The institute founded by Abel at San Cugat was the fourth. Bioethics had emerged as a significant area of scholarship and practical engagement.

Abel summarized the objectives of his Institute as: (1) analyzing the philosophical and theological problems that hinder the progress of bioethics and affect society in general and its systems of values in particular; (2) publishing this information in specialty journals; (3) promoting interdisciplinary dialogue between scientists and humanists; (4) encouraging a dialogue between Christians and members of other religious faiths concerning the foundation of the scientific, philosophic, and theological aspects of ethics as applied to the health sciences; (5) offering a service to society in general and to the Church in particular, focused on the goal of defining criteria for equality ingredient in the life and the dignity of the human person; and (6) participating in the work of bioethics committees established under both civil and ecclesial authority (Abel 1999, p. 13). An intellectual and practical field of engagement with medicine and biomedical sciences had taken shape in America and had successfully been exported to Europe.

In the fall of 1975, the three members of the Borja Institute organized their first research project, which was directed to identifying previously unrecognized ethical problems in perinatal care. A review and intensive study of the first 100 clinical records, beginning on January 1, 1975, produced by the obstetrical service at Barcelona's Hospital of Saint John of God formed the focus of the research. To Abel's, Cuyas', and Escude's great surprise, the study disclosed a great number of totally heretofore unrecognized ethical problems ingredient in the care given the mothers and their babies whose records were studied. This study served as the basis for the first Spanish—indeed European—clinical conference on bioethics, which Abel, Cuyas, and Escude held in December, 1975. Titled "From Clinical Records to Ethical Decision Making," and with invitations extended to more than 50 academicians representing the disciplines of obstetrics, demography, philosophy, and theology, the conference proved to have considerable influence. The earlier influence of the Kennedy Institute reached to the formation of Spanish and European bioethics.

In early 1976, the Fathers—Abel, Cuyas, and Escude—established a Committee for Health Care Ethics at the Hospital Sant Joan de Deu in Barcelona. They incorporated into the work of this committee, which they named the Committee for Health Care Ethics, the members and the work of the hospital's Committee on Family and Therapeutic Counseling, which Dr. Louis Campos, the Chief of the Obstetrical Service at the hospital, had founded in 1974. After consideration of all aspects of a clinical problem(s), the committee explored what would count as the most reasonable paths for ethical decision-making in difficult clinical obstetrical cases that presented conflicting values of often equal weight. The result was that the influence of André Hellegers and the Kennedy Institute took root in Europe. The labor of a European physician who shaped bioethics in America had now come to shape bioethics in Europe. By creating the Borja Institute in Barcelona, the first bioethics institute in Europe, and the Committee for Health Care Ethics at the St John of God Hospital in Barcelona, the first hospital clinical ethics committee in Europe, Dr. Francesc Abel, S.J., the student of Hellegers, stepped beyond—but nevertheless affirmed his continuity with—the moral commitments sustaining his understanding of Christianity. Abel, who had learned bioethics and the functioning of bioethics institutes from André Hellegers, the Dutch-born obstetrician, philosopher, and theologian, had now successfully transferred the vision of this new field to Europe and to European bioethics.

The members of the Borja Institute collaborated actively in creating the International Study Group in Bioethics with independent status as an activity of the International Federation of Catholic Universities. Abel has thoroughly summarized the objectives of this group in his study of the emergence of bioethics (Abel 1999, p. 108). In 1983, Abel also quietly led the initiative (though he has always given credit to Jean Francois Malherbe, Father Eduoard Bone, S.J., and Maurice de Wachter) to organize the European Association of Centers of Medical Ethics (AECEM). Jean Francois Malherbe, with Fr. Eduoard Bone, S.J., who was then Executive Director of the International Federation of Catholic Universities, established a Center for Medical Ethics at Louvain-le-neuve, Belgium in 1978. Maurice de Wachter was the Director of the Dutch Institute of Medical Ethics, created in 1980 in Maastricht, the Netherlands, by the then Dutch Minister of Health, Dr. Louis Stuyt. Abel was the first bioethics scholar in Europe publicly to call for an organization of the developing centers of bioethics. He also suggested its name: The European Association of Centers of Medical Ethics. The result was an important moment that brought together centers for the study of bioethics with roots primarily in the Roman Catholic tradition, even when their predominant language was secular and post-traditional.

Abel had been introduced to the emerging field of bioethical decision-making on a hospital clinical service under the tutelage of Hellegers, who was an extraordinary and superbly gifted teacher. Hellegers, the European turned American, taught Abel well. His European pupil, Francesc Abel, S.J., was also brilliant, and as Hellegers' student was very well prepared to absorb Hellegers' views and import them for the purposes of creating the field of European bioethics.[5] Through the initial and continued work of Francesc Abel, André Hellegers came to exert a major influence on the development of bioethics in Europe. In a sense, he might be considered the "grandfather" of European bioethics. To acknowledge this fact, to honor a great teacher, and to recognize Hellegers as the founder (together with Daniel Callahan) of the academic discipline of bioethics, Abel, who in 1977 had registered the Borja

Institute of Bioethics in the Official Register of Foundations in the Generalitat of Catalonia, decided to change its name to the Hellegers Institute of Bioethics. His Jesuit superiors, however, preferred to retain Borja in the name of the Institute at San Cugat to honor the second Father General of the Society of Jesus ("the Jesuits").

Paul Schotsmans, professor of philosophy and Director of the Institute of Bioethics at the University of Leuven, who published a comprehensive history of the European Society for Philosophy of Medicine and Health Care, also documents the salient role of Abel, as well as of the background influence of Hellegers (Schotsmans 2005). In August, 1987, Hank ten Have, then professor of medical ethics at the University of Nijmegen, Holland (and later the first Director of the Bioethics Institute of UNESCO) had with others founded the Society. Against the backdrop of this history, Schotsmans gave a glowing tribute to Francesc Abel for his role in bringing bioethics to Europe, in organizing the first bioethics institute in Europe, in organizing the first hospital Bioethics Committee in Europe, and in organizing the European Association of Centers of Medical Ethics. In the introduction to his paper, Schotsmans acknowledges the connection with Hellegers. Schotsmans writes:

> Describing bioethics in Europe is impossible without honoring the founding fathers of bioethics. Several eminent bio-ethicists have to be mentioned, like Edouard Bone, S.J., (Brussels, Belgium), Maurice de Wachter (Montreal, Canada and Maastricht, Netherlands), Richard Nicholson (London, UK), Nicole Lery (Lyon, France), Patrick Vespieren (Paris, France) and—even more than all the others—Francesc Abel (Barcelona, Spain). On the occasion of his election as a full member of the Royal Academy of Medicine of Catalonia, F. Abel, S.J., started with a description of the early beginnings of bioethics: 'Biomedical advances and new technologies caused such bewilderment—not to say fear—that doctors and biologists understandably became interested in clarifying concepts such as what is good, who has the authority to decide what is good and what is not good, and on what is this authority based. They also began searching for ethical decision-making criteria, which could be gradually applicable. A group of doctors and researchers at the Johns Hopkins Hospital in Baltimore, Maryland, began meeting almost spontaneously to discuss these and similar questions under the leadership of Dr. André Hellegers (originally from Holland). At the same time and with the same purpose, other groups of university professors began meeting in Hastings-on-Hudson, a small township in Garrison, New York, and at the University of Wisconsin Medical School in Madison. Similar meetings no doubt took place in other parts of the United States [...] I arrived at the Kennedy Institute at the beginnings of 1972'.... The Kennedy Institute, the Hastings Center, Barcelona's Institute Borja, and Montreal's Institute of Bioethics were the four leading institutes in the earliest days of bioethics, and the Hospital de Sant Joan de Deu's Committee for Health Care Ethics was the first in Spain and probably, all of Europe (Schotsmans 2005, p. 38, quoting Abel 1999, pp. 17–18).

All of this was the result of the consequences of the choices Hellegers had made in re-coining the term bioethics and in shaping the field at the Center for Bioethics of the Kennedy Institute of Ethics.

The result of these decisions of particular persons, especially Hellegers and Abel, given their background and abilities, led to the emergence of a European bioethical movement that had its proximate roots in Catalonia and its distant roots in the Kennedy Institute. What arose was an appreciation of bioethics set within a view of the capacities of reason to lay out a moral vision to guide bioethics. Schotsmans goes on to say:

> This makes clear that bioethics in Europe started mainly in the South [and among Roman Catholic thinkers]. On the worldwide level, however, the 'start' of bioethics must be sought in the U. S. A.... The creation of Barcelona's Institut Borja de Bioethica is certainly one of the earliest developments in European bioethics. At the same time, fortunately enough, the European dimension of the bioethical debate was stimulated by the creation of the European Association of Centers of Medical Ethics (EACME). F. Abel played an eminent role in this organization. The Barcelona Institute (with F. Abel) developed an international research and communication network (Schotsmans 2005, p. 38 ff.).

The result was a sense of direction for bioethics that bore the imprint of the choices of particular men, André Hellegers and Francesc Abel. They brought to bioethics a faith in moral-philosophical reflection with roots in reflections on natural law and the bond between faith and reason. It is no accident that these figures were Roman Catholic.

4.6 The Roots of Bioethics

The story of the emergence of bioethics is complex. There are, for example, parallel stories that could be told to complete the account offered here. These accounts would look to the roles played by early journals such as the *Journal of Medicine and Philosophy* (1976) and the Philosophy and Medicine book series (1975) unconnected with André Hellegers, the influence of the department of the history and philosophy of medicine at the University of Texas Medical Branch at Galveston, as well as to the influence of the Society for Health and Human Values (1968), especially through an allied entity, the Institute on Human Values in Medicine (1971). In the Society initially, there was also a predominantly Protestant influence through an associated interest-group, Ministers in Medical Education (1971). By 1975, there was an Institute for the Medical Humanities at the University of Texas Medical Branch (UTMB, Galveston) that had its roots in the commitments of the president of UTMB, Truman Blocker, M.D. Here H. Tristram Engelhardt, Jr., was teaching bioethics and had already in 1974 concluded a one-month intensive course in bioethics supported by the National Endowment for the Humanities. Engelhardt had been influenced by his early engagement as an associate editor of the *Encyclopedia of Bioethics*, which was being edited by Warren Reich, in addition to bringing an independent set of scholarly concerns as a philosopher and historian of medicine. From humanists, physicians, and philosophers, there were independent springs of interest in ethics and medicine usually set within concerns that nested bioethics with philosophy and the humanities more broadly. The American Philosophical Association, for instance, had by 1973 established a Committee on Philosophy and Medicine. The Institute for Medical Humanities at UTMB in Galveston had for its part influenced the creation in 1975 of the Instituto de Humanidades Medicas in La Plata, Argentina. Much was happening in the late 1960s and early 1970s.

There were as well other influential personalities, including Patrick Romanell (1912–2002) and Edmund Pellegrino in the United States, as well as Jose Alberto Mainetti in Argentina and Pedro Laín-Entralgo (1908–2001) in Spain, as well as many others, most of whom are noted in Jonsen's book (Jonsen 1998). These persons also shaped the forces that made it possible for bioethics to come rapidly into existence and take root. These persons did not make the initial decisive choice to deploy the term 'bioethics.' It was the term 'bioethics' with all its ambiguities that conveyed a heuristic unity to the moral reflections on medicine and the biomedical sciences, thus achieving an explicit intention of André Hellegers. Hellegers had had the foresight to envision and name a new field. Images and captivating terms can have a power that transcends strict conceptual borders. This was surely the case with the re-coinage of bioethics by André Hellegers and his associates and their creation of the Center for Bioethics at the Kennedy Institute. By acting at the right time and in the right place, the re-coining of the term and the creation of the Kennedy Institute of Ethics at Georgetown University were like a nidus in hot water. Things dramatically changed and the field of bioethics took shape.

The history of bioethics in the United States cannot be understood apart from the choices made by Hellegers and Abel. So, too, with bioethics in Europe. The Borja Institute has continued to participate with other European centers of bioethics in joint scientific investigations of various bioethical questions. In many of these endeavors, if not all, the Borja Institute for Bioethics has been the leader and Francesc Abel, S.J., its central tireless planner, organizer, host, and leader. All of this had been shaped by the influence of Hellegers. Over the intervening years, the field has surely often developed positions quite different from those embraced by either Hellegers or his student Abel. One might think of Engelhardt's *The Foundations of Bioethics* (1996) and *The Foundations of Christian Bioethics* (2000), as well as Alora and Lumitao's *Beyond a Western Bioethics* (2001), Hoshino's *Japanese and Western Bioethics* (1997) and Qiu's *Bioethics: Asian Perspectives* (2004). This ever-growing and ever more heterogeneous literature still engages the term bioethics. A field has come into existence with a focus and direction of its own. The imprint and influence of André Hellegers and the Center for Bioethics of the Kennedy Institute remains.

Notes

1. It must be acknowledged that Sargent Shriver claims to have been the person who created the term anew in 1971 (Personal letter to H. Tristram Engelhardt, Jr., 26 January 2001). In any event, it was Hellegers who effectively deployed the term.
2. This work initially appeared serialized in three parts in the June 16, 23, and 30 issues of the *New Yorker Magazine* in 1962.
3. The Kennedy Foundation funded some ten fellowships for study in academic disciplines. These contributed to the further understanding of mental retardation, a subject in which the adult children of Joseph Patrick and Rose Fitzgerald Kennedy (led by Eunice Kennedy Shriver) were exceedingly interested because their oldest sister had such a condition.

4. Professor Ramsey's lectures to the faculties of the Yale Medical School and the Divinity School in 1969 were later published as a book (Ramsey 1970). This book proved to be immensely popular and established Ramsey as an influential medical ethicist. Professor Ramsey spent the academic year 1972–1973 as a special observer in the department of obstetrics at Georgetown Medical School and Hospital, during which he not only experienced the practice of obstetrics, but also came seriously to consider the ethical issues it raised. The sessions that Professor Ramsey led were immensely popular with the obstetrical staff members. Professor Ramsey, as a Visiting Scholar at the Kennedy Institute of Ethics, played a large role in the Institute's first 2 years.
5. Abel reviewed these activities in his book *Bioethics: Originally, Presently, and in the Future* (Abel 2000). Unfortunately, Fr. Abel's book has not yet been translated into English from the Spanish. Thus, this excellent and comprehensive record of the founding of bioethics institutes in Europe and the history of the growth of this academic discipline in Europe cannot be studied unless the scholar knows Spanish.

References

Abel i Fabre, Francesc. 2000. *Bioethica: Origenes, presente y future*. Madrid: MAPFRE.

Abel, Francesc, S.J. 1999. Bioethical dialogue in the perspective of the third millennium. Unpublished paper delivered in Barcelona, May 9.

Alora, A.T., and J. Lumitao. 2001. *Beyond a western bioethics*. Washington, DC: Georgetown University Press.

Carson, R. 1962. *Silent spring*. New York: Houghton Mifflin.

De Chardin, T. 1971a. *Human energy*. New York: Harcourt Brace Jovanovich.

De Chardin, T. 1971b. *L'Activation de l'Energie*. New York: Harcourt Brace Jovanovich.

Ehrlich, P. 1968. *The population bomb*. New York: Ballantine Books.

Engelhardt Jr., H.T. 1996. *The foundations of bioethics*, 2nd ed. New York: Oxford University Press.

Engelhardt Jr., H.T. 2000. *The foundations of Christian bioethics*. Salem: Scrivener Publishing.

Harvey, J.C. 2004. André Hellegers and Carroll House: Architect and blueprint for the Kennedy Institute of Ethics. *Kennedy Institute of Ethics Journal* 14: 199–204.

Hoshino, K. 1997. *Japanese and Western bioethics*. Dordrecht: Springer.

Jahr, F. 1927. Bio-Ethik: Eine Umschau über die ethischen Beziehungen des Menschen zu Tier und Pflanze. *Kosmos: Handweiser für Naturfreunde* 24(1): 2–4.

Jonsen, A.R. 1998. *The birth of bioethics*. New York: Oxford University Press.

Leopold, A. 1949. *A sand county almanac: With other essays on conservation from Round River*. Oxford: Oxford University Press.

Lolas, F. 1999. *Bioethics: Moral dialog in life sciences*. Santiago de Chile: Editorial Universitaria.

McClory, R. 1997. *Turning point: Papal birth control commission 20 pack*. Peabody: Crossroad Publishing.

Mead, M. 1957. Toward more vivid utopias. *Science* 126(3280): 957–961.

Potter, V.R. 1949. *Global bioethics – Building on Leopold's legacy*. East Lansing: Michigan State University Press.

Potter, V.R. 1970. Bioethics: The science of survival. *Perspectives in Biology and Medicine* 14: 127–153.

Potter, V.R. 1971. *Bioethics: Bridge to the future*. Englewood Cliffs: Prentice Hall.

Qiu, R. 2004. *Bioethics: Asian perspectives*. Dordrecht: Springer.

Ramsey, P. 1970. *The patient as person: Explorations in medical ethics*. New Haven: Yale University Press.

Reich, W. 1995. The word 'bioethics': The struggle over its earliest meanings. *Kennedy Institute of Ethics Journal* 5: 19–34.
Reich, W. 1999. The "wider" view: André Hellegers' passionate, integrating intellect and the creation of bioethics. *Kennedy Institute of Ethics Journal* 9: 25–51.
Schotsmans, P.T. 2005. Integration of bio-ethical principles and requirements into European Union statutes, regulations and policies. *Acta Bioethica* 11: 37–46.

Chapter 5
Bioethics as a Liberal Roman Catholic Heresy: Critical Reflections on the Founding of Bioethics

H. Tristram Engelhardt, Jr.

5.1 The Moral-Theological Roots of a Secular Movement

A great wonder of recent cultural history is the birth of bioethics.[1] Like the Protestant Reformation, once begun bioethics[2] spread at a remarkable, seemingly unstoppable pace. Its missionaries were soon in Europe, the Pacific Rim, and South America. The only plausible account of how this could have happened so quickly and broadly is that bioethics appeared to satisfy an important cluster of significant cultural needs. On the one hand, bioethics promised to be a source of important moral direction. On the other hand, bioethics supported a collage of attractive socio-political agendas for legal and public policy change. At its inception, bioethics was biopolitics. In this essay, I offer an account of the complex phenomenon of bioethics' emergence, as well as of the goals this movement supported. This account is a history of the field that compasses a personal memoir. Some of what I present relies on my recollections of the intellectual and cultural niche in which bioethics took shape. These recollections are set within the public history of the emergence of bioethics. My goal is to afford future historians of bioethics a more complete view of the texture of the concerns that framed the emergence of bioethics. This essay is offered as both a primary and secondary source for the history of bioethics.

In one regard, I do not depart from accounts I have given previously (Engelhardt 2002, 2003). As elsewhere, I acknowledge that bioethics emerged to fill an intellectual vacuum produced by (1) the deflation of medicine as a quasi-guild,[3] (2) the secularization of American society,[4] and (3) the erosion of the cultural standing of traditional authority figures[5] in a period in which medicine and the biomedical

H.T. Engelhardt, Jr., M.D., Ph.D. (✉)
Department of Philosophy, Rice University,
6100 S. Main Street, MS-14, Houston, TX 77005-1892, USA
e-mail: htengelh@rice.edu

Baylor College of Medicine (Emeritus), Houston, TX, USA

J.R. Garrett et al. (eds.), *The Development of Bioethics in the United States*,
Philosophy and Medicine 115, DOI 10.1007/978-94-007-4011-2_5,

sciences posed questions regarding (1) old but morally problematic medical procedures that had become safer (e.g., abortion[6]), (2) old distinctions that appeared to need greater clarity (e.g., the definition of death; President's Commission 1981), (3) the increasing cost of health care (e.g., problems regarding the provision of health care resources to all in need) that appeared to call out for moral limits to be set (President's Commission 1983), and (4) seemingly novel questions posed by new scientific and technological developments (e.g., cloning and genetic engineering; Recombinant DNA Research 1976, 1978; National Academy 1977; Medical Research Council 1977). All of this I accept as the lineaments of the circumstances in which bioethics came into existence. In this study, I give primary focus to a further and significant set of influences associated with Roman Catholicism, some of which are recognized by John Collins Harvey (2013), to which I have previously given insufficient attention. Their significance has been largely overlooked.

The emergence of bioethics cannot be adequately understood apart from a recognition of the role played by Roman Catholic moral-theological and moral-philosophical assumptions, as well as by the consequences for Roman Catholicism of the Second Vatican Council. First, I will argue that bioethics proceeded on assumptions regarding the capacities of moral philosophy, which assumptions had shaped Roman Catholicism as this religion emerged between the late eighth and the end of the thirteenth century.[7] These intellectual commitments regarding the bond between reason and faith provided a cultural resource that gave bioethics an intellectual plausibility as bioethics took shape at Georgetown University in the 1970s. There was a religiously-grounded faith that moral philosophy could warrant a common normative morality. Second, I will argue that the intellectual and in particular the moral-theological chaos that washed over the Roman Catholic ecclesial community in the period after Vatican II also contributed an existential urgency that led to the emergence of bioethics. During this period of uncertainty, bioethics emerged to give guidance, grounded in a faith in reason that in some circumstances had direct roots in the Middle Ages, and in other cases indirect roots through the Enlightenment, which had affirmed the powers of moral reason without faith. Among other things, I will argue that the development of bioethics depended crucially on the cultural assumptions sustained by, as well as in, an existential urgency present at the Kennedy Institute of Ethics. The Kennedy Institute within a Roman Catholic, indeed a Jesuit academic institution, supported the confidence that bioethicists could disclose a canonical morality able to guide medicine and the biomedical sciences. The result was a remarkably successful post-traditional moral and political movement, grounded in and motivated by theological dissent and liberal-democratic political aspirations.

5.2 After Vatican II: Bioethics

John Collins Harvey's account in this volume of the influence of Roman Catholic thinkers on the genesis of bioethics, as well as of the special concerns supported by the cultural milieu that followed Vatican II, contributes importantly to the history of bioethics by addressing a crucial dimension of the emergence of bioethics largely overlooked (Harvey 2013). Harvey appreciates the central role of Roman Catholic intellectuals such as André Hellegers (1926–1979) and Sargent Shriver (1915–2011), as well as the dynamics within the Roman Catholic community of the 1970s for the genesis of bioethics. Crucial to the development of bioethics, as Harvey recognizes, were the contributions of Sargent Shriver as a moral and indirectly financial supporter of the Kennedy Institute, and of André Hellegers as a person who had a faith in the bond between faith and reason. Moreover, both recognized the heuristic ability of the term "bioethics" to focus a cluster of cultural concerns. Shriver and Hellegers understood that the term bioethics would sell the importance of the moral-political movement they were creating. They also saw the cultural and political promise of the nascent field. Further, as Harvey's account indicates, the emergence of bioethics was tied to a cluster of concerns and conditions within Roman Catholicism of the time, which were important for both Shriver and Hellegers.

Shriver and Hellegers were from the liberal wing of both the Democratic party and Roman Catholicism. Intellectually, they embraced the secular promises of Roman Catholic thought regarding issues of social justice. Given the background Roman Catholic assumptions regarding the power of reason to disclose the character of the moral life independent of faith, Shriver and Hellegers were convinced that moral philosophers could by sound rational argument establish a canonical moral vision to guide bioethics. They also assumed that this moral vision comported in general with their own moral and political intuitions. They were primarily practical men oriented toward policy, who on the basis of an intellectual foundation that others promised presumed that they could proceed with a reliable sense of moral direction. Bioethics would articulate the moral foundation they sought. I recall luncheons with André Hellegers in the "Tombs" of the restaurant 1789[8] across from the old car barn where the Kennedy Institute was located in the Georgetown area of Washington, where the discussions addressed both broad and particular issues. With martinis and within a cloud of cigarette smoke (his smoking and my passively absorbing the smoke), we would discuss a collage of ideas and possible policies, all of which he was certain could be guided by a fabric of natural-law arguments and general philosophical arguments developed within the new field of bioethics. So, too, during drinks and meals at the home of Eunice and Sargent Shriver, a similar conviction prevailed.

In all of this, it is important to recognize the turmoil and passions of the times that provided the framing context and conveyed an urgency. Although its assumptions about the capacities of moral philosophy remained and were powerfully guiding, the post-Vatican II (1962–1965) Roman Catholicism in the early 1970s was itself entering into a crisis of manpower, spirituality, and theology. From being an ecclesial community that *de jure* and *de facto* had successfully contained dissent, it became a community that *de facto* if not *de jure* nurtured dissent. John XXIII's call for *aggiornamiento* had among other things brought into Roman Catholicism the turmoil and dissent of the time. As a result of the forces set loose by Vatican II, traditional moorings were brought into question both within and outside of Roman Catholicism. An indication of the depth of this crisis is given by the drop by 41% between 1965 and 1970 in the number of seminarians in the United States, from 48,992 to 28,819 (Jones 2003, p. 27). Established pieties were undermined. Friday abstinence had been largely abandoned. A new form of the Mass was established (usually banally celebrated, if not celebrated in various experimental variations), which incorporated numerous novelties, including priests no longer praying *ad orientem*. The secularizing character of the times, combined with a strong faith in reason's capacity to articulate and defend a common morality, tended to bring Roman Catholicism's faith in its tradition ever more into question in favor of a new moral vision defended through the resources of philosophy. All could and should be rethought. In the face of all of this, moral philosophy was still held to be able in the face of moral pluralism nevertheless to scry a common moral vision.[9] The result was not a unity of moral vision, a new Pentecost, but a new Tower of Babel, a crisis in theological, especially moral-theological vision, combined with an augmented centrality for moral philosophy.

This was the milieu born of Vatican II in which André Hellegers developed his mature intellectual perspective. LeRoy Walters acknowledges, "[w]e should not forget the probable impact of Vatican II on the young André Hellegers" (Walters 2003, p. 222). In the debates during and following Vatican II, André Hellegers saw himself as, and was regarded as, both a Roman Catholic intellectual and a dissenter. Hellegers reflected the post-traditional spirit of the times. Walters captures well this salience of intellectual dissent.

> In André's own life and work, there are several striking examples of his courage and willingness to protect dissent. The first two Catholic theologians whom André invited to the Kennedy Institute were Warren Reich and Charles Curran. Both had been active in the attempt to reform Catholic moral theology in the late 1960s, and Warren Reich had been a visible and active supporter of Charles Curran when Catholic University and the Vatican had tried to set limits on the scope of his theological inquiries. Warren was the first long-term Catholic scholar at the Institute, and Charles Curran our first visiting scholar. In 1974 and 1975, André invited Bernard Häring, another burr under the Vatican's saddle, to join us as a visiting scholar (Walters 2003, p. 228).

Walters recognizes the *Zeitgeist*: there was the feeling that moral and theological reflection had been set loose from the artificial constraints of what had been a canonical, guiding, moral and theological tradition. Among other things, Hellegers regarded himself as being in a position to change Roman Catholic doctrine on the basis of what he held could be established through empirical science and rational moral reflection. For example, Hellegers, who served on a Roman Catholic commission that had recommended revising the doctrine on contraception (McClory 1995), rejected the view that the doctrinal tradition in Roman Catholicism had a status of its own that could be neither brought into question nor changed.

> Had the encyclical stated that the data, advanced by the commission, were wrong or irrelevant, or were insufficient to warrant a change in teaching, that would have been one thing. It is quite another thing to imply that agreement with past conclusions is the *sine qua non* for acceptance of a study. Such wording pronounced the scientific method of inquiry irrelevant to Roman Catholic theology (Hellegers 1969, p. 217).

Hellegers reacted adversely to Pope Paul VI's *Humanae Vitae* (1968). In the moral-theological cultural niche within which bioethics took shape, Hellegers was supportive of the view that reflections in bioethics could lead to the substantive development of Roman Catholic doctrine. Although there was a view that there was a common morality, there was also at the same time a desire to revise the *de facto* common morality. For example, there was the goal to revise the Roman Catholic moral position on contraception in favor of a revised morality that accepted contraception and that should then be regarded as the common morality.

This spirit of dissent and state of moral-theological unclarity shaped the background within which bioethics took shape. At the beginning of the 1970s, Paul Ramsey saw that this theological chaos had despoiled the West of a traditional source of medical-moral guidance and had created a moral vacuum that would favor whoever would offer plausible direction. Guidance was sought, although a coherent medical morality was by no means self-evidently available. "Due to the uncertainties in Roman Catholic moral theology since Vatican Council II, even the traditional medical ethics courses in schools under Catholic auspices are undergoing vast changes, abandonment, or severe crisis" (Ramsey 1970, p. xvi). To engage a Kuhnian metaphor (Kuhn 1962), Roman Catholicism, as well as much of Western Christianity, and with it much of the culture of the West, entered into a moral paradigm crisis. There was unclarity as to what moral standards should guide behavior in general and decisions in medicine and the biomedical sciences in particular. There was deep controversy regarding what norms should guide health care policy.

Richard McCormick, S.J., a professor at Georgetown University, member of the Center for Bioethics of the Kennedy Institute, and future director of the Kennedy Institute, recognized the depth of the chaos in moral theology and the extent to which that chaos had contributed to confusion in the moral culture of the 1970s. The turmoil within Roman Catholicism as the historically most influential and numerically dominant Christian religion of the West, generally de-stabilized the surrounding culture. This chaos McCormick associated with the consequences of Vatican II.

> The Second Vatican Council, after speaking of the renewal of theological disciplines through livelier contact with the mystery of Christ and the history of salvation, remarked simply: 'special attention needs to be given to the development of moral theology.' During the past 6 or 7 years moral theology has experienced this special attention so unremittingly, some would say, that the Christianity has been crushed right out of it (McCormick 1981, p. 423).

In all of this, Richard McCormick also played the role of a gentle dissenter. McCormick said to LeRoy Walters in the early 1980s, "LeRoy, I'm sort of a sacred cow in Catholic theology" (Walters 2003, p. 228). Much appeared open to revision. Points of orientation had been brought into question, and there remained an urgency to find moral guidance.

This state of affairs engendered dramatic changes. Up until the mid-1960s, Roman Catholicism had possessed and engaged a substantial and sophisticated literature in medical moral theology in what was termed the manualist tradition.[10] Roman Catholic medical-moral theology had proceeded with a sense of certainty. Abruptly, this certainty came to an end. Among other things, the post-Vatican II theological chaos terminated a three-centuries-old manualist tradition, along with its manuals on medical ethics. As Ramsey observed,

> [t]he day is past when one could write a manual on medical ethics. Such books by Roman Catholic moralists are not to be criticized for being deductive. They were not; rather they were commendable attempts to deal with concrete cases. These manuals were written with the conviction that moral reasoning can encompass hard cases, that ethical deliberation need not remain highfaluting but can "subsume" concrete situations under the illuminating power of human moral reason (Ramsey 1970, p. xvi-xvii).

After Vatican II, Roman Catholicism's traditional moral-theological research project collapsed nearly overnight, leaving Roman Catholic physicians and moralists without clear guidance. The literature and reflections of three centuries suddenly seemed out of date, overly Scholastic, and irrelevant within the emerging post-Vatican-II thought-style.

Despite this state of affairs, Sargent Shriver, André Hellegers, John Collins Harvey, and Francesc Abel remained confident that moral-philosophical reflection within bioethics could give guidance. Without self-consciously appreciating the consequences of their engagement, they secularized their religious commitments, transforming what had been religious concerns into the substance of a secular moral movement. The development of bioethics was part of a larger secular, moral, and cultural transformation that had been and was occurring in Western civilization. With respect to bioethics, this involved the transformation of a liberal religious-theological movement into a liberal political-cultural movement with a special focus on health law and policy. As to the general changes, as Vattimo appreciates, religious commitments and concerns were recast as moral concerns. As Vattimo puts it,

> [t]o embrace the destiny of modernity and of the West means mainly to recognize the profoundly Christian meaning of secularization. I return to the observation…, namely that the lay space of modern liberalism is far more religious than liberalism and Christian thought are willing to recognize (Vattimo 2002, p. 98).

Vattimo recognizes that the transformation of Western Christian self-understandings from the late eighteenth to the mid-twentieth century, had produced a post-traditional, secular force that is reshaping the culture of the times. A consequence is that the Christianity of the main-line Christianities and especially of Roman Catholicism saw itself even more appropriately expressed in secular moral commitments.

A Christian moral vision that had possessed a strong metaphysical anchorage had been rendered into a secular moral, post-metaphysical phenomenon in which the moral concerns no longer were the province of a particular ecclesial community. Once this transformation occurred, a transformation encouraged by a strong faith in moral philosophy, traditional religious moral understandings were recast in terms of a commitment to social reform. Under the circumstances, it did not make sense for a Christian bioethics to maintain what Vattimo characterizes as "a strong identity" (2002, p. 98). This emerging secularity and recasting of Christian commitments into general secular commitments transformed medical moral-theological reflections into a secular bioethics. In this light, as Vattimo argues, "…Christianity's vocation consists in deepening its own physiognomy as source and condition for the possibility of secularity" (Vattimo 2002, p. 98). The result was that the focus and reflections of those involved changed from having a religious character and took on a secular character that gave voice to bioethics as a moral and political movement.

Where religion had been focused on converting the world, bioethics focused on changing public policy and law.

In his account, Vattimo underestimates the depth of the rupture that occurred as much of Western Christianity secularized itself. Indeed, it is the case that there are many ways in which the contemporary secular culture of the West is a development out of the culture of Western Christianity. However, the development also involved a foundational metamorphosis, which Vattimo chooses to ignore. It was not just that Christendom fell into ruins but that a profoundly different dominant culture took shape.[11] Karl Löwith captures well the qualitative character of the changes involved and not acknowledged by Vattimo. "[T]he break with tradition at the end of the eighteenth century ... produced the revolutionary character of modern history and of our modern historical thinking" (Löwith 1949, p. 193). The depth of the rupture was a function of the profound secularization that followed and that led to the establishment of a post-Christian, post-traditional dominant culture. The new assumptions supported a radically different worldview.

> If the universe is neither eternal and divine, as it was for the ancients, nor transient but created, as it is for the Christians, there remains [after Christendom] only one aspect: the sheer contingency of its mere "existence." The post-Christian world is a creation without creator, and a *saeculum* (in the ecclesiastical sense of this term) turned secular for lack of religious perspective (Löwith 1949, pp. 201–202).

It was in this radically post-Christian cultural context, where moral concerns with formerly religious roots found secular expressions, that bioethics emerged. Löwith recognizes the radical character of the cultural break. However, as Vattimo appreciates, this secularization allowed energies once given to religious concerns to power moral and political movements.

The result was that these developments undermined a wide range of traditional cultural structures, such that a place was made, a need generated, and energies liberated for a secular bioethics that was foundationally different from the medical-moral theological reflections in Roman Catholicism that had preceded it. Nevertheless, this secular bioethics could draw on displaced religious zeal and concerns so as to sustain a new moral movement. Its ecumenical openness favored the collaboration of dissenters not just within Roman Catholicism, but from all religions as well as from the fully post-religious. Vatican II had accelerated and deepened the secularization of Western culture, especially in a crucial cultural niche where it was possible to fashion the field of bioethics. Bioethics emerged as a form of Roman Catholic dissidence.

5.3 Bioethics: Post-religious Zeal

These cultural changes made a new form of Roman Catholic institutions not just possible but inviting. Bioethics took shape in a "think tank" established and directed by liberal Roman Catholics within a Roman Catholic institution, which was itself in the process of secularization. Unlike the Brookings Institute (1916–) or the Institute of Society, Ethics, and the Life Sciences, alias the Hastings Center (1969–), the Kennedy Institute and its Center for Bioethics were lodged within a major university that aspired to be considered among the Ivy League colleges. It pursued this goal in part by discounting and marginalizing its formerly Roman Catholic sectarian character. This transformation often occurred through engaging a significantly post-traditional version of Roman Catholicism's commitment to natural law and social justice. As a result, not only could the Kennedy Institute draw on the resources of a prominent liberal-democratic Roman Catholic faculty, as well as the salience and ideological commitment of the flagship, and likely most liberal, of America's Jesuit universities, but the university was at peace with the theological dissidents present in the Kennedy Institute, as well as with the post-traditional character of the Kennedy Institute and the bioethics it supported.

Many of those who supported the founding of the Center for Bioethics and the spread of bioethics as an intellectual and social movement, even if not Roman Catholic, were impelled by post-Vatican-II commitments to reshape Roman Catholicism as well as the Western Christian traditionalist elements of American and European culture through a better appreciation of the claims of a secular view of individual autonomy and social justice. The commitments were post-traditional. Sargent Shriver, along with Richard McCormick, S.J., who followed André Hellegers as director of the Kennedy Institute, felt at home, for example, with inviting the dissident Roman Catholic theologian Hans Küng on a bioethics missionary trip of the Kennedy Institute to China in 1979. Given that I was generally a traditionalist in religious matters, my presence at the Kennedy Institute as the Rosemary Kennedy Professor of the Philosophy of Medicine was a puzzle to me and perhaps to them as well. On the one hand, I recognized that there were no secular sound rational arguments to sustain as canonical the moral content of either Roman Catholic moral theology or the content of what was becoming the received standard vision of bioethics. On the other hand, I knew God lived, and my sympathies were with liturgical conservatism. As a consequence, during the Kennedy Institute visit to China, Richard McCormick attempted to conduct a discussion with the Patriotic Roman Catholic Church of China that was in schism from the Vatican, and that had preserved the Tridentine Mass. At the end of the visit, the members of the Patriotic Church asked my opinion of their reconciling with Rome. I replied: "Keep the Tridentine Mass, the rest are in chaos."

The Kennedy Institute sustained a precarious balance between the presence of dissident theologians along with ideas drawn from Roman Catholicism (e.g., natural law and social justice) and a vigorous post-Christian, post-traditional commitment to bioethics as an agent of change engaged on behalf of liberty, equality, and a secular appreciation of social justice. The Kennedy Center for Bioethics as an institution was thus at the heart of the culture wars. Over against the Christendom that had emerged with St. Constantine the Great, Equal-to-the-Apostles (272–337), and that had created a Christian Europe and Americas, there was now the liberalism born of the Enlightenment and of the French Revolution. Old Rome, with its synthesis of faith and reason, had framed Roman Catholicism as it emerged as a distinct religion in the early thirteenth century. Now, moral philosophy was engaged in exorcizing the residual influences of a traditional Christian ethos from the dominant culture. After over half a millennium, especially after the Thirty Years' War (1618–1648) and the Civil War (1642–1649), the Western Christian confidence in reason had given issue to Paris and her children. Paris, taken as the metaphor for the laicist Europe that emerged after the French Revolution, was now a cultural force transforming America and much of Roman Catholicism. Paris now stood successfully over against Constantinople: Christendom was in ruins. Bioethics promised to be a guide out of the ruins.

The spirit of Paris and the Enlightenment post-Vatican II with which Roman Catholicism had become allied were at tension with the pragmatic spirit of common law, which derived its substance from evolving precedents rather than from abstract principles of justice and equality. Common law could understand governance within an American formal-right constitution that eschewed all reference to human dignity, human rights, and social justice. Bioethics, although made in America, generally embraced the moral, social, and legal visions of the Continent, visions congenial to Hellegers and Shriver. Hellegers, who was born in Belgium, and Shriver, who had been American ambassador to France, were at home with Roman Catholicism's Continental view of solidarity and social justice. Bioethics, a late child of the Enlightenment and of the French Revolution, grew out of the cultural assumptions and the forces that formed a post-traditional, post-Christian moral vision. It was shaped by a culture influenced by the Frankfurt School and the student protests of 1968 in Europe. Bioethics was of the general questioning of tradition in the United States in the wake of Woodstock, the Civil Rights movement, the anti-Vietnam War protests, and the sexual revolution that marked the late 1960s and the 1970s. This complex cultural turmoil, which both influenced and was influenced by the turmoil in the Roman Catholic church as a result of Vatican II, nurtured a genre of intellectual and theological dissent that supported scholars who approached traditional mores with a hermeneutic of suspicion. As already noted, in this intellectual milieu, bioethics drew on a late flowering of the Western Christian synthesis of faith and reason, which allowed bioethics to experience itself as a movement driven by a social urgency impelled by post-traditional forces. The expectation was that moral reflections would guide a final liberation from unwarranted constraints imposed by tradition, including Christian tradition. In this cultural context, bioethics found itself nested within a Roman Catholic academic institution whose identity was also in question.

Georgetown University, a Jesuit university that, as already noted, was on a substantial trajectory towards its self-secularization, provided an institutional setting congruent with the aspirations of many of the thinkers of post-Vatican II American and European Roman Catholicism, as well as persons who had been nurtured within other denominations. In this institutional setting, it seemed quite possible to give birth to a new field of moral reflection. In this context, it was anticipated that bioethics could substitute for the medical-moral theological manualist tradition that with Vatican II had been abruptly terminated. In this environment, the institutional affirmation of the role of the dissenter against tradition supported the dynamic of the development of bioethics. As Warren Reich, a former Roman Catholic priest, recalls,

> [t]hus, when I came to Georgetown as a "dissenter," I felt radically disconnected from my past academic pursuits and had no clear vision of my professional or intellectual future. Furthermore, after the 1960s I felt that I, together with countless others throughout the world, was experiencing the decisive end of one cultural, moral, and social era and the beginning of another, the contours of which were not yet defined. Thirty-one years later, I see that my serendipitous situation of being suspended between cultures in 1971 was precisely the requisite spiritual and intellectual condition for the task of trying to absorb and articulate the contours, meanings, and normative issues of a new social, intellectual, and political reality that was rapidly taking shape before our very eyes, the future of which we did not know (Reich 2003, p. 166).

With the sweet taste of dissent, bioethics allowed the fashioning of a post-religious, post-traditional moral discourse, which promised to allow the re-articulation of moral-theological concerns regarding health care and the biomedical sciences in a fashion that would allow the founders of bioethics to lodge the field within their commitments to autonomy, equality, and the realization of social justice.

These developments, along with the institutions and ethos they endorsed, accomplished an immanent displacement of the transcendent. What would in a theological context have been driven by a religious zeal was pursued with passion for the realization of a particular immanent moral vision, along with the social, legal, and policy reforms it authorized. In most cases, this moral vision affirmed an anti-traditional view of autonomous individual choice, an endorsement of egalitarian ideals, and the support for social democratic views of redistributive justice. With regard to individual autonomy, this moral vision did not affirm actual choice, but involved an account of individual choice understood and reformed in terms of ideal choices that were to be made with full knowledge of all material facts, uninfluenced by the manipulation of others (e.g., undistorted by the influence of a husband or a physician), and in terms of what were considered to be the authentic values of the individual, undistorted by a false consciousness (e.g., by an affirmative vision of a patriarchal authoritarian family structure). This pursuit of autonomy licensed a gentle manipulation on behalf of a particular vision of autonomy and its importance. This goal was pursued against the *de facto* free choices of the patient (e.g., the account endorsed persuading the patient to state the patient's wishes in the absence of possibly manipulating authority figures, such as a husband). Actual choice and autonomy were to be critically re-assessed within the framework of a particular ideal of proper choice and autonomy.

The commitments to an egalitarianism embedded in a social-democratic ideology led to the discounting of market solutions to policy questions. What occurred in bioethics was a part of a broader, indeed global, surge of interest in issues of social justice. Such interest had already earlier in the twentieth century expressed itself in the invocation of social justice by the Soviet Union, which, as Hayek noted, quoting Andre Sakharov, led to "millions of men in Russia" becoming "the victims of a terror that 'attempts to conceal itself behind the slogan of social justice'" (Hayek 1976, p. 66). In the developing world, such commitments favored the persistence of poverty in the absence of effective markets.[12] In Roman Catholicism, appeals to social justice were engaged to recast if not outright nullify traditional moral commitments. Again, to quote Hayek,

> [social justice] seems in particular to have been embraced by a large section of the clergy of all Christian denominations, who, while increasingly losing their faith in a supernatural revelation, appear to have sought a refuge and consolation in a new 'social' religion which substitutes a temporal for a celestial promise of justice, and who hope that they can thus continue their striving to do good (Hayek 1976, p. 66).

The general point is that the bioethics that emerged in the 1970s and early 1980s was not simply an academic attempt to puzzle through conceptual and moral controversies raised by medicine and the biomedical sciences. It was not simply an intellectual undertaking, but a moral, indeed political movement. There were seemingly paradoxical elements. In the face of a *de facto* moral pluralism that could in principle not be set aside by sound rational argument (Engelhardt 2006), a single common morality was asserted and affirmed for which bioethics was to be its exegete in matters bearing on the biomedical sciences and health care. Of course, such counterfactual assertions strengthened the plausibility of the movement.

The religious roots of bioethics and their importance for the early institutional emergence of bioethics is reflected in the circumstance that of the eight associate editors of the first edition of *The Encyclopedia of Bioethics* (Reich 1978), which by its publication helped to establish the field under the rubric "bioethics," seven had studied within faculties of divinity, theology, or religious studies. They had displaced their prior religious zeal to a secular arena. The editor, Warren Reich, was a former Roman Catholic priest, as was Albert Jonsen. Early members of the Kennedy Institute such as Beauchamp and Childress (who resisted the term 'bioethics' and instead engaged the rubric "biomedical ethics") also had backgrounds in theology. Much of the discourse involved a secularized, moral-theological idiom. Indeed, among the associate editors of *The Encyclopedia of Bioethics*, I was the sole person innocent of anything but a few compulsory courses I had been required to take in theology during my first 2 years in college. My background was instead in the history and philosophy of medicine, as well as in work regarding Kant, Hegel, Husserl, and more generally phenomenology (Engelhardt 1973; Schutz and Luckmann 1973).

5.4 A Personal Confession: Libertarian by Default

As a consequence, I tended to approach bioethics with fewer politically-driven conclusions in mind. My engagements in bioethics were motivated primarily by an interest in solving conceptual puzzles. One such puzzle was my amazement that I could not secure by sound rational argument the lineaments of traditional, Western Christian morality regarding such issues as abortion, infanticide, and euthanasia. I found no foundation for a canonical, content-full, secular morality. Moreover, it became clear that moral pluralism was real and intractable to secular sound rational argument. *De facto*, there was no common morality. There were instead substantive disagreements concerning the morality of abortion, sexual relations outside of the marriage of a man and a woman, capital punishment, and physician-assisted suicide, not to mention the nature of *jus ad bello* and *jus in bello*. Different moralities affirmed different orderings of cardinal human values and of right-making conditions. There was no moral consensus. Nor was it clear what amount of agreement or consensus with regard to which moral issues should carry moral-epistemological weight. I was not a metaphysical-moral skeptic. I did not deny the existence of moral truth. However, I was a secular moral-epistemological skeptic. I knew that secular moral reflection was not able to establish a particular morality as canonical.

This position was not one I had pursued as a goal, but the very opposite. I had begun by exploring the capacities of moral philosophy and had ended with attempts to determine what, if anything, could be salvaged from the Enlightenment project of disclosing a universal secular morality that should unite all persons as such. The results were inescapable and not to my liking. I had surely hoped for much more from moral philosophy. The first edition of *The Foundations of Bioethics* (Engelhardt 1986) begins with a lament concerning our state of affairs and a recognition of the limits of the secular moral-philosophical project.

> Much that one would like to prove regarding the nature of the good life and/or the authority to enforce it appears unavailable. The life of reason, unlike the life of belief or special grace, appears impoverished.... My intent has not been to defend that ethic [i.e., the sparse ethic available, given the inability through sound rational argument to establish a canonical, content-full morality] but rather the very opposite. I have endeavored to find grounds for establishing by reason a particular view of the good life and securing by general rational arguments the authority for its establishment. To my dismay and sorrow, such have not been available (Engelhardt 1986, p. viii).

I was led in my moral-philosophical reflections, which focused on matters bioethical as a heuristic, to conclusions I did not want to embrace regarding the limits of moral-philosophical reflection. My conclusion was that secular, moral-philosophical reflection could only secure a purely procedural morality grounded in permission. This morality could account for the foundation of such practices as the market, which can bind moral strangers as such.

To repeat this point regarding my attitude about the infeasibility of the Enlightenment's secular moral-philosophical project, the project of identifying and warranting a common canonical content-full morality: it was not a circumstance that I celebrated. This is a point often overlooked in commentaries on my work: the conclusions I reached were not those I sought or desired. I had been raised Roman Catholic and had taken for granted that moral-philosophical arguments could be developed to justify the traditional morality of Roman Catholicism. These assumptions regarding the capacities of reason lay at the foundations of Roman Catholicism's emergence as a separate denomination at the beginning of the second millennium. I assumed sound rational argument could establish a canonical, content-full morality. I had also assumed, along with Socrates and Roman Catholics of at least the twentieth century, that God affirmed the good, the right, and the virtuous because they were good, right, and virtuous, and that the good, the right, and the virtuous were not such because God so affirmed them. As I began to study Kant, his version of the Enlightenment project appeared plausible, in that it was a continuation of the aspirations for moral philosophy born of the Middle Ages. However, this project, especially Kant's project, proved unfeasible. My recognition of the limits of secular moral rationality led in the end to my abandonment of the moral-theological positions of Roman Catholicism and of the moral-philosophical positions of the Enlightenment. In the end, I was brought to embrace the Church of the first millennium (Engelhardt 2000), which has had a quite different relationship with philosophy (Bradshaw 2007).

Here a corollary point warrants particular stress. I was not even a libertarian in secular political theory because I was in my heart committed to that position, but because the position proved unavoidable for secular political theory, given the limits of secular moral arguments. Absent an agreement concerning God and His demands, and absent a canonical, content-full morality or view of justice established by conclusive secular sound rational argument, one is left by default, in secular contexts at least, with that authority for common projects that is conveyed by those agreeing to that common undertaking. I was puzzled by the unfounded secular moral certainties of my colleagues in bioethics. This state of affairs led me to assess the influence of prior moral and political commitments on the appreciation of "facts," an assessment that I pursued through an exploration of controversy theory (Engelhardt and Caplan 1987) and the history of ideas. As I first visited the Center for Bioethics at Georgetown University in about 1974, as I began my work as an associate editor for the first edition of *The Encyclopedia of Bioethics*, I had the dispositions of a puzzle-driven philosopher shaped by my work with philosophers such as John Findlay (1903–1987), Marjorie Grene (1910–2009), Charles Hartshorne (1897–2000), Klaus Hartmann (1925–1991), and Gottfried Martin (1901–1973). I had come from an intellectual milieu quite different from that of my ex-theological and ex-ministerial colleagues. I had intellectual interests much more than a moral and political cause.

5.5 The First Decade: Creating a Moral, Social, and Political Movement

The first decade of the development of bioethics was characterized by a plurality of intellectual visions, as well as by a diversity of institutions that competed with the Kennedy Institute. Each institution with its own intellectual vision sought to command the future. Already in 1969, the Hastings Center had brought together broadly influential reflections on medical ethics, as well as a nucleus of scholars examining the intersection of science, ethics, and public policy. The Hastings Center maintained a broad compass and eschewed a narrow focus on bioethics. Yet, in great measure it supported and through the *Hastings Center Report* still supports the general field of bioethics. At the same time, there were various coalitions of interest in the humanities. These in various ways were late expressions of the Third and New Humanisms. At the end of the nineteenth century and the beginning of the twentieth century, there had been a growing concern that the new sciences and technologies should be placed within a recognition of the values that ought to frame human life. These concerns pressed for a better appreciation of the relationship between the culture of science and technology and the culture of the humanities. Abraham Flexner (1866–1959), well known for reshaping American medical education (Flexner 1910), argued in his 1928 Taylorian Lecture for a humanism that would critically reassess cultural commitments. "[T]he assessment of values, insofar as human beings are affected, constitutes the unique burden of humanism" (Flexner 1928, p. 12). Then C.P. Snow (1905–1980) in his 1959 Rede Lecture further developed these views (Snow 1956). Snow's Rede Lecture, published as an article, became the basis of a book that supported the need for a broader humanistic vision (Snow 1962, 1964). These publications reflected views advanced by Matthew Arnold (1822–1888) in his Rede Lecture (Arnold 1882).

The concern to establish a secular view of the human condition led to the creation of the National Endowment for the Humanities through the Humanities Act of 1965. The desire better to appreciate the force and significance of the new sciences and technologies came to have a special focus on health care and the biomedical sciences. The Society for Health and Human Values, which had come into existence in 1968 and then created the Institute on Human Values in Medicine, supported programs in the humanities, including programs in the history of medicine (Engelhardt 1975), one of which I joined in 1972, transforming it into a program in the history and philosophy of medicine. Now some four decades later it is difficult adequately to capture the texture of the ambiguities, passions, and institutional competition of the time. There was a widespread sense that a field needed take shape that could guide medicine and the biomedical sciences. However, aside from Hellegers and Shriver, no one had given name or focus to the field. In 1971 on a plane I serendipitously met William Knisely, vice-chancellor of the University of Texas Medical Schools. He encouraged me to apply to the University of Texas Medical Branch in Galveston. After reading a *Time* article, I had learned of the establishment of the Center for Bioethics at Georgetown University and had written without receiving a reply. In the summer of 1972 I found myself on Galveston Island, where one of my ancestors had landed in December, 1834. In my youth, I had often wondered why anyone would live in that strange town of the Gulf littoral.

Programs in the medical humanities, biomedical ethics, and bioethics were *de facto* mutually supportive, although in the early years I was asked to choose being associated with the Hastings Center or with the Institute on Human Values in Medicine. The Institute, an entity associated with the Society for Health and Human and Values, organized missionary educational groups in which I participated, compassing philosophers, historians of medicine, theologians, and others committed to bringing medical schools to establish teaching positions in the medical humanities (Pellegrino and McElhinney 1982). Many current programs in bioethics and the medical humanities came into existence under the influence of zealous supporters of the cause of establishing the medical humanities in medical education. There was on the part of some the view that the medical humanities were the proper framing intellectual context within which bioethics was to be understood. In particular, Edmund Pellegrino, whose writings have proved widely influential (Engelhardt and Jotterand 2008), wished to lodge bioethics within the broader context of the philosophy of medicine, and the latter within the medical humanities. Although bioethics is surely one of the medical humanities and even though bioethics can be understood as part of the philosophy of medicine, bioethics has continued to flourish *sui generis*. Bioethics has drawn on a wide range of cultural resources to form a moral and ideological movement of its own.

Bioethics proved within the first decade and a half of its existence to have created a robust interaction between a genre of academic scholarship and the provision of a clinical service. On the one hand, bioethics became established in the Academy as a recognizable scholarly endeavor productive of a literature appearing through well-regarded publishers and peer-reviewed journals. On the other hand, bioethics produced service personnel able to provide targeted legal and policy advice in hospitals and on Institutional Review Boards (IRBs), mediate disputes among physicians, nurses, families, and patients, support risk management, and clarify the nature of moral and legal puzzles bearing on the care of patients, all the while nesting these services within the authority of bioethics as an academic, intellectually recognized field. In the face of salient and intractable moral pluralism, bioethics as a clinical practice has continued to function as if there were a common morality, as if there were a consensus regarding moral issues. Such a false assumption supports the interests of bioethicists by conveying the impression that bioethicists are experts regarding which morality should guide concrete moral choices in particular circumstances. However, in the face of intractable moral pluralism, the aspiration to identify a common morality and a global bioethics fails (Engelhardt 2006). Such does not exist. This false consciousness nevertheless serves policy-makers interested in conveying a moral authority to their enactments, as well as supporting the interests of bioethicists in lending moral authority to their opinions.

Beyond such self-serving claims that bioethicists often make for themselves, there remain underlying cultural roots of the genesis of bioethics. In particular, bioethics as a moral and political movement is the secular expression of the immanent displacement of what had been transcendent concerns. Christians, especially Roman Catholics, transformed what would have been expressed in a religious and theological zeal into the moral force for a social and political movement. For the Roman Catholics who played a cardinal role in the development of bioethics, it was initially their rejection of such Roman Catholic doctrines as the immorality of contraception that turned them into dissenters who then re-directed their interests, including their interests in social justice and in Enlightenment views of human autonomy and human dignity, to bioethical issues generally. Other points of dissent then became of interest and a secular movement was formed. Their "heretical dissent" led to the fashioning of a moral and political movement at a crucial historical juncture when the endeavor could fill a moral vacuum produced by the deligitimization of traditional authority figures and the collapse of traditional moral constraints. To find direction, appeal was made to consensus, which appeal was to an effective sociopolitical coalition advanced under a secular form of a theological appeal to a *consensus fidei*. This complex understanding took root and developed under the rubric of bioethics. In the end, bioethics drew to itself much of the interest that had existed in the medical humanities.

At the beginning, it could perhaps have turned out otherwise. In the early years, there was resistance to bioethics in many quarters. Some in the Academy were doubtful of bioethics' intellectual depth. Others, especially in medical schools, were skeptical of its staying power, holding that bioethics was likely a fad whose time would soon pass. However, prominent physicians failed to defend the integrity of medical ethics over against the expansionist claims of bioethics. Physicians did not defend the claim that the clinical experience of physicians conveys a moral knowledge of its own. Traditionalists within Roman Catholicism failed to maintain the medical-moral manualist tradition. With the cultural field generally deserted, bioethicists quickly claimed the high ground—the Roman Catholic dissenters had become secular intellectuals, gathering others to their cause, thus creating a new academic and clinical field. Bioethics grew and took root both in humanities faculties and in schools of medicine. Eventually, this success led to the establishment of independent doctoral programs, as well as to the recognition of bioethics as appropriately incorporated as a part of graduate education in philosophy. Bioethics came to be established in various clinical and policy settings, and eventually hospital ethics committees (HECs) were established. With time, clinical ethicists became a regular feature of large hospitals. What had been the brainchild of Sargent Shriver and André Hellegers grew into an internationally accepted fusion of academic and clinical fields, which still maintains some of its initial character as a moral and political movement.

5.6 Bioethics: From Religious Inspiration to Secular Reality

Within two decades of the emergence of bioethics, its socio-historical roots in the religious and cultural turmoil of the 1960s and 1970s were obscured by the general recognition of bioethics as an academic and clinical field. The first generation, who were largely Roman Catholics and Protestants with significant personal academic histories rooted in theology and ministerial studies, were being replaced or at least joined by a second generation often educated in philosophy programs. The Ministers in Medical Education, an association within the Society for Health and Human Values, was no longer present, as the American Society for Bioethics and the Humanities entered the twenty-first century, which in 1998 came into existence as an amalgam of the American Association of Bioethics, the Society for Bioethics Consultation, and the Society for Health and Human Values. The religious dissent of many of the founders, especially of those who were Roman Catholic dissenters and prominent advocates of a more liberal Roman Catholic theology, had become part of a liberal socio-political movement that generally supported the policy changes imposed on health care in the United States in 2010 by the Obama regime. What had initially been a moral-theological movement within Roman Catholicism led by Francesc Abel, Charles Curran, André Hellegers, Sargent Shriver, and Warren Reich became a complex socio-political movement with substantial academic and clinical roots.

From the National Commission for the Protection of Human Subjects of Biomedical and Behavioral Research to the Commission for the Study of Bioethical Issues of the Obama administration, bioethical reflections aspired to influencing law and public policy. Bioethics was not simply an academic movement or a clinical endeavor, but also a political movement. As already noted, in the light of bioethics as biopolitics, one can understand the ubiquitous appeals to consensus (i.e., why would any amount of agreement carry moral weight in the face of moral pluralism, and in the face of unclarity as to who should count as moral or bioethical experts?) as in fact an appeal to the existence of an effective political majority or effective coalition. The political character of bioethics was underscored by the threat of demonstrations at the American Society of Bioethics and the Humanities against Leon Kass, the then-chairman of the President's Council on Bioethics, when he was invited to visit the annual meeting in 2005. The more conservative commitments of George W. Bush's bioethics experts offended many in the movement's more liberal mainstream. Different perspectives sought to establish their own bioethics and thus influence biopolitics.

The partisan character of bioethics and the thin borders between scholarly reflections and ideological commitments would not have surprised André Hellegers. He knew of the connection between bioethics and politics through his dealings with the Kennedy family who supported the nascent Kennedy Institute. He also knew of the connection between scholarly reflections and Vatican quasi-political maneuvering. Bioethics had come to compass all of this in an effective socio-political movement with intellectual and academic trappings. What had initially been a quasi-heretical movement of theological dissidents in Roman Catholicism had become mainstream biopolitics. Bioethics had achieved a culturally, academically, and politically recognized existence as a complex hybrid of moral and social concerns. Bioethics had proved able to sustain itself as an academic field and a collage of clinical practices that was expressed as a quasi-political movement aimed at refashioning law and public policy.

Notes

1. There are a number of accounts of the emergence of bioethics. These include Abel (2001), Jonsen (1998), Rothman (1991), and Stevens (2000).
2. The term 'bioethics' appears first to have been coined by Fritz Jahr (1927) and then again independently employed by Van Rensselaer Potter (1949). Potter's use in the 1970s (Potter 1970, 1971) is frequently and mistakenly taken to be the first use of the term. Either André Hellegers (Reich 1999) or Sargent Shriver (personal letter to H. Tristram Engelhardt, Jr., 26 January 2001) independently recoined or re-engaged the term as it is now used.
3. The medical profession was recast from a quasi-guild into a trade through a number of United States Supreme Court holdings. See, for example, *The United States of America, Appellants, v. The American Medical Association, A Corporation; The Medical Society of the District of Columbia, A Corporation et al.*, 317 U.S. 519 (1943); and *American Medical Assoc. v. Federal Trade Comm'n*, 638 F.2d 443 (2d Cir. 1980). See also Krause (1996) and Starr (1982).
4. The first Amendment to the compact styled the Constitution of the United States did not prohibit the establishment of a religion at non-federal levels. This Amendment by itself only

prohibits the establishment of a national, not a state, county, or city religion. "Congress shall make no law respecting an establishment of religion, or prohibiting the free exercise thereof; or abridging the freedom of speech, or of the press; or the right of the people peaceably to assemble, and to petition the government for a redress of grievances." For court holdings recognizing the predominantly Christian and more broadly acknowledging the religious character of the American people, see, for example, *Church of the Holy Trinity v. United States*, 143 US 457 (1892) at 470; *United States v. Macintosh*, 283 US 605 (1931) at 625; *Tessim Zorach v. Andrew G. Clauson et al.*, 343 US 306, 96 L ed 954, 72 S Ct 679 (1952) at 313. The ethos of America was Christian, indeed, predominantly Protestant Christian. "Evidence that Protestant Christianity [was] the functional common religion of [American] society would overwhelm us if we sought it out" (Wilson 1986, p. 113). In fact, Huntington advances the view that the American ethos is rooted in Protestant Christianity (Huntington 1994). It was only in the twentieth century that America's dominant culture was secularized as the First Amendment was imposed on the states. See, for example, *Roy R. Torcaso v. Clayton K. Watkins*, 367 US 488, 6 L ed 2d 982, 81 S Ct 1680 (1961). These court cases were followed by Supreme Court rulings which established a secular bioethics, as with its ruling on abortion (*Roe v. Wade*, 410 US 113 [1973]), which developed against the background of other holdings, particularly regarding access to contraception. See *Griswold v. Connecticut*, 381 US 479, 85 S Ct 1678, 14 L Ed 2d 510 (1965) and *Eisenstadt v. Baird*, 405 US 438, 92 S Ct 1029, 31 L Ed 2d 349 (1972).

5. *Canterbury v. Spence*, 464 F. 2d 772, 789 (D.C. Cir. 1972); *Cobbs v. Grant*, 8 Cal. 3.d 229, 246; 502 P.2d 1, 12; 104 Cal. Rptr. 505, 516 (Calif. 1972); and *Sard v. Hardy*, 397 A. 2d 1014, 1020 (Md. 1977).
6. The founding of the National Commission for the Protection of Human Subjects of Biomedical and Behavioral Research was driven in part by the issue of abortion. Although the U.S. Supreme Court case had established a constitutional prohibition of any legislation that categorically forbade abortion, it was still possible to prohibit research on human fetal tissue. As a consequence, the Commission first addressed research on fetuses (National Commission 1975).
7. Roman Catholicism as a separate denomination was engendered by a number of decisive political and theological developments that occurred from the eighth to the thirteenth centuries. Many of these changes were tied to the influence of the Franks on the Western Roman empire. A crucial step occurred through the donation in A.D. 756 to the papacy of the territory of the papal states, leading the papacy to its first step in assimilating kingship to priesthood. Then in 800 through Charles the Great's being crowned emperor by Pope Leo III, Christendom was politically divided and the office of Western emperor was set in tension to the growing claims of the pope, who was by then not just patriarch but also a temporal sovereign.
A new Christianity had taken shape not just with its own ecclesiology (i.e., universal papal jurisdiction and subsequently papal infallibility) and its own dogmas (e.g., the *filioque*, purgatory, indulgences, no second marriage after divorce due to sexual impurity, and the immaculate conception), but its own understanding of theology, namely, as a scholarly academic discipline rather than as the fruit of holiness. The political decisions of King Peppin of the Franks to create the papal states, as well as of Charles, king of the Franks, to become emperor, established a new political environment in which this new view of theology could take root. Between the Council of Frankfurt (794) with its affirmation of the *filioque* and Hildebrand's (pope Gregory VII, 1073–1085) assertion of universal spiritual and temporal papal authority and his enforcement of priestly celibacy, a Western Christianity had emerged that was quite different from that of the Roman Orthodoxy that had once united the territory of the Roman Empire, and that persisted in the East. Crucial for the subsequent history of the West and its dominant bioethics was the circumstance that this new religion embraced a theology that was no longer realized through a life of holiness but rather in academic study for which moral philosophy played a central role. Theology in the primary sense was no longer achieved through an ascetic life crowned by illumination from God. Theology was instead ever more pursued as a scholarly academic practice guided by the changing norms of philosophy and coming to possess an ever more independent life in the new universities of the West. By the fourteenth century, this change in theology was condemned by the Ninth Ecumenical Council, the fifth council held in Constantinople (1341, 1347, 1351).

8. André Hellegers held forth so often in the Tombs of the restaurant 1789 that after his death a placard was erected on the wall to mark the site where he sat at table and drank.
9. Charles Curran among others generally reduced Roman Catholic moral theology to moral philosophy, with moral theology differing mainly in terms of which scholarly work was cited.

    Obviously a personal acknowledgement of Jesus as Lord affects at least the consciousness of the individual and his thematic reflection on his consciousness, but the Christian and the explicitly non-Christian can and do arrive at the same ethical conclusions and can and do share the same general ethical attitudes, dispositions and goals. Thus, explicit Christians do not have a monopoly on such proximate ethical attitudes, goals and dispositions as self-sacrificing love, freedom, hope, concern for the neighbor in need or even the realization that one finds his life only in losing it. The explicitly Christian consciousness does affect the judgment of the Christian and the way in which he makes his ethical judgments, but non-Christians can and do arrive at the same ethical conclusions and also embrace and treasure even the loftiest of proximate motives, virtues, and goals which Christians in the past have wrongly claimed only for themselves. This is the precise sense in which I deny the existence of a distinctively Christian ethic; namely, non-Christians can and do arrive at the same ethical conclusions and prize the same proximate dispositions, goals and attitudes as Christians (Curran 1976, p 20).

10. As one entered the mid-twentieth century, there already existed a considerable Roman Catholic medical-moral-theological literature. In the United States, this literature was in continuity with the manualist tradition. It was also strengthened through the creation of directives to American Roman Catholic hospitals, which were published as a code by the Catholic Health Association. The Catholic Health Association had developed out of a movement initiated in June, 1915, led by Charles B. Moulinier, S.J. Rev. Michael P. Bourke was later responsible for the creation of a code of ethics for the Diocese of Detroit in 1920. In 1921 a surgical code was adopted by the Catholic Hospital Association. Finally, a document with the title "Code of Ethics – 1948" was drafted and first published in 1949. An account of this history is provided by Griese (1987, pp. 1–19). A literature then developed exploring the implications of this code and the issues it addressed. See, for example, Kelly (1958). This work was initially published as pamphlets: Kelly (1949, 1950, 1951, 1953, 1954). See also *Ethical* (1971). The substance and general character of the reflections on the code were a part of a considerable and often quite sophisticated moral-theological literature, which as already noted was set within the over-300-years Roman Catholic tradition of producing manuals to guide moral decisions. This literature included: Bonnar (1944), Bouscaren (1933), Capellmann (1882), Coppens (1897), Ficarra (1951), Finney (1922), Flood (1953–1954), Hayes et al. (1964), Healy (1956), Kenny (1952), La Rochelle and Fink (1944), McFadden (1946a, b), O'Donnell (1956) and Sanford (1905). As mentioned, after Vatican II this literature fell into desuetude.
11. Western Christendom through various catastrophic events fell into ruins. The great Western schism in the late fourteenth and early fifteenth centuries with popes in Rome, Avignon, and Pisa contributed to the decline. There was even a Pope John XXIII (1410–1415), rendering for some the John XXIII of the twentieth century the so-called Pope John XXIII. However, among the most significant events was the Protestant Reformation, which led to the bloodshed and destruction of the Thirty Years' War (1618–1648) and the Civil War (1642–1649). The latter broke the power of the Western Christian Empire, undermining Christendom as a regnant moral and political vision.

    The Treaties of Westphalia finally sealed the relinquishment by statesmen of a noble and ancient concept, a concept which had dominated the Middle Ages: that there existed among the baptized people of Europe a bond stronger than all their motives for wrangling—a spiritual bond, the concept of Christendom. Since the fourteenth century, and especially during the fifteenth, this concept has been steadily disintegrating.... The Thirty Years' War proved beyond a shadow of a doubt that the last states to defend the idea of a united Christian Europe were invoking that principle while in fact they aimed

> at maintaining or imposing their own supremacy. It was at Münster and Osnabrück that Christendom was buried. The tragedy was that nothing could replace it; and twentieth-century Europe is still bleeding in consequence (Daniel-Rops 1965, vol. 1, pp. 200–201).

With the French Revolution and the formal end of the empire on August 6, 1806, a new vision of morality and polity became fully dominant.

12. Commitments to social justice have proved especially harmful to the developing world. See, for example, Aiyar (2009).

References

Abel i Fabre, F. 2001. *Bioética: Orígenes, presente y futuro*. Madrid: MAPFRE.
Aiyar, S.S.A. 2009. Socialism kills: The human cost of delayed economic reform in India. *Center for Global Liberty and Prosperity* 4(October 21): 1–12.
Arnold, M. 1882. *Literature and science*. Cambridge: Cambridge University Press.
Bonnar, A. 1944. *The catholic doctor*. London: Burns Oates & Washbourne.
Bouscaren, T.L. 1933. *Ethics of ectopic operations*. Chicago: Loyola University Press.
Bradshaw, D. 2007. *Aristotle east and west*. New York: Cambridge University Press.
Capellmann, C.F.N. 1882. *Pastoral medicine*. Trans. William Dassel. New York: F. Pustet, (orig. 1877).
Catholic Hospital Association. 1971. *Ethical and religious directives for Catholic health facilities*. St. Louis: Catholic Hospital Association.
Coppens, C. 1897. *Moral principles and medical practice*, 3rd ed. New York: Benziger Brothers.
Council, Medical Research. 1977. *Guidelines for the handling of recombinant DNA molecules and animal viruses and cells*. Ottawa: Medical Research Council.
Curran, C. 1976. *Catholic moral theology in dialogue*. Notre Dame: University of Notre Dame Press.
Daniel-Rops, H. 1965. *The Church in the seventeenth century (Le grand siècle des âmes* [1963]), 2 vols. Trans. J.J. Buckingham. Garden City: Doubleday.
Engelhardt Jr., H.T. 1973. *Mind-body: A categorial relation*. The Hague: Martinus Nijhoff.
Engelhardt Jr., H.T. 1975. The history and philosophy of medicine: A report on a postgraduate seminar on the humanities in medicine. *Clio Medica* 19(September): 57–63.
Engelhardt Jr., H.T. 1986. *The foundations of bioethics*. New York: Oxford University Press.
Engelhardt Jr., H.T. 2000. *The foundations of Christian bioethics*. Salem: Scrivener Publishing.
Engelhardt Jr., H.T. 2002. The ordination of bioethicists as secular moral experts. *Social Philosophy and Policy* 19: 59–82.
Engelhardt Jr., H.T. 2003. The foundations of bioethics: Rethinking the meaning of morality. In *The story of bioethics*, ed. J. Walter and E. Klein, 91–109. Washington, DC: Georgetown University Press.
Engelhardt Jr., H.T. (ed.). 2006. *Global bioethics: The collapse of consensus*. Salem: Scrivener Publishing.
Engelhardt Jr., H.T., and A. Caplan (eds.). 1987. *Scientific controversies: A study in the resolution and closure of disputes concerning science and technology*. New York: Cambridge University Press.
Engelhardt Jr., H.T., and F. Jotterand (eds.). 2008. *The philosophy of medicine reborn: A Pellegrino reader*. Notre Dame: University of Notre Dame Press.
Ficarra, B.J. 1951. *Newer ethical problems in medicine and surgery*. Westminster: Newman Press.
Finney, P.A. 1922. *Moral problems in hospital practice*, 2nd ed. St. Louis: B. Herder.
Flexner, A. 1910. *Medical education in the United States and Canada*. A report to the Carnegie Foundation for the Advancement of Teaching, Bulletin No. 4. New York: Carnegie Foundation.
Flexner, A. 1928. *The burden of humanism*. Oxford: Clarendon.

Flood, P. 1953–1954. *New problems in medical ethics*, 2 vols. Trans. Malachy Gerald Carroll. Westminster: Newman Press.
Griese, O. 1987. *Catholic identity in health care: Principles and practice*. Braintree: Pope John Center.
Harvey, J.C. 2013. André Hellegers, the Kennedy Institute, and the development of bioethics: The American–European connection. In *The development of bioethics in the United States*, ed. J.R. Garrett, F. Jotterand, and D.C. Ralston, 37–54. Dordrecht: Springer.
Hayek, F. 1976. *Law, legislation and liberty, vol. 2: The mirage of social justice*. Chicago: University of Chicago Press.
Hayes, E., P. Hayes, and D. Kelly. 1964. *Moral principles of nursing*. New York: Macmillan.
Healy, E.F. 1956. *Medical ethics*. Chicago: Loyola University Press.
Hellegers, A. 1969. A scientist's analysis. In *Contraception: Authority and dissent*, ed. C.E. Curran, 216–239. New York: Herder and Herder.
Huntington, Samuel P. 1994. *The clash of civilizations and the remaking of world order*. New York: Simon & Schuster.
Jahr, F. 1927. Bio-Ethik: Eine Umschau über die ethischen Beziehungen des Menschen zu Tier und Pflanze. *Kosmos: Handweiser für Naturfreunde* 24: 2–4.
Jones, K.C. 2003. *Index of leading Catholic indicators*. St. Louis: Oriens Publishing.
Jonsen, A.R. 1998. *The birth of bioethics*. New York: Oxford University Press.
Kelly, G.S.J. 1949. *Medico-moral problems, Part I*. St. Louis: Catholic Hospital Association.
Kelly, G.S.J. 1950. *Medico-moral problems, Part II*. St. Louis: Catholic Hospital Association.
Kelly, G.S.J. 1951. *Medico-moral problems, Part III*. St. Louis: Catholic Hospital Association.
Kelly, G.S.J. 1953. *Medico-moral problems, Part IV*. St. Louis: Catholic Hospital Association.
Kelly, G.S.J. 1954. *Medico-moral problems, Part V*. St. Louis: Catholic Hospital Association.
Kelly, G.S.J. 1958. *Medico-moral problems*. St. Louis: Catholic Hospital Association.
Kenny, J.P. 1952. *Principles of medical ethics*. Westminster: Newman Press.
Krause, E. 1996. *Death of the guilds*. New Haven: Yale University Press.
Kuhn, T. 1962. *The structure of scientific revolutions*. Chicago: University of Chicago Press.
La Rochelle, S. A., and C. T. Fink. 1944. *Handbook of medical ethics*. Trans. M.E. Poupore. Westminster: Newman Book Shop.
Löwith, K. 1949. *Meaning in history*. Chicago: University of Chicago Press.
McClory, Robert. 1995. *Turning point: The inside story of the papal birth control commission*. New York: Crossroads.
McCormick, R. 1981. *Notes on moral theology, 1965 through 1980*. Washington, DC: University Press of America.
McFadden, C.J. 1946a. *Medical ethics for nurses*. Philadelphia: Davis.
McFadden, C.J. 1946b. *Medical ethics*. Philadelphia: Davis.
National Academy of Sciences. 1977. *Academy forum: Research with recombinant DNA*. Washington, DC: National Academy of Sciences.
National Commission for the Protection of Human Subjects of Biomedical and Behavioral Research. 1975. *Research on the fetus*. Washington, DC: H.E.W.
O'Donnell, T.J. 1956. *Morals in medicine*. Westminster: Newman Press.
Paul VI, Pope. 1968. *Humanae vitae*. New York: Pauline Books and Media.
Pellegrino, E.D., and T. McElhinney. 1982. *Teaching ethics, the humanities, and human values in medical schools*. Washington, DC: Society for Health and Human Values.
Potter, V.R. 1949. *Global bioethics – Building on Leopold's legacy*. East Lansing: Michigan State University Press.
Potter, V.R. 1970. Bioethics: The science of survival. *Perspectives in Biology and Medicine* 14: 127–173.
Potter, V.R. 1971. *Bioethics: Bridge to the future*. Englewood Cliffs: Prentice Hall.
President's Commission for the Study of Ethical Problems in Medicine and Biomedical and Behavioral Research. 1981. *Defining death*. Washington, DC: U.S. Government Printing Office.
President's Commission for the Study of Ethical Problems in Medicine and Biomedical and Behavioral Research. 1983. *Securing access to health care*. Washington, DC: U.S. Government Printing Office.

Ramsey, P. 1970. *The patient as person*. New Haven: Yale University Press.
Recombinant DNA Research. 1976. *Document relating to "NIH guidelines for research involving recombinant DNA molecules" Feb. 1975–June 1976*. Publ. no. (NIH) 76–1138. Washington, DC: H.E.W.
Recombinant DNA Research. 1978. *Document relating to "NIH guidelines for research involving recombinant DNA molecules"*. Publ. no. (NIH) 78–1139. Washington, DC: H.E.W.
Reich, W. (ed.). 1978. *Encyclopedia of bioethics*. New York: Macmillan Free Press.
Reich, W. 1999. The "wider" view: André Hellegers' passionate, integrating intellect and the creation of bioethics. *Kennedy Institute of Ethics Journal* 9: 25–51.
Reich, W. 2003. Shaping and mirroring the field: The *encyclopedia of bioethics*. In *The story of bioethics*, ed. J. Walter and E. Klein, 165–196. Washington, DC: Georgetown University Press.
Rothman, D. 1991. *Strangers at the bedside*. New York: Basic Books.
Sanford, A. 1905. *Pastoral medicine: A handbook for the Catholic clergy*. New York: Joseph Wagner.
Schutz, A., and T. Luckmann. 1973. *The structures of the life-world*. Trans. R.M. Zaner and H.T. Engelhardt Jr. Evanston: Northwestern University Press.
Snow, C.P. 1956. The two cultures. *New Statesman,* October 6, 413–414.
Snow, C.P. 1962. *The two cultures and the scientific revolution*. Cambridge: Cambridge University Press.
Snow, C.P. 1964. *The two cultures: And a second look*. Cambridge: Cambridge University Press.
Starr, P. 1982. *The social transformation of American medicine*. New York: Basic Books.
Stevens, M.L.T. 2000. *Bioethics in America: Origins and cultural politics*. Baltimore: Johns Hopkins University Press.
Vattimo, G. 2002. *After Christianity*. Trans. Luca D'Isanto. New York: Columbia University Press.
Walters, L. 2003. The birth and youth of the Kennedy Institute of Ethics. In *The story of bioethics*, ed. J. Walter and E. Klein, 215–231. Washington, DC: Georgetown University Press.
Wilson, J. 1986. Common religion in American society. In *Civil religion and political theology*, ed. L.S. Rouner, 111–124. Notre Dame: University of Notre Dame Press.

Part II
The Nature of Bioethics: Cultural and Philosophical Analysis

Chapter 6
A Corrective for Bioethical Malaise: Revisiting the Cultural Influences That Shaped the Identity of Bioethics

Warren T. Reich

This is an essay on (1) the recurrent bioethics malaise that is traceable notably to an expanding reductionism in the methods and scope of bioethics; (2) how both the malaise that nestles in the community of bioethicists and the reductionism that is its cause distort the meaning of bioethics as it has been shaped by cultural factors; and (3) how a corrective might be achieved by taking seriously a consideration of three sorts of cultural forces that shaped bioethics from its origins.

The tendency towards the dual reductionism to which I am alluding deprives bioethics of the enrichment it could still derive from the cultural forces that have had a profound influence on ways of perceiving the relationship of science and technology to society, and on the humanistic, moral, and ethical dimensions of the life sciences and health care. Those three cultural forces are: (1) the force of cultural conflict as analyzed by some intellectuals who launched a worldwide debate on "the two cultures" of science and humanities; (2) the culture of the 1960s that precipitated the rapid coalescence of the elements ingredient in a new and challenging area of inquiry known as bioethics; and (3) the expansive culture of Western intellectual inquiry since the Renaissance, which accounts for the diverse approaches to bioethics.

6.1 The Malaise of Bioethics; The Reductionism That Feeds It

Something is being lost in the field of bioethics. The unique interest and sense of discovery that characterized its early years – precipitated by a new vision of the life sciences, their importance for human health and well-being, and our discovery of the need to create a dynamic interface between the life sciences and the humanities

W.T. Reich, S.T.D. (✉)
Department of Theology, Georgetown University,
120 New North, Box 571135, Washington, DC 20057, USA
e-mail: wreich01@georgetown.edu

J.R. Garrett et al. (eds.), *The Development of Bioethics in the United States*, Philosophy and Medicine 115, DOI 10.1007/978-94-007-4011-2_6,

(Reich 1999) – have been gradually declining during recent decades. Sensational, innovative hi-tech issues such as in vitro fertilization, mind control, artificial organs, and abusive biomedical research that had been approved by the medical community and governmental agencies deservedly attracted much attention from scholars and journalists from the start. Those issues jump-started the field very rapidly, and assured the allocation of extensive funding for the early research and educational programs in bioethics, in hopes of resolving public outcry and moral debate as well as apparent threats to established values.

Understandably – but unfortunately – under these pressures the early stages of bioethics tended to veer too strongly in the direction of policy-making bioethics, with the result that a number of influential scholars tended to restrict bioethics to – or at least over-emphasize – that sort of issue (Reich 1990). These developments had a big impact on younger scholars, who felt that those influential bioethics leaders were setting an example for the way they should proceed. A significantly influential number of publications, bioethical institutes, and bioethics educational programs were modeled excessively after this quest for socially applicable regulations, even in the clinical setting, with the result that major dimensions of bioethics were placed in the shadows (Evans 2002).

Among the dimensions that tended to be relegated to the margins, especially in the first two decades of bioethics history, were the pressing quotidian problems that, in the long run, are often more important than the hi-tech issues; ethical issues regarding the health professions beyond physicians (nursing, medical social work, etc.); religious bioethics that had lost its voice by being excluded from public bioethics debates; ethical methodologies that took seriously human experience (e.g., suffering and compassion); the role of images that tend to shape moral behaviors (e.g., images of the mentally deficient child, or of the physician); the role of narrative (literary, historical, etc.) in constructing an ethic for issues in health and the life sciences; and bioethical issues and methodologies advanced by scholars in other countries (Reich 1997).

One effect of this mainstream vs. marginal split has been that a number of the scholars who devoted themselves to the issues and methods that had been relegated to the fringes of mainstream bioethics became, for all intents and purposes, bioethics Outsiders (Swinton 2004).[1]

Had more effort been devoted to achieving a better balance of topics, scope, and method in the field – a goal that was probably unrealistic at the time, considering the many forces that were militating against it – the field might have achieved not only more energy and long-term balance, but less ennui.

Initially, I was a bit startled to hear, over the past several decades, competent and even leading bioethicists complaining that bioethics had become *boring*. I realized, however, that this ennui had also affected me. While spending a year at the National Humanities Center in 1982–1983, I realized that the routine, minimalist methods that had come to dominate the field not only repelled me, but produced a sort of malaise – a negative intellectual, moral, and emotional experience that pushed me in new intellectual directions. Some 12 years later, I was more convinced of the seriousness of the problem when my long-time colleague Richard A. McCormick, S.J., analyzed this phenomenon.

By 1994 McCormick felt compelled to write about "The Malaise in Bioethics," which he claimed had been influencing bioethics for 20 years – i.e., since only about 3 years after the word "bioethics" was coined, and less than 10 years since sporadic bioethical investigations had begun in the 1960s! Here is McCormick's analysis of the problem:

> It is admitted by a significant number of scholars that there is a pervasive dissatisfaction with the status quo of bioethical reflection in the United States. A sense of malaise is unmistakably present and hovers over the subject like a dark cloud. Most of the dissatisfaction points to the "principles approach as the regnant paradigm." It is said to be a dilemma-oriented, problem-solving, deductive, rationalistic, individualistic and rights-focused enterprise. By contrast there are repeated calls for less reliance on principlism, proximity to the lived-in existence, inclusion of the imagistic, symbolic, emotional dimensions, less absolutism, greater emphasis on the self as related to the other.... I agree that there is malaise. So much of what is written seems tired and predictable, only marginally relevant (McCormick 1994, p. 149).

It is noteworthy that McCormick then argues, not for a renewed culture of *bioethics*, but for a renewed culture of the *professions*. I acknowledge that malaise in the health professions, which has been well documented, is a serious problem, and that it also affects bioethics negatively (Wear and Kuczewski 2004; Famer and Campo 2004). But the approach that I am taking is that in order to correct the malaise of bioethics, which is experienced and acknowledged by bioethicists, the needed response is to examine and articulate for a new era the meaning of bioethics, to understand the cultural factors that have influenced it and continue to influence it, and to take steps to renew the underpinnings, relationships, and activities of bioethics itself. This chapter is but a very modest and incomplete attempt to move in the direction of that process by analyzing some causes of this malaise and to propose a mitigation of that problem by connecting bioethics with the cultural factors that have shaped bioethics and either helped or hindered its progress.

I am convinced that a major underlying problem causing this malaise has been the tendency towards reductionism in defining the scope and determining the methods of bioethics.[2] What are those reductionist developments?

We have witnessed a reduction of the moral issues judged relevant for bioethical inquiry; the reduction of bioethics to an exercise in applied moral philosophy, which in turn assured a substantial neglect of interdisciplinarity; a reduction of methodology to a narrow, rationalist system of moral philosophy that excludes cultural influences and methods embracing sentiment and imagination; more generally, the reduction of bioethics to a freestanding discipline with few or no connections to its surrounding culture; and especially, the reduction of bioethics to medical ethics and, finally, at least in large part, to physician ethics.[3] The narrower the focus, the method, and the intellectual breadth, the more one risks converting ethics to a mechanical routine, thus producing boredom and malaise. This reductionism (of topic, method, culture, and bioethics itself) and the resulting boredom have undoubtedly been factors in causing what seems to be a reduced interest in what excited earlier scholars in bioethics and gave vitality to the field: a large range of interlocked, interdisciplinary problems that share roots in the life sciences and humanities and that invited a dialogical system of ethics that kept this interdisciplinarity alive. Some of the expansive, creative inquiry into the larger issues seems to have diminished, at least in some quarters, due to devoting more and more attention to the more narrowly defined issues that do, indeed, plead for clarification.

6.2 What Was/Is Bioethics? The Gradual Reductionism of Bioethics

I now turn my attention to the reductionism of and in bioethics. My views on this crucial question are, of course, influenced by my own history and experience in the field. My experience, in addition to establishing and directing an educational program in bioethics and medical humanities in a medical faculty and university hospital for many years, has largely been devoted to the planning and editing of two editions (nine volumes) of an encyclopedia for the field of bioethics. I believe that project was the first time anyone had ever drawn up a comprehensive, detailed plan of, rationale for, and definition of, the entire field of bioethics (which, in 1971, had not yet developed).

Forty years ago, I spent the best part of a year sketching out the parameters, contents, and methods for the field we were calling bioethics, even though the term bioethics had been used for the first time in the title of a book – at least in the English language – only a few months earlier. The author, Van Rensselaer Potter, advocated the creation of a field of bioethics as a bridge to the future (Potter 1971; Reich 1994, 1995a). I was sorting out a complex web of topics, issues, methods, events, ideas and concepts, history, purposes and goals, disciplines, values, methods for deriving normative standards, and the meaning, scope, and kinds of methodologies that we might gather into the future field of bioethics. It was an exciting task I had set for myself[4] – the task of determining what bioethics should become – on the basis of the elements that had already emerged, those that were just beyond the horizon, and the old and new cultural developments that were shaping them. What cluster of issues was bioethics beginning to confront, and what types of issues would it be likely to address? To which disciplines and areas of knowledge should bioethics create links in order to address those issues properly? What sort of moral methodologies already existed in all the major cultures of the world? How have they responded to moral issues in health and life, and how would they be likely to respond to the newer problems of the life sciences? Briefly put: what intellectual resources would bioethics require, and to which publics would it be likely to address itself (Reich 2003a)?

Before selecting and presenting my plan to the Associate Editors of the *Encyclopedia*, who would critique my plan and reshape the portion of the encyclopedia they would be responsible for, I had had the benefit of advice from scholars from many fields of learning, from resident and visiting scholars at Georgetown University's then-new Center for Bioethics (now called the Kennedy Institute of Ethics), as well as from encyclopedists.[5] I read the available literature (quite sparse), and then attempted to reach into all the corners of the life sciences and the biomedical and ethical worlds to expose all the relevant problems and issues of all the life sciences that invited interaction with the humanities as well as careful ethical scrutiny, and to identify, as well as I could, which topics and issues would require cultural, historical, scientific, philosophical, religious, theological, social, psychological, ethical and/or legal underpinnings.[6] I developed 27 "editions" of the plan for the *Encyclopedia*. Only then did I define the field of bioethics.

I concluded that bioethics is the ethics of the life sciences, with special (though not exclusive) attention to human health. (It would embrace the health and well-being of non-human animals; public health; environmental issues ingredient in a comprehensive bioethics; etc.) My definition of bioethics, in the first edition of the *Encyclopedia of Bioethics*, was as follows: bioethics "is the systematic study of human conduct in the area of *the life sciences and health care*, insofar as this conduct is examined in the light of moral values and principles" (Reich 1978, p. xix, italics in original). I used the word "principles" in its traditional meaning as *any* of the *sources* of bioethics – ideas, scientific paradigms and data, moral images, values, virtues, norms, laws, etc. – not in the sense in which "principlists" subsequently used the term, although their usage became the dominant and still-current meaning of "principle" in the context of ethics in the United States. In the second edition, I changed the formulation of the definition, but retained its essential ingredients: bioethics "can be defined as *the systematic study of the moral dimensions* – including moral vision, decisions, conduct, and policies – of *the life sciences and health care,* employing a variety of ethical methodologies in an interdisciplinary setting" (Reich 1995b, p. xxi, italics in original).

The future of bioethics will depend on whether bioethics is viewed as nothing more than medical ethics (which is a rather narrow field, inasmuch as it is rooted in the experience, responsibilities, and ethical rules of physicians), or whether it will be pursued as "the ethics of the life sciences." Furthermore, the future of bioethics malaise depends, according to the analysis offered above, on whether bioethics is subjected to ongoing reductionism.

As noted above, currently there is pressure to reduce bioethics to medical ethics. Consider, for example, statements made recently by Albert Jonsen that have major implications for the question of reductionism in bioethics. His comments are highly significant for several reasons: the context in which they appear and the deservedly high regard in which Jonsen's work in the history of bioethics is held (Jonsen 1998). The Jonsen article that I cite is not just another article on the history of bioethics. It appears in a reference work that has a very distinctive purpose, *The Cambridge World History of Medical Ethics*, edited by Robert Baker and Laurence McCullough (2009) – the first-ever comprehensive, global history of medical ethics, one of whose major purposes is to probe and clarify the historiography of medical ethics. Articles on bioethics in this work have to be read in the context of the purpose of this work, including consideration of how bioethics relates to medical ethics (and the historiography thereof).[7] Consequently, it is important to assess Jonsen's analysis of bioethics in reference to medical ethics, i.e., against the background of the historiography of medical ethics. The introduction to Jonsen's article reveals, implicitly, a struggle between medical ethics and the ethics of the life sciences, in the following way. First, he describes Van Rensselaer Potter's idea of bioethics with words that are quite impenetrable, considering the practical message Potter was conveying. Jonsen's complete summary of Potter's notion of bioethics is the following: he says Potter proposed a "vision of a new conjunction of scientific knowledge and moral appreciation of the converging evolutionary understanding of humans in nature" (Jonsen 2009, p. 477). What is missing from Jonsen's account is the clarity *and passion* with which Potter was explicitly writing, regarding a very urgent question: *survival*, a health (and medical) problem of great import. He was an oncology scientist who saw cancer being spread by environmental factors, with extensive lethal potential not only for individual patients, but for entire populations and for civilization itself (Carson 1962). He regarded it as a global health problem that required a global ethic. He was a life scientist, who knew – as did a coterie of life scientists in the 1960s, who functioned as modern, scientifically-informed moral prophets – that ethical problems of public health, personal health, and the life sciences encountered by his specialization demanded interdisciplinary solutions and an overriding ethic that would provide the vision those solutions required (Potter 1971). His was an important message for his era, on the cusp between the 1960s and the 1970s. It is not entirely clear why Potter was not taken more seriously by moral philosophers in his own country, but I believe there are at least three reasons for the neglect: because Potter spoke as prophet more than as philosopher (see the analysis of modes of discourse below, where I discuss three cultures of the Western world); because Potter was not particularly astute in philosophy; and because he lacked resources for interdisciplinary research (a factor that he acknowledged). So in his arena of cancer research – the global causes and effects of cancer, where so many major problems are encountered, and often suppressed by, great social and political powers – Potter's real ethical message was effectively buried, partly through the neglect of those of us who were his peers. Perhaps U.S. bioethicists were not prepared to take seriously the ethics of health in the broad ethical and scientific context of an environment that makes people terribly ill and kills them. Alternatively, perhaps many U.S. bioethicists simply wanted to restrict their ethical interests to clinical issues and publicly controversial issues of biomedical technologies. I think those two were the alternatives that Potter was trying to teach us. I think he was trying to shake us into acknowledging that the health problems, the scientific problems, and the ethical problems were bigger than we thought. Now we can say: too bad he was not more effective. One wonders whether his vision will remain on the margins – a vision that regarded bioethics as comprising issues in the life sciences, understood ethically in the context of the culture of the life sciences (Reich 1994, 1995a, 1996).[8]

Jonsen misrepresents the definition of bioethics that I provided in the first edition of *The Encyclopedia of Bioethics*, and in so doing he misrepresents the entire approach to bioethics that permeates the "Introduction" and accounts for the plan of the entire four-volume work. The definition that he attributes to me is this: "Warren Reich … defined bioethics [in the *Encyclopedia of Bioethics*] as 'the study of the ethical dimensions of medicine and the biological sciences.'" He cites pages xix–xx, but I cannot find that wording on those pages. What is objectionable about that definition is that he places "the ethical dimensions of medicine" in first place (bioethics is "the study of the ethical dimensions of medicine....") and makes no mention of the main theme that I emphasize over and over again in the "Introduction": that bioethics is the ethics of the life sciences (which includes medical ethics, but also many other special areas of the ethics of the life sciences). Medical ethics dominates Jonsen's definition of bioethics, which he attributes to me; but medical ethics decidedly does not dominate the *Encyclopedia*'s definition. The *Encyclopedia*'s definition of bioethics is found on the page (xix) mentioned by Jonsen, and it is featured very prominently: "Bioethics ... can be defined as the systematic study of human conduct in the area of the life sciences and health care, insofar as this conduct is examined in the light of moral values and principles" (Reich 1978, xix). Furthermore, it is a definition that has been adopted over and over again by authors all over the world. By contrast, Jonsen's writings, including this article, take medical ethics as the starting point and major interpretive tool for bioethics. For example, in this article he takes Thomas Percival's *Medical Ethics* as his starting point for explaining the meaning of bioethics.

Jonsen offers a third example of an understanding of bioethics that I believe distorts its authors' understanding of bioethics. I am speaking of how Jonsen explains the meaning that the Hastings Center assigns to bioethics. He offers two sentences regarding the Hastings Center's mission, mentioning "medical practice" and "crucial issues that arise within medicine," without ever mentioning the ethics of the life sciences which has featured so prominently in the approach taken by the Hastings Center. Actually, the Hastings Center, since its founding in 1969 and for many years afterwards, featured the term "life sciences" very prominently in its name ("Institute of Society, Ethics and the Life Sciences: The Hastings Center"). They discontinued that longer title, but in recent personal communications both Thomas Murray (Director and CEO of the Center) and Daniel Callahan (a founder and long-time Director of the Center) assured me that they have not changed their understanding of bioethics ("Ethics and the Life Sciences"); and, in fact, they are conducting a project that does not deal with medicine at all.

6.3 Culture and Bioethics

Reflecting on a conference held in Ottawa in 1988, George Weisz, professor of social and historical studies of medicine at McGill University in Ottawa, criticized the failure of American bioethics to be attentive to cultural influences (Weisz 1990, p. 3). He commented that in 1984, Renée Fox and Judith Swazey developed a sophisticated critique of American bioethics based on its failure to recognize the social and cultural sources and implications of its own thought. They characterized as culturally specific and intellectually problematic the uncritical emphasis in American bioethics on the individual and his rights (as opposed to the web of human relationships that engender mutual obligations and interdependence), the techniques of rational abstraction that uproot issues from their concrete human reality, and the assumption that these techniques and values have universal applicability.[9]

It may be useful to consider three cultural factors that gave rise to bioethics in the 1960s and that continue to shape bioethics today: (1) the "two cultures" debates of the 1960s; (2) the origins of bioethics in the culture of the 1960s; and (3) three cultures of the Western world that shape bioethics today. Perhaps this exercise will place us in a better position to give direction to the bioethics of the future.

6.3.1 The Two Cultures and Bioethics

If one is concerned about the ways in which cultural influences of the 1960s affected the rise, scope, and trajectory of bioethics, one good place to start is with the famous lecture "The Two Cultures and the Scientific Revolution," which C.P. Snow, British scientist and novelist, delivered in 1959 (Snow 1960). His lecture appeared in altered form (in response to criticisms, it included a "third culture") in a book that Snow published in the 1960s – a book that reached a vast international readership, was frequently discussed by scholars who were pioneering the field of bioethics in 1970–1971, and has been regarded as one of the 100 most important books published since World War II (Snow 1963).

In his 1959 lecture, Snow argued that scientists and humanists (in the latter category he spoke of "literary intellectuals") were isolated from one another; and between them "he claimed to find a profound mutual suspicion and incomprehension, which in turn had damaging prospects for applying technology to the alleviation of the world's problems (Collini 1963, p. viii)." Although Snow was not a "systematic thinker," preferring instead the territory of "the Big Idea" (Collini 1963, p. viii), his essay sparked intense international interest, dialogue, and a spate of publications on bridging these two worlds. While some philosophers, theologians, literary intellectuals and scientists informally debated the existence of the two cultures in the 1960s and early 1970s, others argued that an additional detriment would be created by the chasm that Snow described: that the seedling efforts in bioethics would be ineffective without building a bridge between the two cultures. If Snow's thesis was to be taken seriously in the origins of bioethics, that bridge would have to be built not simply between ethics and medicine, but between the humanities and the life sciences.

How to build that bridge? Snow proposed a "third culture" – a group that could bridge the gap between scientists and humanists. One could say that two of the most prominent and effective "third culture" initiatives in the US, in the midst of what was then called the "biological revolution," were the early conferences on ethics and science that involved leading scientists, physicians, theologians, philosophers, and legal scholars in the 1960s and 1970s; and the highly-influential consultations held at the Hastings Institute for Society, Ethics, and the Life Sciences in Hastings-on-the-Hudson, NY.

Where do "the two cultures" stand today? Because a number of very influential publications in the first generation of bioethics emphasized the establishment of a *politically powerful ethic for* biomedicine (and for U.S. government laws, regulations, and other policies pertaining to biomedicine), the problem regarding the cultures involved in *the meeting* of science and the humanities raised by Snow (however awkwardly and not always convincingly) continues to exist as a problematic issue for bioethics.

There are, however, at least two positive, mitigating points to make regarding this contemporary problem. First, there is a cadre of remarkably effective, contemporary, "second-generation" bioethics scholars who, in their own very substantial work, bridge the gap between science and humanities in a thorough and credible way that has rarely, if ever, been seen before. Second, in the 50 years since Snow delivered his lecture, there have been significant advances in the analysis of the dissimilarities and similarities in the two cultures. One of the world's most distinguished bioethicists, Onora O'Neill, in her lecture entitled "The Two Cultures Fifty Years On," makes the observation that whereas Snow had judged the two cultures wholly different in approach and achievements, "... if we consider the approaches and methods actually used by inquiry in the humanities and in the natural sciences we find many similarities. In both domains inquiry relies on interpretation and inference, makes and seeks to support empirical truth claims and deploys and defends normative assumptions." We need to keep in mind not only these common elements, but "the diversity of work undertaken in either culture" (O'Neill 2010).

An example of the cultural dimensions of the encounter between science and humanities at the present time can be found in Japan, where scholars are struggling *against* the dominance of one outside culture and *in favor of* the *inclusion* of interests raised by their own culture.

When American bioethics was imported into Japan from the Kennedy Institute of Ethics in the 1980s, it was presented as a rights-based ethic which arose from the U.S. civil rights movements in the 1960s. This restrictive understanding of bioethics was used to protect Japanese patients' rights (Tsuchiya 1998). A group of Tokyo professors from a variety of disciplines, all of them expert in bioethics, convinced that the "Washington approach" to bioethics simply did not match the culture and ethical needs of Japan, formed a multi-year research project called the Project on Historical and Meta-Scientific Study of Bioethics, under the direction of Professor Yoshihiko Komatsu.

This research group was convinced that the original edition of the *Encyclopedia of Bioethics* (Reich 1978) had important epistemic value for the field as a whole in the twenty-first century. It should be noted that most of the members of this research project had served as scholarly translators for the first edition of the *Encyclopedia* which, more than anything else, had established bioethics as a legitimate discipline in Japan. Hence they were working in familiar territory.

A major portion of their research project was to conduct a meticulous study of the *Encyclopedia*, to determine what its intellectual structure had to say about the nature, scope, and methods of the field. When I arrived in Tokyo to deliver lectures to the research group in 2007 (Morimoto 2010) they informed me that their research had already led them to several conclusions – foremost among them that the approach to bioethics should not be restricted to that of standard applied moral philosophy. Instead, they found a useful approach to the field as a whole in the intellectual design of the *Encyclopedia of Bioethics*.

Their analysis of the *Encyclopedia* led them to the conclusion that the humanities clearly provided the framework for bioethics: the relevant ideas,[10] moral traditions, ethical methods, cultural/religious aspects, etc. Within that framework are found the ethical topics presented as broadly as possible – as problem areas – along with an appropriate range of ethical methodologies. Information and interpretation from the life sciences (not just medicine) were presented directly in the context of the humanistic/ethical analysis of a given problem area, thus enabling a dialogue between the two cultures. The Japanese scholars found this approach liberating, for it allowed them to determine which ideas, which topics in the humanities, which life science problems, which disciplines, which approaches and methodologies would present a credible resource for bioethics in the Japanese culture, with careful attention to other cultures. Their underlying concern was the role of culture in bridging the "gap" between humanities (including ethics) and the life sciences.

6.3.2 The Origins of Bioethics in the Culture of the 1960s

My own first experience of the force of the culture of the 1960s began when I went to Europe to study for a doctorate in moral theology at the Gregorian University in Rome. The most exciting intellectual experience of my life occurred in 1959–1960, when scholarly papers, offering a bold new approach to theology (anticipating the gathering of the Second Vatican Council) were circulated underground in Rome, authored by theologians such as Eduard Schillebeeckz, Yves Congar, and Hans Küng, countered by rebuttals from Vatican conservatives such as Cardinal Ottaviani. One of the Roman papers denounced the "transalpine fog" that had drifted across the Alps down to Rome. This Roman denunciation of theologians in France, Germany, Belgium, and the Netherlands – parts of which were all contained in ancient "transalpine Gaul" – were reminiscent of what Caesar said about the transalpine culture almost 2000 years earlier, in his memoirs *The Gallic Wars*: "for as the temper of the Gauls is impetuous and ready to undertake wars, so their mind is weak, and by no means resolute in enduring calamities (Caesar 2006, 50 B.C.)." I felt I was immersed in a firmly planted cultural conflict that spanned the centuries. On the theological level, I believed I was experiencing the most tumultuous intellectual upheaval in almost 400 years – i.e., since the closing of the Council of Trent (the Council of the Counter-Reformation) in 1563.

After studying in a cisalpine setting (Rome), though under the guidance of a transalpine professor, I accepted the offer of a scholarship to study in a transalpine (German) University. While at the University of Würzburg, I followed the advice of a former professor of mine, the distinguished patristics scholar Johannes Quasten (1900–1987), to visit another world-famous patrologist, Berthold Altaner (1885–1964), whose (only slightly jocular) advice was: "Now you must forget all that you were taught in Rome!" Certainly, I was moving back and forth between two cultures that had long been in conflict; and I was to continue experiencing this cultural shift for the next 50 years, starting with the cultural upheavals that my contemporaries and I were to encounter in the 1960s.

After I began teaching ethics at the graduate level in 1962, and throughout the 1960s, I felt that something unique was happening in the U.S. culture, as well as in the culture of many other countries, which was having an effect on the public ethics of our society. I knew I had to change my approach to theological ethics – first, by immersing myself in and interpreting the social issues, values, and politics, as well as the music and the literature of the era, and then determining their relevance for new insights into ethics and the teaching of ethics. I began sensing that our moral and intellectual culture was undergoing a profound change as a result of social and cultural transitions.

I am now convinced that the culture of the 1960s played a significant role in giving shape to bioethics, and that the bioethics that emerged initially in the 1960s and more decisively around 1970–1971 was not just one phase in the history of applied moral philosophy, as some scholars claim. On the contrary, bioethics, at the time that it emerged, was a new, interdisciplinary phenomenon that had unique intellectual, moral, and social characteristics.

In what way did the culture of the 1960s give rise to bioethics? The culture of the 1960s was incisively described and analyzed from a sociological perspective in a classic work of that era, Theodore Roszak's *The Making of a Counter Culture* (Roszak 1969). Roszak explained that the U.S. counterculture of the 1960s – with its opposition to the older generation's Viet Nam War and its advocacy of love rather than war as a social force for change; its abhorrence of racial injustice and the disenfranchisement of black youth; its radical discontent with the dishonest values of the older generation; its tragic loss of hope with the assassinations of President John Kennedy, his brother Senator Robert Kennedy, and Martin Luther King, Jr.; its advocacy of a new sense of community (which was not dispelled through the failure of many of the communes that were established in that era); and its proposals for a new ethos for a new society – requires attention even today for the force it exerted on basic approaches to social issues.

The rejection of the moral authority of the previous generation and the calling into question of the moral authority (or the absoluteness of the authority) of major social institutions including church and state were major characteristics of the 1960s counterculture. A good example of how this part of the 1960s ethos was portrayed is found in the film "The Graduate," directed by Mike Nichols and starring Dustin Hoffman as the young man who allows himself to be seduced by an older woman, portrayed by Anne Bancroft. The Dustin Hoffman character (Ben Braddock) is a recent college graduate; and the fact that the seductive Anne Bancroft character (Mrs. Robinson) is a close friend of his parents has the effect of flaunting the dishonest values of the older generation. The self-centered moral power taken on by the older generation is accentuated by Mrs. Robinson's rage when Ben Braddock falls in love with her daughter Elaine. The film is not just a relic from the past; it continues to be a poignant reminder of the ethos of the 1960s.

Mention of "The Graduate" points to the fact that the counterculture of the 1960s was a youth culture, which raises the question whether the youth culture of the 1960s can be generalized into a broader culture. Probably the majority of youth – and certainly the majority of the population – were not opposed to the Viet Nam War and did not espouse most of the elements of the youth culture. In fact, the alienated youth of the era were rejected, not only by hostile fellow-citizens, but even by their fathers and mothers. Yet the counterculture of which Roszak wrote became, in a sense, the defining culture of the 1960s, because what was most important in the counterculture, as in any culture – its symbols, its mood, its music, its deep sentiments, its activities that altered a society's assumptions and attitudes, and above all its captivating description of a new society in which technology, including biomedical technology, would be made subservient to, rather than master of, an inspiring social vision – assumed an incredibly influential and durable force that still has the capacity of affecting social and political outlook.

The mood of the 1960s was captured in one of the most moving songs of the era, "American Pie" by Don McLean.[11] This song, like the songs of Pete Seeger, Bob Dylan and others, has the effect of making the experience of the 1960s present today. It is a song about "the day the music died," when "something touched me deep inside." It speaks of "the generation lost in space," which has "no time left to start again." Hope seems to have disappeared: "not a word was spoken, the church bells all were broken." And yet there was a turning-point – the poignant sense of loss and outrage in the face of the violence of the 1960s was accompanied by new tears and dreams for a new future, as in these words: "In the streets the children screamed, the lovers cried and poets dreamed."

The counterculture of the 1960s was advocating an ethos in which an art of living could be developed on a new foundation – one in which men and women could face an era influenced by science and technology, but in which they could be detached from some of the more harmful characteristics of a technocratic life. It is not surprising that some of the leading philosophers of the twentieth century detected and explored precisely this sort of issue; the technocratic society.[12]

The ethic of the anticipated era would include spiritual-religious elements, but with major changes – it must rely on a broader range of religious traditions than the religions that dominated the contemporary culture. The new ethos could include normative ethical approaches rooted at least partly in familiar religious insights of authoritative religions, but it must place far greater emphasis on spiritual ideals, attitudes, and shared experiences that have the potential for being life-changing and that can help form a life in which the authentic spirituality of the individual can reach out and help create a new and better world. The ethos of the future that constituted part of the vision of the 1960s would be rational, but it would not authenticate a rationalistic ethic that looks only to classical conceptual arrangements of the past, to the neglect of the importance of ongoing human experience, including social and spiritual experience.

Bioethics has, in fact, taken major steps in these directions (Reich 2003b). But the vision of the 1960s would have called for a new vision of the relationship between secular ethics, everyday moral experience, spirituality, and religion – a relationship that is still being developed some 40 years later.

Those, then, are some of the general characteristics of the longing for a new society, the moral vision that we experienced in the 1960s. But now it is important to take a closer look at how the "technocracy" of the 1960s influenced the emerging bioethics.

In the 1960s America, like much of the rest of Western society, was experiencing an enormous force of alienation, especially alienation from society, its values, and the authorities of its traditional institutions. Roszak explains that much of that alienation was due to the technocracy that controlled so much of our society. The technocracy was made up of the social structures, values, and industrial-political priorities that constantly increased the ascendancy of technology as the savior of all human ills. This attitude was the result of the Enlightenment combined with the Scientific Revolution.

Characteristic of the technocratic force in society was the "objective consciousness" associated with it. This objective consciousness, for all practical purposes, was viewed as excluding or at least minimizing the importance of all higher ideals and values. All human problems were reduced to practical problems, and technocracy would be the savior (Marcuse 1964).

This technocratic force was felt at the intersection of science and society. The result was that when bioethics arose, the question would be whether it, too, would be the instrument of a technocratic society, or whether it would promote ideals and images of human life independent of technology. This question still remains a burning issue: what really happened in the new field of bioethics regarding the superiority of technology, and where does that issue stand today? It is disturbing for me to realize that the same question still exists today, for it can be accurately shown that much of the bioethics of today is oriented towards solving social and public policy problems with a form of rationality that is designed to fit into the presuppositions and needs of the civil society and the public organizations that fund bioethics and that anticipate receiving from bioethics the answers that will fit most suitably into the established social structures. Thus, I believe that an examination of the ethical implications of the technocracy that Roszak described in the 1960s should be a reason for once again examining the influence of technocratic culture on the role, methods, and activities of the field of bioethics. For I believe that the technological issue at the basis of modern society was not a passing phenomenon, but one that has deep roots in today's postmodern culture. This phenomenon continues to challenge us to scrutinize how the "problem-solving," policy-oriented bioethics of today really functions in the context of the values and vision that should characterize a morally-aware society.

At the very least, an examination of the socio-cultural origins of bioethics in the 1960s should make us realize that bioethics did not arise simply as a response to a set of biomedical moral problems, but as a result of an enormous moral upheaval in our society, an awareness of the power of medical technocracy over our lives, and a healthy skepticism as to whether the power-oriented religious and civil authorities could solve those problems without contributions from all available intellectual and moral resources.

6.3.3 *Three Cultures of the Western World That Shape Bioethics Today*

I now want to turn my attention to an assessment of where bioethics is today. To do that, I will again use the concept of culture, but in a different way, for now I want to suggest that those who investigate and propose bioethics, understood as an intellectual and social enterprise undertaken in the Western world, have the opportunity of using any of three major paradigms, which can be called three cultures of the Western world. John O'Malley (2004), whose masterful work on the cultures of the West I am using,[13] explains that culture means a large, self-validating configuration "of symbols, values, temperaments, patterns of thinking, feeling, and behaving, and patterns of discourse" (p. 5). Culture, he continues, deals with form as much as content.

6.3.3.1 Culture No. 1

In O'Malley's analysis, Culture No. 1 is *the prophetic culture*, a culture exemplified by the Protestant reformer Martin Luther. O'Malley says of Luther: "He made courage in fighting for systemic change perhaps the most distinctively Western of all the virtues" (pp. 9–10). The person who functions within Culture No. 1 declares the need for a major change in society, church, or state, and typically the argument will be expressed in stark terms such as the unquestionable will of God or, in a secularized world, a basic human right.

A good bioethical example of Culture No. 1 is found in an article on the ethics of clinical research published in (1966) by Dr. Henry Beecher, Professor of Research in Anaesthesia at Harvard University. In this article Beecher assumed the role of *the prophetic outsider* by presenting, in a major medical journal, 21 examples of clinical investigations in which investigators endangered "the health or the life of their subjects" without informing them of the risks or obtaining their permission. While denouncing the scientific abuses of his colleagues, including colleagues at his own university, Beecher advocated shifting the power of decision regarding the ethics of research from the physician-researcher to the patient by way of the principle of informed consent, a principle that was widely ignored in the 1960s in research universities and governmental agencies. Speaking like the prophets of old, Beecher called for the medical profession to show more respect for their vulnerable patients. He became a founder of the bioethical sub-field of research ethics.

Bioethics cannot rely *solely* on prophetic statements designed to shake up a corrupt society or profession, but sometimes we find the prophetic voice mixed with the voice of another culture. For example, the article on "Medical Ethics Under National Socialism" by Fritz Redlich (1978) in the first edition of the *Encyclopedia of Bioethics* is an historic article of great importance. Though published 30 years after the Nuremberg Trials, it was one of the very few articles published on this topic in the U.S. after World War II, prior to the 1980s. His article was also important because he was Jewish and Austrian, earned his M.D. degree at the University of Vienna while Hitler was being treated there, and emigrated to the U.S. in the late 1930s. He was a very accomplished professor of Psychiatry and Dean of Medicine at Yale University, who wrote a book on Hitler's physical and psychological infirmities (Redlich 1998). His bioethics article on medical ethics under National Socialism was remarkable because of the way he exposed the medical abuses of the Nazi era, while using a variety of historical and systematic methods.

There is continuing need for bioethicists to be prophets in the sense that they must be prepared to be countercultural, ready to challenge the moral priorities of the powerful institutions – the state, the church, the professions, and the military. This prophetic attitude of Western culture has deep roots in the prophetic religious tradition of the Jewish and Christian religions.

6.3.3.2 Culture No. 2

Culture No. 2 is *the academic/professional culture*. This is the culture of Plato and Aristotle and other thinkers of the School of Athens and the School of Rome. Platonism and Neoplatonism "infiltrated into the thought patterns of the West from antiquity into modern times to such an extent that they became ... almost indistinguishable" (O'Malley 2004, p. 10). After Aristotle "emerged with startling brilliance ... in the High Middle Ages," Plato and Aristotle stimulated a certain style of learning and discourse that was characterized by an analytical, questioning, rigorous, and relentless style (O'Malley 2004, p. 11). Employing argumentation that relied on close reasoning, this culture developed most strongly in the universities of the Middle Ages and has continued to be manifested in the universities and professions of today. "It finds its most appropriate home in the classroom, the laboratory, the library, the think tank, the research institute, the ... meetings of learned societies" (O'Malley 2004, p. 13). Although there is not a straight line of descent from its Greek and Roman origins, it is no wonder that Culture No. 2 came to be the dominant culture for the articulation of bioethics in the form of abstract principles in the United States and other countries of the Western world. It is not surprising that Culture No. 2 dominated U.S. bioethics in the first decades of the flourishing of bioethics. After theological ethics had been discarded by the bioethics community that served public discourse – in spite of the fact that (or perhaps because) it alone had been the major developer and teacher of "medical morality" over a period of at least 300 years – principle-based bioethics, developed under the influence of Culture No. 2, was ideal in many ways. It could be relied upon to quickly fill the vacancy created by the withdrawal of theological ethicists from the scene and the lack of preparation of many philosophers in the area of ethics. The principles of Culture No. 2 provided a "lingua franca" that rather easily spanned the gap between philosopher, physician, nurse, and at least some families. As criticism of the abstractness, impersonal character, and rejection of sentiment that characterized this bioethics method mounted, some philosophers started combining this method with a more satisfying method, such as virtue ethics, feminist ethics, and narrative ethics.

Under the influence of major thinkers such as Descartes, Kant, and Mill, Culture No. 2 produced a phenomenon known as contemporary Anglo-American applied moral philosophy, which existed, but was not being broadly pursued by philosophers, when bioethics was launched. In the U.S. bioethics came to be most strongly associated with Culture No. 2 when some philosophers, principally at Georgetown University, adopted just four principles as an adequate foundation for bioethics – the principles of autonomy; justice (it would have been useful if a "thick" notion of justice had been a bioethics principle – here, the initiators of the four bioethics principles were actually restricting themselves to *distributive* justice); non-maleficence; and beneficence (which, without benevolence, was excessively minimalist and rationalist). This approach is exemplified in the most widely used textbook for biomedical ethics, *Principles of Biomedical Ethics*, by Tom Beauchamp and James Childress (1979), which has been published in six editions.

The principle-based, Culture No. 2 approach to bioethics prefers abstract, universal principles like autonomy and beneficence, which it typically sets in opposition to each other. It is suspicious of the role of emotions and imagination in bioethics; and it has been strongly influenced by a few minimalist attributes of modern scientific inquiry, such as precision of reasoning and certitude of conclusions.

Some philosophers functioning within Culture No. 2 claim that bioethics is merely a continuation of applied moral philosophy, but that claim fails to see how the framework of applied ethics mistakenly assumes the adequacy of a few moral principles that are applied, with little interpretive variation, to an enormous range of moral problems in all sorts of settings, disciplines, and professions. Bioethics needs to draw on the rigors of Culture No. 2, but not be restricted to it.

6.3.3.3 Culture No. 3

Culture No. 3, *the humanistic culture*, is in stark contrast to Culture No. 2, which leads people to rely too heavily on abstract ethical principles in the context of an ethic of duty. The historic culture known as the humanistic culture moves us *to understand and interpret our experience and our moral vision*. Over 30 years ago, Daniel Callahan said that to accomplish the task of rethinking the role and methodology of bioethics, it is essential to realize that ethical analysis and prescription rarely "lead the way in social and cultural change." Instead, he explained, "the more normal role [of ethics] is to try *to interpret and structure the flux of behavior, experience, and intuition, and from there to develop general moral visions and goals* together with specific principles and rules to exemplify them." In order for ethics to carry out its task effectively, "in the first place, *it must correctly grasp and understand the realities of culture and society*…." (Callahan 1980, pp. 1228–1233, italics in original).

Culture No. 3 arose in Florence in the fourteenth, fifteenth, and sixteenth centuries, when intellectual and artistic expressions were associated with the broad range of humanity, in what today is called the humanistic approach to learning. Cicero, Virgil, St. Ambrose, and St. Augustine were trained in the skills and ideals of a tradition that was based more on literature than on the ancient Greek philosophers. Their education was heavily weighted towards poetry, drama, history, and rhetoric – an education that would eventually be called a humanistic education. This tradition dominated the intellectual life of the Western world before the universities appeared; and it was this tradition that the humanists of the Renaissance reinstated.

In today's splintered world, as we search for a credible foundation for ethics, including bioethics, it should be no surprise if some scholars use Culture No. 3 to find the solution. On the North American scene, they have included William F. May, Katherine Montgomery, Howard Brody, S. Kay Toombs, Richard Zaner, Rosemarie Tong, Eva Feder Kittay, and more generally, Martha Nussbaum. For example, in the 1980s and 1990s there was a powerful move to base bioethics in an ethic of care (Fry 1989, Carse 1991) as proposed by philosophers who drew on a contemporary description of the care-experiences of women (Benhabib 1987) who were socially-advantaged, principally white women living in the U.S., neglecting the care-experiences of women of other color and race (Townes 1998). Culture No. 3 reminds bioethics scholars that the breadth of this inquiry that is so fundamental to bioethics must extend even further. For to grasp what care is about we must return to those who have described and portrayed precisely the idea (and experience) of care itself in the great literary and rhetorical traditions of Horace, Virgil, Seneca, Augustine, Dante, Petrarch, Goethe, and Sudermann, as well as modern philosophers, including Martin Heidegger (Pattison 2000) and Michael Slote (Slote 2007), along with theologians, psychologists, and literary figures who have been strongly influenced by Culture No. 3.[14] Thus, we should not be surprised if we discover that it is only through the recovery of ancient, medieval, and modern literary texts and their models of care, humanity, mercy, compassion, and justice that we can turn bioethics into a believable, twenty-first century discipline.

This literary culture continued as a major influence into the twentieth century, largely because the humanists created the humanistic secondary school, the Liceo, the Lycée, the Gymnasium, the Public School in England, the Grammar School, or the Latin School; and the same culture also entered the Arts faculty of the university (O'Malley 2004).

Within the discipline of bioethics, a minority of scholars has favored Culture No. 3 over Culture No. 2. They have refused to rely too heavily on the clear-cut definitions of Culture No. 2, preferring instead the rich layers of meaning and the arguments of heart and imagination found in Culture No. 3. In that culture, we see the development of "Narrative Bioethics," which employs literature, literary criticism, and a philosophy of literature to develop the vision and insight, as well as specific behavioral images that provide a moral guide – and in this sense can be normative for conduct in the area of health and health care. A phenomenological approach to the philosophy and ethics of health and health care is also nested in Culture No. 3, as is the field now known as Literature and Medicine. Many works in the latter category, clustered in Culture No. 3, manifest a close connection between literary culture and the ethics of medical and health care (Rabuzzi and Daly 1989; Bremen 1993).

More specifically, a narrative ethic makes use of philosophical tools that have been developed especially by German scholars, namely "image ethics" (*Bildethik*) and "model ethics" (*Modellethik*). Literary sources for these approaches are innumerable, but a dominant example arose in the 1960s, when the novels of Hermann Hesse (1877–1962) – published in English translation, in paperback, at the height of the 1960s – were consumed by hordes of readers, especially university students. Hesse appealed to the youth of the 1960s counterculture movement because he spoke to their culture, especially their interest in searching for self-knowledge and spiritual experience through contact with Eastern religions, in spite of the initial chagrin of relatives and friends, who might have been adherents of mainline traditional, American religions. Especially popular was the quest-for-enlightenment theme that the youth found in Hesse's *Siddhartha* (Hesse 1951) and his *Journey to the East* (Hesse 1968a). A very different theme appealed to a wide audience: the life-journeys of two men in Hesse's *Narcissus and Goldmund* (Hesse 1968b). Narcissus was a fifteenth-century monk who was oriented to the interior life, with its intellectual and spiritual interests, while his friend, the young Goldmund, who was attending the monastery school, was outward-oriented and chose to wander and find himself through experiences in the world. This story provided a vivid example of the potential for an image ethic by way of the contrasting personal images of the art of living manifested in the two major characters.

I believe the youth of the 1960s precipitated a major spiritual and religious shift in American culture. Their interest in Eastern religions with its meditation may have been on the fringe in the 1960s, but today that spiritual culture provides a real option which is growing enormously and changing the character of our society.

6.4 Conclusion

By mid-twentieth century – but beginning much earlier – the sciences were coalescing; the humanities, including ethics, were coalescing; and scientific and humanistic cultures were expanding, separating, and coalescing. Those developments – coalescence and interaction – produced the culture which was the energy that propelled us into bioethics. C.P. Snow – in his novel, less-than-brilliant way – helped us to see that, even if he initially was arguing for the exact opposite.

At this point, it is illusory to regard science, the humanities, and ethics as separate; and, in addition, it is illusory to cut up the life sciences into small sections, each examined separately from an ethical perspective. Of course, some individuals will want to specialize in one aspect of ethics and the life sciences; but the vision of the life sciences joined with humanities provides a framework that offers a unifying character to our ethical initiatives.

The "life sciences" is a modern conceptualization, and making "life" a major object of concern is also a modern development. Our common concern is life, and what leads us to focus on a core concept like life is evolution. More particularly, what urges us to join life-issues together and to contribute to *the meeting* of the life sciences and the humanities is evolution. Lewis Thomas, in *The Lives of a Cell*, calls evolution "our most powerful story, equivalent … to a universal myth" (Thomas 1978, p. 122).

There seem to be indications that evolution is already becoming the narrative that unites our science-humanities discourse.

I believe this vision – of life, the life sciences, and ethics-within-humanities – can propel us with enthusiasm and a broader vision into the next era of bioethics. Without malaise.

Notes

1. The experience and concept of the "Outsider" is important for understanding contemporary bioethics, not to mention many other aspects of the intellectual and academic worlds. The classical source for this concept is the (somewhat dated yet richly constructed) work by Colin Wilson that was widely studied in the 1960s (Wilson 1956). I have recently discussed the Outsider category and its place in bioethics (Reich 2007, 2009).
2. The professional setting of many *clinical* bioethics consultants creates a different source of ennui or boredom: the constant repetition of the same sorts of clinical questions. That repetitious situation is, of course, typical of many caring professions. In fact, some clinical bioethicists specialize in applying the tools of their profession to the boredom, burnout, and difficult work situations that create profound problems for many health professionals and bioethicists.
3. Thomas Percival (1740–1804), who coined the term medical ethics in 1803, is increasingly regarded historiographically – mistakenly, I believe – as the father not only of physician ethics, but of all of medical ethics and even of the much broader field of bioethics (Percival 1803).
4. Initially, I did not know that I was undertaking a task that had never before been done – the creation of an encyclopedia that would report on a field of learning that was in the process of being established (Personal communication from Robert Lewis Collison, author of Collison, 1964).
5. Two editors of encyclopedias to whom I am indebted for their guidance were David L. Sills, editor of the *International Encyclopedia of the Social Sciences* (Sills 1968) and Philip Wiener, Editor-in-Chief of the *Dictionary of the History of Ideas: Studies of Selected Pivotal Ideas* (Wiener 1968).
6. Albert Jonsen mistakenly records the beginning of this major project as occurring in 1972. It began in 1971. I spent the academic year 1971–1972 developing a draft (through 27 revisions) of the topics that represented what I perceived as the future scope and methods of bioethics, selecting and inviting the Associate Editors, and writing and obtaining a major grant from the National Endowment for the Humanities to make the project possible.
7. As a matter of disclosure, I should state that I served as an advisor to *The Cambridge World History of Medical Ethics* project and was the final reviewer engaged by Cambridge University Press prior to publication. Nonetheless, in this chapter I am speaking neither for the editors of this work nor for the publisher, but solely as an individual scholar in the field.
8. To understand how Potter's thought evolved on the relationship between global concerns and clinical concerns in his understanding of bioethics, it is necessary to read all his works, not just his first article and first book in which he launched his ideas.
9. For a more recent, more thorough, and more critical analysis of U.S. bioethics and its history by the same authors, see Fox and Swazey (2008).
10. Some of the ideas covered in the *Encyclopedia of Bioethics*, 1st ed., had never been written on before as moral topics, e.g., "Care." This topic, nonetheless, was included because (1) "care" had, for several thousand years, been an important idea in the history of ideas in the Western world; and (2) we regarded this and other comparable ideas as an undeveloped but necessary resource for *the future* of bioethics.
11. See the text as performed by Don McLean; copyright 1971, Capital Records, Inc.: http://www.lyrics007.com/Don%20McLean%20Lyrics/American%20Pie%20Lyrics.html. Accessed 6 Dec 2007.

12. Martin Heidegger's philosophy of technology and his exploration of a new post-technological way of thinking, which was developed for the same era, comes to mind in this context. See Pattison (2000).
13. I omit the fourth of O'Malley's four cultures, because that culture, dealing with art, has less direct relevance for the verbally-oriented, intellectual and social pursuits that are characteristic of bioethics. It does, however, enhance Culture No. 3, as advanced in this essay.
14. For more on Culture No. 3, see O'Malley, op. cit., pp. 14ff.

References

Baker, R.B., and L.B. McCullough (eds.). 2009. *The Cambridge world history of medical ethics*. Cambridge/New York: Cambridge University Press.

Beauchamp, T.L., and J.F. Childress. 1979. *Principles of biomedical ethics*. New York: Oxford University Press.

Beecher, H.K. 1966. Ethics and clinical research. *The New England Journal of Medicine* 274(24): 1354–1360.

Benhabib, S. 1987. The generalized and the concrete other: The Kohlberg-Gilligan controversy and moral theory. In *Women and moral theory*, ed. E.F. Kittay and D.T. Meyers, 37–55. Totowa: Rowman & Littlefield.

Bremen, B.A. 1993. *William Carlos Williams and the diagnostics of culture*. New York: Oxford University Press.

Caesar, C.J. 2006, 50 B.C. "De Bello Gallico" and other commentaries. Trans. W. A. Macdevitt. With an Introduction by Thomas de Quincey. First published 1915. Project Gutenberg Ebook #10657 (2006): http://www.gutenberg.org/cache/epub/10657/pg10657.html. Accessed 12 Sept 2010.

Callahan, D. 1980. Shattuck lecture – Contemporary biomedical ethics. *The New England Journal of Medicine* 302(22): 1228–1233, at p. 1228.

Carse, A.L. 1991. The 'voice of care': Implications for bioethical education. *The Journal of Medicine and Philosophy* 16: 5–28.

Carson, R. 1962. *Silent spring*. Boston/Cambridge, MA: Houghton Mifflin/Riverside Press.

Collini, S. 1963. Introduction. In *The two cultures and a second look*, ed. C.P. Snow vii–lxii. New York: New American Library.

Collison, R.L. 1964. *Encyclopedias: Their history throughout the age: A bibliographical guide with extensive historical notes to the general encyclopedias issued throughout the world from 350 B.C. to the present day*. New York: Hafner.

Evans, J.H. 2002. *Playing God? Human genetic engineering and the rationalization of public bioethical debate*. Chicago: University of Chicago Press.

Farmer, P., and N.G. Campo. 2004. New malaise: Bioethics and human rights in the global era. *The Journal of Law, Medicine & Ethics* 32(2): 243–251.

Fox, R.C., and J.P. Swazey. 2008. *Observing bioethics*. New York: Oxford University Press.

Fry, S.T. 1989. The role of caring in a theory of nursing ethics. *Hypatia* 4(2): 88–103.

Hesse, H. 1951. *Siddhartha*. New York: New Directions Books.

Hesse, H. 1968a. *Journey to the East*. New York: Farrar, Straus & Giroux.

Hesse, H. 1968b. *Narcissus and goldmund*. New York: Farrar, Straus and Giroux.

Jonsen, A.R. 1998. *The birth of bioethics*. New York: Oxford University Press.

Jonsen, A.R. 2009. The discourses of bioethics in the United States. In *The Cambridge world history of medical ethics*, ed. R.B. Baker and L.B. McCullough, 477–485. Cambridge/New York: Cambridge University Press.

Marcuse, H. 1964. *One-dimensional man: Studies in the ideology of advanced industrial society*. London/New York: Routledge.

McCormick, R.A. 1994. The malaise in bioethics. In *Corrective vision: Explorations in moral theology*, 149–164. Kansas City: Sheed & Ward.

Morimoto, N. 2010. History of bioethics – Approach from cultural background; from the lecture of Warren T. Reich. In *Toward construction of meta-bioethics: Redefining bioethics*, ed. Y. Komatsu and C. Kagawa, 109–134. Tokyo: NTT Publishing.

O'Malley, J.W. 2004. *Four cultures of the West*. Cambridge, MA: Harvard University Press/ Belknap Press.

O'Neill, O. 2010. *The two cultures fifty years on*. A lecture delivered November 4, 2010, sponsored by the Birkbeck Institute for the Humanities, University of London, at Senate House, London. http://backdoorbroadcasting.net/2010/11/onora-oneill-the-two-cultures-fifty-years-on/. Accessed Jan 14, 2011.

Pattison, G. 2000. *Routledge philosophy guidebook to the later Heidegger*. New York: Routledge.

Percival, T. 1803. *Medical ethics; or a code of institutes and precepts, adapted to the professional interests of physicians and surgeons*. Manchester: S. Russell.

Potter, V.R. 1971. *Bioethics: Bridge to the future*. Englewood Cliffs: Prentice-Hall.

Rabuzzi, K.A., and R.W. Daly (eds.). 1989. *The cultures of medicine*. Baltimore: Johns Hopkins University Press.

Redlich, F. 1978. Medical ethics under national socialism. In *Encyclopedia of bioethics*, ed. W.T. Reich, 1015–1020. New York: Macmillan, Free Press.

Redlich, F. 1998. *Hitler: Diagnosis of a destructive prophet*. New York: Oxford University Press.

Reich, W.T. 1978. "Preface" (xi–xiv) and "Introduction," (xv–xxii). In *Encyclopedia of bioethics*, vol. 4, 1st ed, ed. W.T. Reich. New York: Macmillan/Free Press.

Reich, W.T. 1990. La Bioetica negli Stati Uniti: Orientamenti e Tendenze. In *Vent'anni di Bioetica: Protagonisti, idee, istituzioni*, ed. C. Viafora, 141–175. Padua: Fondazione Lanza and Gregoriana Libreria Editrice.

Reich, W.T. 1994. The word 'bioethics': Its birth and the legacies of those who shaped it. *Kennedy Institute of Ethics Journal* 4(4): 319–335.

Reich, W.T. 1995a. The word 'bioethics': The struggle over its earliest meanings. *Kennedy Institute of Ethics Journal* 5(1): 19–34.

Reich, W.T. ed. 1995b. *Encyclopedia of bioethics*. Rev. ed. 5 vols. New York: Simon & Schuster/ Macmillan.

Reich, W.T. 1995c. Introduction to the revised edition. In *Encyclopedia of bioethics*, ed. W.T. Reich, xix–xxxii. Rev. ed. 5 vols. New York: Simon & Schuster/Macmillan.

Reich, W.T. 1996. Preface. In *History of bioethics: International perspectives*, ed. R. Dell'Oro and C. Viafora, 3–6. San Francisco: International Scholars Publications.

Reich, W.T. 1997. Origine, Sviluppo e Futuro della Bioetica [Origins, development, and future of bioethics]. In *Bilancio di 25 Anni de Bioetica: Un Rapporto dai Pionieri* [An assessment of 25 years of bioethics: A report from the pioneers], ed. G. Russo, 25–35. Leumann: Editrice Elle Di Ci.

Reich, W.T. 1999. The 'wider view': André Hellegers's passionate, integrating intellect and the creation of bioethics. *Kennedy Institute of Ethics Journal* 9(1): 25–51.

Reich, W.T. 2003a. Shaping and mirroring the field: The encyclopedia of bioethics. In *The state of bioethics: From seminal works to contemporary explorations*, ed. J.K. Walter and E.P. Klein, 165–196. Washington, DC: Georgetown University Press.

Reich, W.T. 2003b. Dai principi alle persone: Evoluzione (necessaria) della bioetica [From principles to persons: A necessary evolution of bioethics]. *Janus* 8: 8–19.

Reich, W.T. 2007. La Bioetica: Il Punto di Vista dell'Outsider. *Medicina e Morale* 3: 533–553.

Reich, W.T. 2009. Bioethics and the Outsider: The Misfit-Ethicist, Torture, and Laicità. In *Yearbook of philosophical hermeneutics, the dialogue – Das Gespräch. Il Dialogo, 1/2009: Autonomy of reason – Autonomie der Vernunft*, ed. R. Dottori, 224–241. – *Proceedings of the V Meeting Italian-American Philosophy*. Berlin/London: Lit Verlag Dr. W. Hopf.

Roszak, T. 1969. *The making of a counter culture: Reflections on the technocratic society and its youthful opposition*. Garden City: Doubleday (Also republished with a new Introduction, Berkeley: University of California Press, 1995).

Sills, D.L. (ed.). 1968. *International encyclopedia of the social sciences*. New York: Macmillan.
Slote, M. 2007. *The ethics of care and empathy*. New York/London: Routledge.
Snow, C.P. 1960. *The two cultures and the scientific revolution. The Rede Lecture 1959*. New York: Cambridge University Press.
Snow, C.P. 1963. *The two cultures; and, a second look*. New York: New American Library.
Swinton, J. (ed.). 2004. *Critical reflections on Stanley Hauerwas' theology of disability: Disabling society, enabling theology*. Binghamton: Haworth Pastoral Press.
Thomas, L. 1978. *The lives of a cell: Notes of a biology watcher*. New York: Penguin.
Townes, E. 1988. *Breaking the fine rain of death: African American health issues and a womanist ethic of care*. New York: Continuum.
Tsuchiya, T. 1998. How bioethics was introduced in Japan. Presented at the Fourth World Congress of Bioethics, 6 November 1998, Tokyo, Japan. http://www.lit.osaka-cu.ac.jp/user/tsuchiya/gyoseki/presentation/IAB4.html. Accessed Sep 3, 2010.
Wear, D., and M. Kuczewski. 2004. The professionalism movement: Can we pause? *The American Journal of Bioethics* 4(2): 1–10.
Weisz, G. (ed.). 1990. *Social science perspectives on medical ethics*. Dordrecht: Kluwer Academic Publishers.
Wiener, P. (ed.). 1968. *Dictionary of the history of ideas: Studies of selected pivotal ideas*. New York: Scribner.
Wilson, C. 1956. *The outsider*. New York: Gollancz.

Chapter 7
American Biopolitics

George J. Annas

7.1 Introduction

Bioethics and biopolitics have been linked in the United States at least since the Civil War—about a century before either term was coined. In the Civil War the biopolitical issue was ownership and control of the bodies and health of people we called slaves and the government's support of the institution of slavery. In contemporary America, biopolitics encompasses the wide range of measures the government takes to try to make people more healthy, what we usually refer to simply as public health. And in post 9/11 America, where public health has de facto become part of homeland security, biopolitics has expanded to include terrorism prevention measures that implicate the body politic and the bodies of Americans—most spectacularly exemplified in proposals to subject all Americans who want to fly on commercial airlines to whole body scanning.

Michel Foucault is generally credited with inventing the concept of biopolitics, and he used the term primarily to designate the use of government power (biopower) not just to make a life or death decision with regard to a subject, as kings used to do (e.g., executions and torture), but to manage entire populations, primarily through surveillance and control of the health-related aspects of life[1] (Rose 2007, p. 54). More specifically, Foucault defined biopolitcs as "the way in which attempts have, since the eighteenth century, been made to rationalize the problems posed for governmental practice by phenomena characteristic of a group of living beings constituted as a population: health, hygiene, natality, longevity, races…" (Macey 1993, p. 358; Foucault 2003, pp. 242–250).

G.J. Annas, J.D., M.P.H. (✉)
Department of Health Law, Bioethics and Human Rights, School of Public Health, Boston University, 715 Albany Street, Boston, MA 02118, USA
e-mail: annasgj@bu.edu

J.R. Garrett et al. (eds.), *The Development of Bioethics in the United States*, Philosophy and Medicine 115, DOI 10.1007/978-94-007-4011-2_7,

Using biopower as an aspect of state power over its people reached its pinnacle in Nazi Germany with the Holocaust. Hitler had laid out his biopolitical agenda for the Jews in *Mein Kampf*, where he characterized Jews in Germany as racial parasites living off of and contaminating the blood of the "German national body" (Hitler 1925/1999, pp. 300–329). National Socialism, of course, sought to use state power to control all aspects of the health of the nation and its citizens. The dark side of state control reached its depths in Auschwitz where Foucault's biopolitics, the power to *make live*, intersected with the government's historical power to *make die* (Agamben 2002; Esposito 2008).

The American breed of biopolitics, sometimes termed liberal biopolitics, seeks (at least since World War II) instead to limit the power of the state over individual lives and health. Although any aspect of governmental control over the health of its citizens necessarily involves biopolitics, in the United States we almost never use this term. Instead, we usually speak simply of politics. And in the realm of population health, the term we use is public health, defined as the actions that governments (state and federal) take to help ensure a healthy population. The two terms are not synonymous, and biopolitics is broader than public health—nonetheless, to understand biopolitics in the United States it is necessary to understand the evolution and contemporary practice of public health. Similarly, although bioethics is often used simply as another word for medical ethics, it was originally invoked to encourage the merger of the entire field of biology with the humanities as one response to C.P. Snow's critique of *The Two Cultures*. Nonetheless, bioethics cannot be understood without understanding its primary component, medical ethics, and biopolitics cannot be understood without understanding public health and its contemporary companion, human rights. For the purposes of this overview of biopolitics American-style (as well as its relationship to bioethics), I adopt the American public health model of biopolitics, as well as the original concept of bioethics—which is broad enough to encompass not only the medical sciences, but also public health and ecology.

America has been defined by rights since the Declaration of Independence and the adoption of the US Constitution, two of the most significant products of the Enlightenment, and the Constitutional rights of citizens have both defined and limited the state's biopower from the beginning of the country. Two major threads in the fabric of American public health illustrate what I take to be unique individual rights characteristics of biopolitics (and bioethics) American style: the pervasiveness of Mill's harm principle in limiting the legitimacy of government's public health interventions and the authority of the US Supreme Court to set limits on government-imposed public health measures (based on an understanding of individual rights protected by the US Constitution); and the continuing political attempts by the federal government to adopt a national (public) health program.[2] Since the first has undergirded American biopolitics from its inception, I will devote most of this chapter to it.

7.2 John Stuart Mill and the US Supreme Court

Little useful can be said about American biopolitics (or American bioethics for that matter) without at least some understanding of America itself. For John Stuart Mill, understanding America began with reading Alex de Tocqueville's *Democracy in America* (1990). We need not agree with all the conclusions of Tocqueville to recognize that even today American democracy remains a work in progress, and that because of this so does American biopolitics.

In his autobiography, Mill tells us that he relied heavily on Tocqueville's "remarkable work" not only to better appreciate the "excellences of democracy" in America but also, and perhaps more importantly, its majoritarian "infirmities." Mill believed that these infirmities, if unchecked by active political involvement of individual citizens, "could degenerate into the only despotism of which, in the modern world, there is a real danger—the absolute rule of the head of the executive over the congregation of isolated individuals, all equals but all slaves" (Mill 1964, p. 143). For Mill, equality was necessary for a functioning democracy, but not sufficient: liberty was central. And in America, liberty is protected by law, especially Constitutional law. As Abraham Lincoln put it, perhaps intentionally overstating his case:

> Let every American, every lover of liberty, every well wisher to his posterity, swear by the blood of the Revolution, never to violate … the laws of the country… Let reverence for the laws be breathed by every American mother to the lisping babe that prattles on her lap—let it be taught in schools, in seminaries, and in colleges;—let it be written in Primers, spelling books, and in Almanacs … in short, let it become the *political religion* of the nation (emphasis in original) (Gopnik 2009, p. 40).

Lincoln and our country were, of course, grievously tested in the context of the central biopolitical issue of our country's first century: slavery. That issue was itself inflamed by the branch of government that has become the bioethical/legal/biopolitical arbiter in the United States, the Supreme Court. In the infamous Dred Scott case, decided shortly before the Civil War, the Court affirmed that humans could be owned as property, and that the property interest slaveholders possessed in their slaves would be enforced by the government. The freedom and equality of slaves was legally proclaimed only in the midst of the Civil War, and only thereafter enshrined in amendments to the US Constitution. American biopolitics also tracks catastrophic events in the Post-Civil War history of the country, including late nineteenth century epidemics, World War I and the Spanish flu epidemic, World War II, the human rights movement, and the terrorist attacks of 9/11.

Of course Mill did not write about biopolitics per se, and even its primary manifestation, public health, can be said to have been born as a scientific endeavor only 4 years before he published *On Liberty* in 1859. This was when John Snow helped end a cholera epidemic in the Soho neighborhood of London by removing the handle from a water pump for a well that he had identified (by what are now seen as basic epidemiological methods) as the likely source of the disease (Johnson 2006). And since

Mill's death in 1873, medicine and science have changed so much that he would barely recognize them. One major change has been a change in emphasis: from social reform to improve the health of the public to concentration on individual life styles and personal responsibility that bioethics has also been more at home with (Turoldo 2009; Fairchild et al. 2010).

We Americans are healthier than we have ever been, and yet worry more about our personal health and safety than we ever did (Knowles 1976). In our (public) health-crazed society, what can Mill teach us? Mill is probably best known for what he calls "one very simple principle":

> That principle is that the sole end for which mankind [sic] are warranted, individually or collectively, in interfering with the liberty of action of any of their number is self-protection. That the only purpose for which power can be rightfully exercised over any member of a civilized community, against his will, is *to prevent harm to others*. His own good, either physical or mental, is not a sufficient warrant (emphasis supplied) (Mill 1974, p. 68).

This principle, as Mill goes on to argue at some length, is actually far from simple and the rest of his text is spent defining and defending it. Nonetheless, the basic concept is clear: the government has no business interfering with an individual's liberty to decide how to conduct his or her life (or healthcare) unless an individual's actions are likely to be harmful to others, in which case the state has the right to defend its citizens and itself. The Mill doctrine has been applied consistently by US courts, especially the Supreme Court, in a variety of public health contexts, especially vaccination and eugenics. In both of these cases the Court concluded that harm to others justified compromising individual rights, using self-defense language very similar to that currently employed by government officials in our ongoing "war on terror."

7.3 *Jacobson v. Massachusetts* (Smallpox Vaccination)

In the most famous public health case, *Jacobson v. Massachusetts*, decided in 1905 (50 years after the publication of *On Liberty*), the US Supreme Court upheld a Massachusetts law that permitted local health departments to require vaccination against smallpox to protect the community from an epidemic of the disease. The Court relied heavily on Mill's harm principle without citing him directly, noting first that the liberty "secured by the Constitution" was not an "absolute right" that could be exercised in all circumstances without restraint. Instead the Court ruled:

> There are manifold restraints to which every person is necessarily subject for the common good. On any other basis organized society could not exist with safety to its members. Society based on the rule that each one is a law unto himself would soon be confronted with disorder and anarchy. *Real liberty for all could not exist under the operation of a principle which recognizes the right of each individual person to use his own, whether in respect of his person or his property, regardless of the injury that may be done to others* (emphasis supplied).

Later in the opinion the Court observed that when a ship arrived in the United States, and there had been cases of either yellow fever or cholera on board during the voyage, passengers could "in some circumstances" be quarantined against their will even if they themselves did not exhibit any symptoms of the disease, for the sake of the health of the entire community:

> The liberty secured by the 14th Amendment, this court has held, consists in part in the right of a person 'to live and work where he will' ... and yet *he may be compelled*, by force if need be, against his will and without regard to his personal wishes or pecuniary, or even his religious or political convictions, *to take his place in the ranks of the army of his country, and risk the chance of being shot down in its defense* (emphasis supplied).

7.4 *Buck v. Bell* (Sterilization)

Jacobson was an easy case, holding that a state may require healthy adults to accept a safe and effective vaccination when a lethal epidemic endangers the community in which the individual lives for the same reason a state may conscript soldiers—for the defense of the community. Nonetheless, just 20 years later, in *Buck v. Bell*, the Court expanded the "harm principle" beyond all rational bounds. In *Buck* the Court upheld a Virginia law that authorized the involuntary sterilization of "feeble minded" persons held in state institutions on the basis that such sterilizations protected society from the degenerate criminal offspring or imbeciles who would "sap the strength of the state." In the words of Justice Oliver Wendell Holmes, writing for the Court, "The principle that sustains compulsory vaccination is broad enough to cover cutting the Fallopian tubes. Three generations of imbeciles are enough." Carrie Buck's own interests in bodily integrity and procreation were not even considered—only the financial burden her hypothetical children might put on the state.

In *Buck v. Bell* the harm to others exception swallowed up Mill's liberty rule. It was not until after World War II, and the public disclosures of the horrors of Nazi eugenics—that were premised in part on the US eugenics movement—that the Court began to recognize what are now termed "fundamental liberty interests" that could only be interfered with by the government if it could demonstrate a "compelling state interest." And even with such an interest, government interference with fundamental liberties must be by the "least restrictive method" available.

Fundamental liberty interests that have been recognized by the Supreme Court, interests that Mill certainly would also recognize, include the right to choose a marriage partner, the right to use contraceptives, the right to choose to have an abortion, the right to sexual privacy, and the right to refuse any medical treatment. A smallpox epidemic would still be a sufficiently compelling reason for the government to mandate that its citizens take a safe and effective vaccination. On the other hand, involuntary sterilization would likely no longer be permitted since it directly violates a fundamental liberty interest (procreation) and the state has no compelling reason to follow the teachings of a discredited and scientifically-unsound eugenics doctrine. The sterilization of an incompetent person in a state institution remains possible, but would require demonstration to a judge that the procedure is in the person's best interests, and is the least invasive procedure available to accomplish the contraceptive goal (Mariner et al. 2005).

7.5 *Simon v. Sargent* (Seatbelts)

This is not to say that the harm principle has lost its appeal, or that it has been marginalized in public health. In cases in which no fundamental liberty interest is at stake, such as in riding motorcycles, lower courts have been only too willing to apply the financial interests of the state ("harm to society") concept articulated in *Buck v. Bell.* In *Simon v. Sargent*, for example, a federal district court upheld the Massachusetts motorcycle helmet requirement on the basis that the state, not the individual, often winds up footing the bill for injuries sustained in motorcycles accidents. In the court's words, "The thrust of the position [of the plaintiff] is that the police power does not extend to overcoming the right of an individual to incur risks that involve only himself. 'The risk to oneself is no public purpose.' For this proposition he quotes John Stuart Mill (1974/1859), 'The only part of the conduct of anyone, for which he is amenable to society, is that which concerns others.' We may stop right there." But the court continues to elaborate the harm to society it has in mind:

> From the moment of injury, society picks the person up off the highway; delivers him to a municipal hospital and municipal doctors; provides him with unemployment compensation if, after recovery, he cannot replace his lost job, and, if the injury causes permanent disability, may assume the responsibility for his and his family's continued subsistence. We do not understand a state of mind that permits plaintiff to think that only he himself is concerned.

7.6 Commerce and Public Health (Including Obesity and Abortion)

If this is really the test, then Mill as applied to contemporary public health unravels, because everything can be made to connect to everything else. No individual act can be said to have consequences only to the individual if its financial consequences are all that is needed to make it a public matter. The late Chief Justice William Rehnquist addressed a similar problem of setting limits when he refused to conclude that a federal statute that prohibited students from bringing guns to school could be justified based on the Interstate Commerce Clause (which gives Congress authority to regulate activities that affect the US economy). Rehnquist wrote, in *U.S. v. Lopez*, that if the Court accepted the argument that guns in school affect the country's economy because their presence could make students so anxious or nervous that they could not study as well, and as a consequence they become less productive workers, and this hurt the US economy on a global competitive scale, then virtually any activity, including marriage and school curriculum (two traditionally state and local areas of authority), could be regulated by the federal government. Similarly, the 2009–2010 plan of Congress to require individuals to purchase health insurance or pay a fine is being fought by individual states as an infringement on liberty that is beyond the Commerce Clause power of the Congress (Pershing 2010). It seems unlikely, however, that this argument will prevail in a Supreme Court that has recently expanded the scope of the Commerce Clause to include virtually any activity, including growing marijuana in small quantities for medicinal purposes in a flower box at home, that could affect the US economy (Annas 2010).

In *Foucault's Pendulum* (1988), Umberto Eco makes the point that you can see connections between any two objects if you use (or overuse) your imagination, for example, in constructing a conspiracy theory in which "the Templars have something to do with everything." The narrator, Casubon, tells his companion Belbo that "The challenge isn't to find occult links between Debussy and the Templars. Everyone does that. The problem is to find occult links between, for example, cabala and the spark plugs of a car" (Eco 1988, p. 314). If the problem is to find links between a personal activity and public money, the link will always be possible, at least from the government's perspective.

Perhaps the most extreme example comes from American public health's current fascination and horror with obesity. Bad enough, it would seem, that obese individuals have higher health risks for some diseases. But no, it's not bad enough. It has even been suggested that fat people cause global warming. The connection is as straightforward as connecting cabala to a spark plug: it takes more gasoline to transport a fat person than a thin one. Aggregated together, fat Americans use more than a billion gallons extra gasoline because of their weight—and this produces emissions that affect global climate. The author of the study, Sheldon H. Jacobson, is quoted as explaining why the health implications of obesity were not enough: "We felt that beyond public health, being overweight has many other socioeconomic implications". Another researcher (perhaps after reading the motorcycle case) put it more simply, "the individual's decision to lead a sedentary lifestyle will end up costing taxpayers" (Garrow 2007).

Bernard-Henri Levy, the French philosopher who attempted to retrace Tocqueville's journey through America, wound up making some biopolitical observations on America's new public health fixation with obesity in his *American Vertigo*:

> I understand that there is a weight-loss business that is as big as the junk-food business. I understand that the former has an advantage over the latter in being able to rely on the prestigious testimonials of science and medicine. Even better, I understand that inventing obesity-that is to say, claiming first that being fat is a disease; second, that this disease must be treated; and third, that it will never, despite treatment, be completely cured—creates a type of dependence that is at least equal to that produced by the inventors of flavors, fragrances, and packaging that are designed to develop a loyal following among junk-food consumers… Big Brother once again. *No longer a cop but a doctor in everyone's body.* Worse than a doctor, a statistician, imprinting his implacable orders onto the quick of live flesh (emphasis added) (Levy 2006, p. 96).

An outside observer, Levy concluded that obesity is a real problem in America, but obesity (excess) of a different sort, "another brand of obesity":

> A social obesity. An economic, financial, and political obesity. Obesity of cities. Obesity of malls, as in Minneapolis. Obesity of churches… Obesity of parking lots… Obesity of SUVs. Obesity of airports… Obesity of election campaign budgets… Obesity of Hollywood box-office sales… Obesity of large memorials… The bigger it is, the better it is, says America today (Levy 2006, p. 240).

In one of his least flattering observations of America, Levy goes on to describe what he sees as a nation that "has lost control of its own situation—not just alimentary, but mental, cultural, metaphysical; a nation, one feels, that has strayed from, or broken, that secret formula, that code, that prompts the body to stay within its limits and survive." (Levy 2006, p. 241) Could we usefully add, based on the rubric of "cost to society" as a valid measure of the harm principle, "an obesity of public health?"

The challenge to American public health (and thus to American biopolitics) is to concentrate health regulations on controlling things (such as drugs, devices, food and water quality, automobile design, etc.) rather than trying to control the behavior of individuals, which is, both Mill and Foucault knew, a much more tempting target (Glantz et al. 1992). In this regard it can escape no observer's attention that since 1973 and the US Supreme Court's decision in *Roe v. Wade*, that abortion—and the continuing attempt by the government to control the behavior of women and their physicians concerning abortion by using the criminal law—has been a major exception to the general societal belief that the government should not dictate the medical decisions of individuals (Annas 2007, 2009b). The one-time Congressional action to try to keep a feeding tube in a particular patient, Terri Schiavo, is another (Annas 2010; Bishop 2008). On the other hand, Engelhardt may be correct in asserting that, because of the multiple dogmas held by the world's religions, we will never reach political consensus on abortion (Engelhardt 2006).

Support for criminalizing abortion is based, nonetheless I think, on a belief that abortion is not really a medical procedure at all, but is more properly seen as unjustified killing, at least when not done to save the life of the pregnant woman (and perhaps in cases of rape or incest as well). Justice Ruth Bader Ginzberg suggests as much in her dissent in the five-to-four decision of the Court upholding the Congress's authority to make a procedure termed "partial-birth abortion" a crime (*Gonzales v. Carhart*). The majority of the Court not only ruled that Congress had the legal authority to regulate the practice of medicine (something that has historically been left to the individual states), but also that it was reasonable for Congress to conclude that "abortion doctors" could not be trusted either to tell their patients the truth or to act in their medical best interest (Perry 2008).

The majority asserted that giving Congress constitutional authority over medical practice was nothing new, but identified no case in which Congress had ever outlawed a medical procedure. Its reliance on the more than 100-year old *Jacobson v. Massachusetts* case in this regard is especially inapt. As previously noted, *Jacobson* was about mandatory smallpox vaccination during a lethal epidemic. The statute had an exception for "children who present a certificate, signed by a registered physician, that they are unfit subjects for vaccination," and the Court implied that a similar medical exception would be constitutionally required for adults. It is not just abortion regulations that have had a health exception for physicians and their patients—all public health regulations have (Mariner 2003; Mariner et al. 2005).

This bioethics/biopolitics opinion is a wholesale vote of no confidence by the Supreme Court not only in the way physicians are regulated by state medical boards, but also in the ethics of physicians themselves. This is a stark change in the law and in American biopolitics. But it may also be simply a recognition that abortion is unique and abortion (bio)politics may apply only to abortion itself in America. This view would explain why during the 2009–2010 Congressional debate on healthcare insurance reform, the only medical procedure that merited its own special treatment was abortion (Annas 2009b). And it is to the healthcare debate, and its post-World War II human rights aspects, that I now turn.

7.7 The American Right to Health

Although the United States remains the only developed country in the world without a national system that insures all of its citizens access to medical care, the human right to health (which includes, but is much broader than, medical care) has strong American political roots. As I have suggested from both Mill's harm principle and the provisions of the US Constitution, American biopolitics has primarily been concerned with so-called negative rights, those individual rights that restrict government from acting in certain spheres of life, including the right to practice one's religion. The "right to health," on the other hand, is an example of what has historically been denoted a positive right, a right that requires the government to take action and spend money. In the midst of World War II, President Franklin Roosevelt told Congress in his 1944 State of the Union address that it was time to add specific positive rights to the US Constitution because, "We have come to a clear realization of the fact that true individual freedom cannot exist without economic security and independence." FDR called on Congress to adopt a "second Bill of Rights," a bill of economic security, which included "The right to adequate medical care and the opportunity to achieve and enjoy good health."

This proposal was foreshadowed in his more famous "four freedoms" speech in which Roosevelt had argued that positive rights and negative rights were inseparable and interconnected in the real world. He listed four (two positive and two negative): freedom of speech, freedom of religion, freedom from want, and freedom from fear. Roosevelt saw government not primarily as an entity that controlled the life of its citizens individually, or the body politic, but as an entity that protected its vulnerable citizens from the worst aspects of market capitalism, as exemplified in the Great Depression.

Truman continued FDR's fight for the right to health, telling Congress in his 1948 State of the Union address—the same year the Universal Declaration of Human Rights (UDHR) was adopted by the United Nations—"Our first goal is to secure fully the essential human rights of our citizens." Regarding health, Truman continued, "Our ultimate aim must be a comprehensive insurance system to protect all our people equally against insecurity and ill health." In 1965, President Lyndon Johnson traveled to the Truman Library in Independence, Missouri to sign Medicare into law in Truman's presence as a tribute to his dedication to this cause. As American as it is, the right to health is, of course, not an exclusively an American, or even a Western, concept. A 1947 report of UNESCO's philosophers' committee, for example, listed

15 "norms" that the committee had found were widely shared by cultural and religious traditions around the world, including "the right to protection of health," although the word "right" was not found in many traditions (Annas 2009b).

Since the adoption of the UDHR, and later the two major treaties—the International Covenant on Civil and Political Rights, and the International Covenant on Economic, Social and Cultural Rights-we have become accustomed to having human rights declared internationally by treaty, and thereafter promulgated nationally by legislation that enacts specific human rights entitlements. In this framework, a national healthcare plan of the kind proposed by Roosevelt and Truman (and later by Presidents Kennedy and Johnson, and even the much more limited ones by Presidents Clinton and Obama) would be a statutory enactment of America's vision of the right to health. This conclusion is reasonable even though, unlike 150 other countries, the United States has not adopted the International Covenant on Economic, Social and Cultural Rights that contains the right to health. This is because no specific insurance scheme, delivery system, or even benefit package is required by the international right to health. Instead, consistent with basic human rights precepts, a national health plan must be universally accessible, cannot discriminate, and must have special provisions to protect pregnant women and children.

Senator Ted Kennedy predicted, to thunderous applause at the 2008 Democratic National Convention, that Barack Obama would "break the old gridlock and guarantee that every American will have decent, quality health care as a fundamental right and not just as a privilege." The decision of the Obama administration not to champion healthcare as a human right resulted in a very narrow debate that it is difficult even to classify as biopolitical. This is because the debate focused not on human rights and human dignity, but instead on the financing of private health insurance, including eliminating preexisting conditions as a disqualifier for health insurance and a prohibition of rescinding health insurance coverage after a policy-holder gets sick, both under the rubric of "health insurance reform." It is an American biopolitics truism that the only winners of such a narrow healthcare debate, however, will predictably be those with the most effective political lobbyists—the insurance companies, drug companies, hospitals, and (perhaps) physician groups.

Opponents of a real national plan, such as "Medicare for all," had an easier time using false slogans, such as "socialism," and a "government run health care system," that puts a "government bureaucrat between you and your doctor." Opposition forces also created unfounded fear that a national healthcare program (even a public option, known as a "public health plan") would undermine rather than promote human rights by, for example, creating imaginary "death panels" that would encourage euthanasia by arbitrarily deciding to "pull the plug on grandma." And just in case these arguments are insufficient, it is common to see the central biopolitical controversy in America raised again in the healthcare context—the argument that a national health program "would require taxpayers to fund abortions." In fact, neither the House nor Senate versions of health insurance reform were able to garner the majorities they needed until language more demanding than the existing Hyde Amendment was added to reassure at least some supporters that no public funds would be used to pay for abortions not required to save a pregnant woman's life, or because of rape or incest (Annas 2009b).

Senator Kennedy was, I think, correct in seeing human rights as the only language powerful enough to finally break America's biopolitical healthcare reform stalemate. The internationally-recognized right to health (a right that includes the right to equal access to basic healthcare, as well as other pre-conditions for healthy living, including food, water, shelter and clothing), has not been recognized as an enforceable legal right in the United States. Nonetheless, Americans have long recognized access to medical care as a legal right in emergency situations, at least if a sick or injured person can get to a hospital emergency department. Emergency medical care is well understood to be a human right in the United States (and virtually everywhere else in the world). Because of our obsession with medical technology, Americans also have a national program that provides universal access to end-stage renal dialysis, as well as to kidney and heart transplantation. The latter has provided almost a full employment program for bioethicists. The former raises the biopolitical question of how to use politics to expand the legal and moral right to emergency care into the more encompassing right to necessary healthcare.

I have previously suggested that certain characteristics of the American healthcare system mirror similar characteristics of America, and that our healthcare system is not likely to change until our country changes. Four characteristics seem intractable: Americans (and America) are individualistic, technologically-driven, death-denying, and wasteful. Levy concentrated his observations on the last characteristic, although he labeled it as societal "obesity." All four characteristics were on display during the fall 2009 healthcare debate when the U.S. Preventive Services Task Force issued a report, based on a meta-analysis of existing studies of the efficacy of mammography in saving lives, suggesting that annual routine mammograms should no longer be recommended for women in the 40–49 years old age group (Kolata 2009) The study found, among other things, that to prevent one cancer death in this age group it would be necessary to screen 1,904 women every year for this 10-year period. As important, the risk of a false positive (which would be followed up by biopsy, and perhaps treatment) was up to 56% after ten mammography examinations in the 40–49 years old group (Nelson et al. 2009, p. 731). Rather than lead to a reasonable public discussion, the recommendation was immediately denounced. Senator Kay Bailey Hutchinson (R-Texas), for example, said it was an assault on American values: "One life out of 1,904 to be saved, but the choice is not going to be yours. It's going to be someone else that has never met you, that does not know [your] family history. This is not the American way of looking at our health care coverage." (Sack 2009) Kathleen Sebelius, secretary of health and human services (in whose department the report was written), speaking for the Obama administration, was equally and immediately dismissive, saying simply, "our policies remain unchanged." (Sack and Kolata 2009) The question of whether cancer screening was not only wasteful, but could actually do more harm than good, was simply ignored and displaced by individual decisionmaking, death denial, and faith in technology.

Of course the right to health itself does not define the precise content of a medical benefit package—rather the legislative adoption of this right by Congress is a necessary step toward making the definition of a national medical benefit package relevant. As the Supreme Court has made clear on a number of occasions, Congress has the authority to determine the benefit package for both public and private insurers under its Commerce Clause powers. This may be characterized as the ultimate biopolitical (public health) prize in contemporary America, but dysfunctional American characteristics make it a prize we are unlikely to attain in the near future.

American bioethicists will continue to play a critical role in helping to define rights *in* the doctor-patient relationship (as they have with informed consent and even the right to refuse medical care), but should also, I think, take on an expanded role by helping to define the right *to* have a doctor-patient relationship in the first place as a central part of the right to health. In accepting this role, American bioethics will also join American biopolitics, as well as becoming advocates in the human rights realm (Annas 2005).

Resource poor countries have only the obligation to "progressively realize" the right to health. The United States has no such financial excuse. That the world's richest country has yet to establish a right to health is both paradoxical and unconscionable. To paraphrase Eleanor Roosevelt, who said that universal human rights begin "in small places, close to home," it's time to welcome the right to health into its American home (Annas 2009a).

7.8 Biopolitics American Style

American biopolitics, whether referred to simply as politics or public health, is complicated. It is informed by Mill, usually determined by the individual states (which retain the "police" or public health and welfare powers) or Congress (under its Commerce Clause authority) and often approved (or disapproved) by the Supreme Court as the arbiter of liberty and the limits of government action. Mill was almost exclusively concerned with what we now call political and civil rights, and his primary target was government that if left to its own devices would, he believed, become tyrannical and dominate and restrict the liberty, and thus the ability to live a fulfilling life, of its citizens. He was especially concerned with the influence of religions in determining the rules of daily living, and rejected all orthodoxy in this regard.

Mill's concerns have a contemporary echo not only in attempts to legislate morality, especially in the areas of human reproduction and marriage, but also in the public health attempts to legislate healthy living, including anti-obesity measures. To the extent that public health is in danger of becoming a secular religion in which the morality of citizens can be judged by simply looking at the shape of their bodies, Mill's anti-paternalistic sentiments have a direct bearing on contemporary public health practice. The primary aspect of Mill focused on in this chapter can be restated in terms of contemporary constitutional law: The government should only interfere with individual liberty—even to make healthy—when an individual's actions create a significant risk of substantial *physical* harm to others, and even then government restrictions on the individual's liberty should be limited to those necessary to protect others from harm, and implemented by the least restrictive method.

Two additional aspects of American biopolitics are related. First, the Supreme Court (rather than any public official or panel of experts), will determine the limits of the government's authority to regulate both individual and community (public) health, as illustrated primarily in the vaccination and abortion cases. Second, the implementation of any positive right in America, including the right to health, as illustrated by the brief debate over breast cancer screening, will be primarily determined in the political arena where it will require both strong presidential leadership, and Congressional action. In today's Congress, American biopolitics has become much more about private money-making (and spending) than about the health of the public. It's not your grandfather's public health; and it's not Foucault's biopolitics—it's a unique and evolving creature: American biopolitics.

Notes

1. "Biopower is more a perspective than a concept: it brings into view a whole range of more or less rationalized attempts by different authorities to intervene upon the vital characteristics of human existence—human beings, individually and collectively, as living creatures who are born, mature, inhabit a body that can be trained and augmented, and then sicken and die … the term 'biopolitics' [can be usefully defined] to refer to specific strategies brought into view from this perspective, strategies involving contestations over the ways in which human vitality, morbidity, and morality should be problematized, over the desirable level and form of the interventions required, over the knowledge, regimes of authority, and practices of intervention that are desirable, legitimate, and efficacious" (Rose 2007, p. 54).
2. There are at least four other major trends or threads in American public health (biopolitics) that I will not discuss in this chapter, each of which could be the subject of its own chapter: (1) a shift in perspective, from an almost obsessive concentration on local communities to an increasing realization that federal and global actions are needed to protect and improve the health of populations; (2) the developing merger of medicine and public health, exemplified by the expansion of the domain of public health from communicable diseases, like tuberculosis and HIV/AIDS, to chronic ones, like diabetes and hypertension, and the concentration in discussions of health insurance on prevention and wellness programs; (3) the globalization of public health and the rise of international health organizations, perhaps best illustrated by the global swine flu pandemic of 2009 and the role of the World Health Organization in addressing it; and (4) the attempted post 9/ll merger of public health and public safety under the rubric of emergency preparedness. These four threads are discussed in my *Worst Case Bioethics* (2010).

Cases Cited

Buck v. Bell, 274 U.S. 200 (1927).
Gonzales v. Carhart, 550 U.S. 124 (2007).
Jacobson v. Massachusetts, 197 U.S. 11 (1905).
Roe v. Wade, 410 U.S. 113 (1973).
Simon v. Sargent, 346 F. Supp. 277 (1972).
U.S. v. Lopez, 514 U.S. 549 (1995).

References

Agamben, G. 2002. *Remnants of Auschwitz: The witness and the archive*. New York: Zone Books.

Annas, G.J. 2005. *American bioethics: Crossing human rights and health law boundaries*. New York: Oxford University Press.

Annas, G.J. 2007. The supreme court and abortion rights. *The New England Journal of Medicine* 356: 2201–2207.

Annas, G.J. 2009a. The American right to health. *Hastings Center Report*, September/October 3.

Annas, G.J. 2009b. Abortion politics and health insurance reform. *The New England Journal of Medicine* 361: 2589–2591.

Annas, G.J. 2010. *Worst case bioethics: Death, disaster, and public health*. New York: Oxford University Press.

Bishop, J.P. 2008. Biopolitics, Terri Schiavo, and the sovereign subject of death. *Journal of Medical Philosophy* 33: 538–557.

De Tocqueville, A. 1990. *Democracy in America*. New York: Vintage Books.

Eco, U. 1988. *Foucault's pendulum*. New York: Ballantine Books.

Engelhardt Jr., H.T. 2006. Global bioethics: An introduction to the collapse of consensus. In *Global bioethics: The collapse of consensus*, ed. H.Tristram Engelhardt Jr.. Salem: M & M Scrivener Press.

Esposito, R. 2008. *Bios: Biopolitics and philosophy*. Minneapolis: University of Minnesota Press.

Fairchild, A.L., D. Rosner, J. Colgrove, R. Bayer, and L.P. Fried. 2010. The exodus of public health: What history can tell us about the future. *American Journal of Public Health* 100: 54–63.

Foucault, M. 2003. *Society must be defended: Lectures at the College de France, 1975–1976*. Trans. David Macey. New York: Picador.

Garrow, D.J. 2007. Don't assume the worst. *New York Times*, April 21, A25.

Glantz, L.H., W.K. Mariner, and G.J. Annas. 1992. Risky business: Setting public health policy for HIV-infected health care professionals. *The Milbank Quarterly* 70: 43–79.

Gopnik, A. 2009. *Angels and ages: A short book about Darwin, Lincoln, and modern life*. New York: Knopf.

Hitler, A. 1925/1999. *Mein Kampf*. Trans. R. Manheim. New York: Houghton Mifflin.

Johnson, S. 2006. *The ghost map: The story of London's most terrifying epidemic—And how it changes science, cities, and the modern world*. New York: Riverhead Books.

Knowles, J. (ed.). 1976. *Doing better and feeling worse: Health in the United States*. New York: W. W. Norton & Company.

Kolata, G. 2009. Panel urges mammograms at 50, not 40. *New York Times*, November 17, A1.

Levy, B.-H. 2006. *American vertigo: Traveling America in the footsteps of Tocqueville*. New York: Random House.

Macey, D. 1993. *The lives of Michel Foucault: A biography*. New York: Vintage Books.

Mariner, W.K. 2003. Public health law: Past and future visions. *Journal of Health Politics, Policy and Law* 28: 525–552.

Mariner, W.K., G.J. Annas, and L.H. Glantz. 2005. *Jacobson v. Massachusetts*: It's not your great-great-grandfather's public health law. *American Journal of Public Health* 95: 581–590.

Mill, J.S. 1964. *Autobiography of John Stuart Mill*. New York: New American Library.

Mill, J.S. 1974/1859. *On liberty*. London: Penguin. [First pub. 1859].

Nelson, H.D., K. Tyne, A. Naik, C. Bougatsos, B.K. Chan, and L. Humphrey. 2009. Screening for breast cancer: An update for the U.S. Preventive Services Task Force. *Annals of Internal Medicine* 151: 727–737.

Perry, J.E. 2008. Partial birth biopolitics. *DePaul Journal of Health Care Law* 11: 237–247.

Pershing, B. 2010. Some foes of health-care bill hope courts will stop legislation. *Washington Post*, January 3, A3.

Rose, N. 2007. *The politics of life itself: Biomedicine, power and subjectivity in the twenty-first century*. Princeton: Princeton University Press.
Sack, K. 2009. Screening debate reveals culture clash in medicine. *New York Times*, November 20, Al.
Sack, K., and G. Kolata. 2009. Breast cancer screening policy won't change, U.S. officials say. *New York Times*, November 19, Al.
Turoldo, F. 2009. Responsibility as an ethical framework for public health interventions. *American Journal of Public Health* 99: 1197–1202.

Chapter 8
Medicine and Philosophy: The Coming Together of an Odd Couple

Carson Strong

8.1 Introduction

The great bluesman Muddy Waters wrote a song in which he said, "The blues had a baby, and they named it rock and roll." One could apply that metaphor to the topic of the present paper, in which case one might say that medicine and philosophy had a baby, and they named it bioethics. The field of bioethics arose with the coming together of what can be described as an odd couple—the professions of moral philosophy and clinical medicine. The story of bioethics is partly an account of how these two disparate professions joined together, perceived each other, interacted, and worked out an accommodation each with the other. Of course, there were other academic fields playing a role in the creation of bioethics. Some of the earliest bioethicists were theologians and professors of religious studies; and there were lawyers, social scientists, nurses, physicians, and professors of literature, among others. The interaction of medicine and philosophy was only part of the story, but it is worth discussing because it was a significant part. The thesis that I develop in this essay makes two main assertions: the coming together of moral philosophy and medicine changed both of these fields; and the characteristics that bioethics turned out to have were greatly influenced by the ways in which the interaction between moral philosophy and clinical medicine unfolded in the context of medical education. In this paper, I draw in part upon my own experience as a philosopher who was recruited to teach ethics to medical students and residents.

C. Strong, Ph.D. (✉)
Department of Medicine, College of Medicine, University of Tennessee
Health Science Center, 956 Court Avenue, Suite G212, Memphis, TN 38163, USA
e-mail: cstrong@uthsc.edu

J.R. Garrett et al. (eds.), *The Development of Bioethics in the United States*,
Philosophy and Medicine 115, DOI 10.1007/978-94-007-4011-2_8,

8.2 Factors that Brought Philosophers into the Clinical Setting

During the late 1960s and early 1970s, the medical profession in the U.S. began seeking input from philosophers and other humanists, in a search for help in dealing with a number of issues. The medical profession was facing challenges from many directions. At least five main issues are worthy of mention: changes in the practice of medicine had reduced the amount of personal closeness between patients and physicians; the rise of patient autonomy challenged traditional medical paternalism; abuses in medical research involving human subjects had come to light, calling into question the ability of medicine to regulate itself; advances in medical technology were creating new and controversial ethical issues for health care professionals; and there was widespread dissatisfaction with the nature of medical education. A brief review of these problems will be presented.

8.2.1 The Increasing Estrangement Between Patients and Doctors

David J. Rothman (1991) describes an important change in the doctor-patient relationship in the U.S. between 1930 and 1960. He makes a convincing argument that the close personal relationships typical of doctor-patient interactions in the pre-World War II era evolved into an estrangement by the 1960s. In the early 1930s it was common for physician-patient encounters to take place in the patient's home, not in a medical office or hospital. A survey of over 8,000 families in 18 states from 1928 to 1931 reported that 56% of physician-patient interactions involved one or more house calls (Collins 1940). The significance of seeing patients in their own homes was described by Francis Peabody, a professor of medicine at Harvard in the 1920s:

> When the general practitioner goes into the home of a patient, he may know the whole background of the family from past experience; but even when he comes as a stranger he has every opportunity to find out what manner of man his patient is, and what kind of circumstances makes his life... What is spoken of as a "clinical picture" is not just a photograph of a sick man in bed; it is an impressionistic painting of the patient surrounded by his home, his work, his relations, his friends, his joys, sorrows, hopes, and fears (Peabody 1930, pp. 32–33).

Another account, given by a son who accompanied his physician-father on house calls, tells us that the physician was consistently given a warm welcome and was offered tea, cookies, or cakes by people grateful for his presence and anxious to hear what he had to say. His father seemed to know everyone's friends and relatives and would share reminiscences and warm conversations with patients and families (Gerber 1983, p. xiv). In that lower-technology era, detailed and intimate knowledge of the patient's life and surroundings played an important role in the physician's arriving at a diagnosis and planning treatment (Rothman 1991, pp. 117–118).

This closeness fostered trust in the physician, but as time went on, this closeness disappeared. As Rothman puts it, "Practically every development in medicine in the post-World War II period distanced the physician and the hospital from the patient and the community, disrupting personal connections and severing bonds of trust" (1991, p. 127). These factors included the disappearance of the house call. By the early 1960s, home visits accounted for less than 1% of patient-doctor interactions (Friedson 1961, pp. 58–59, 66–67). By bringing patients to their offices, doctors increased their efficiency, seeing more patients in less time and increasing their earnings. This deprived the doctor of a firsthand knowledge of the patient's home environment, with the result that the physician and patient were now more apart. Another factor was the rise of medical specialization and subspecialization. This compartmentalization of medical practice brought it about that, in an increasing percentage of visits, the patient had never met the doctor before. With this estrangement, less attention was being given to the patient's life, background, and family. And the interpersonal skills that enabled a doctor to learn about those things were becoming less characteristic of physicians.

8.2.2 *The Rise of Patient Autonomy*

The origin of today's emphasis on patient autonomy can be found in legal cases that occurred during the early part of the twentieth century. In those cases, patients sued their physicians for performing treatments for which the patients either had not given consent or had not been informed of the associated risks. During that period, these types of cases were new, and the courts were faced with the challenge of articulating what interests of the patient, if any, had been harmed by the physicians' actions. The term "patient autonomy" had not yet come into use, but the courts would eventually formulate a similar concept, referring to it as the right to self-determination.

In one of the earliest cases, *Pratt v. Davis* (1906), Parmelia Davis was a patient afflicted with epilepsy. Her physician, Dr. Pratt, performed two operations upon her that he claimed were necessary. In the first operation, the doctor repaired some lacerations of the uterine cervix, basing his decision to do this on the prevailing medical view that in women there is a relationship between epilepsy and the uterus. There was evidence that Mrs. Davis was competent to give informed consent except during brief periods in which she suffered epileptic attacks. The court ruled that during the periods between her attacks she was competent to give consent. After the first operation, Mrs. Davis told the doctor that she objected to any further operations, and the doctor told her husband that she would have to be brought back to the hospital in 1 or 2 weeks for a little further treatment of minor irritation, indicating that the treatment would be medical and that no surgery would be needed.

When she returned to the hospital, Dr. Pratt had her anesthetized and surgically removed her uterus and ovaries. The patient's sister, Mrs. Howe, testified that the next day she heard her sister tell the doctor that she had been informed by the nurse that she had been operated on and asked him insistently what right he had to operate on her without her consent. Dr. Pratt testified that he did not inform Mrs. Davis that the second operation would take place. In his defense the doctor relied on an alleged implied consent given by Mr. Davis. The doctor testified that Mr. Davis had given him *carte blanche* to use his own discretion in doing what he believed to be medically best for the patient. "I told him," said the doctor, "I hoped I would not have to do anything more than the first work. At the same time I would have to have his free permission to do whatever I thought best for the case or I would not have anything to do with the case, and to that he consented" (p. 178).

The testimony of Mr. Davis indicated that he had withdrawn any implied or express consent he may have given for the second operation. The court ruled that Mr. Davis was not authorized to give proxy consent, anyway, since Mrs. Davis was competent. The court ruled that the operation performed without any consent at all constituted an intentional, unconsented touching and therefore was a battery.

The legal principle upon which the requirement of consent is based was further articulated by Justice Cardozo in a 1914 case, *Schloendorff v. Society of New York Hospitals*. He stated, "Every human being of adult years and sound mind has a right to determine what shall be done with his own body; and a surgeon who performs an operation without his patient's consent commits an assault, for which he is liable in damages" (*Schloendorff* 1914, p. 93).

It was not until 1957 that the courts articulated a requirement that physicians must provide to patients information that is relevant to the decision whether to consent to a proposed treatment. The case was *Salgo v. Leland Stanford Jr. University*. Martin Salgo had undergone a translumbar aortography and subsequently his legs became permanently paralyzed, a rare complication of the procedure. The physicians testified that they had not disclosed any of the risks to the patient. The court stated, "A physician violates his duty to his patient and subjects himself to liability if he withholds any facts which are necessary to form the basis of an intelligent consent by the patient to the proposed treatment" (*Salgo* 1957, p. 181). This new requirement of informed consent raised a number of important theoretical and practical questions: What exactly is the right to self-determination, and what is its moral basis? What does it mean to say that a patient is competent to make decisions? Does the right to informed consent apply to all medical procedures, or are there exceptions? What information does a patient need to be given in order to be adequately informed? Now that informed consent was a legal requirement, these questions called out for answers.

8.2.3 Research Abuses

Medical research came under the microscope when reports emerged of vulnerable subjects being used as a means to further scientific knowledge. This examination of research involving human subjects was stimulated by an article published in 1966 by Henry K. Beecher describing 22 examples of studies that involved a variety of ethically questionable practices, including failing to obtain informed consent, exposing patients to excessive risks, withholding known therapies, and exploiting the availability of mentally impaired subjects. One of Beecher's examples was a 1963 study in which cancer researchers injected live cancer cells into chronically ill patients (Beecher 1966). Within the scientific community, this particular study was considered to be an important exploration of cancer immunology. Previous studies had established that cancer cells injected under the skin would be rejected by healthy subjects in 4–6 weeks and by cancer patients in 6 weeks to several months. To confirm the hypothesis that the slower rate of rejection in cancer patients was due to their cancer and not to the general debility that accompanies any chronic disease, it was necessary to perform the experiment on patients chronically ill with nonmalignant diseases. Live cancer cells were injected into 21 chronically ill non-cancer patients at the Jewish Chronic Disease Hospital in Brooklyn, New York. The subjects were told that a lump would form that would go away in a few weeks. They were not told in clear language that the procedure was a research study unrelated to treatment of their own illness, nor were they informed that the injected substance was cancer cells (Langer 1966).

Perhaps the most notorious example of abuse was the Tuskegee syphilis study, which came to light in 1972. The purpose of the study was to observe the natural history of untreated syphilis. The subjects were approximately 400 African American men with syphilis who lived in and around Tuskegee, Alabama. The study began in 1932, and the men were followed up and observed for four decades. In the early 1950s, penicillin became the accepted treatment for syphilis. However, the subjects were not offered penicillin nor were they told that penicillin was a possible treatment. Many of them later suffered the morbidity and mortality associated with advanced syphilis (Rothman 1982).

Disturbingly, the approaches to research revealed by these studies were common at that time. Although the principle of informed consent had been recognized since the Nuremberg Trials, it was widely accepted within the medical science community that the need for informed consent in a particular study should be left to the conscience of the investigator. Similarly, the determination of whether a study was ethical was left to the judgment of the investigator alone (Jonsen 1998, p. 143). At the National Institutes of Health, an internal survey revealed that only nine of 52 grant-conferring departments had policies concerning the rights of research subjects. In light of this, the NIH Director appointed a committee to review the ethical issues in human medical research, and the committee's report concluded that the judgment of the investigator is not sufficient to determine whether a study is ethically acceptable (Jonsen 1998, p. 143). This new idea did not sit well with many medical researchers, who countered that it would be unethical to allow those who are not experts in the relevant sciences to make judgments about whether a study should proceed. A national debate ensued over whether there should be greater oversight of human subjects research (Jonsen 1998, pp. 145–146).

8.2.4 Issues Raised by New Biomedical Technologies

Rapid advances were occurring in the development of technologies for testing, monitoring, and treating patients. In many areas of medicine, these advancements were creating new and difficult ethical issues. Two examples will suffice to illustrate these developments: the allocation of medical resources for dialysis; and issues involving quality of life and end-of-life care.

8.2.4.1 Allocation of Dialysis Resources

In the 1950s, several American medical centers were constructing experimental artificial kidney machines. Each time a patient was dialyzed with one of these machines, a cannula (tube) had to be inserted into an artery and vein. Because multiple needle sticks cause vessels to sclerose, the machine could be used only a limited number of times for each patient. In 1960 Dr. Belding Scribner, of the University of Washington School of Medicine in Seattle, developed a shunt that permitted a patient to use a kidney machine over and over, for long-term dialysis. The cannulas in the artery and vein were left in place and joined by means of a connecting tube (shunt), permitting blood to run through the shunt without clotting and thereby maintaining the cannulas in functional condition. Dialysis using this method worked well from the start, and the number of patients receiving chronic dialysis began to grow. The King County Medical Society obtained a grant to open the Seattle Artificial Kidney Center in 1962, but it soon became apparent that more patients were being referred for dialysis than could be accommodated in the nine-bed center. The Board of Trustees of the medical society tackled this allocation problem by creating a Medical Advisory Committee and an Admissions and Policy Committee. The former would determine which patients were medically suitable for chronic dialysis, and the latter would decide which of the medically suitable patients would be permitted to receive the treatment (Fox and Swazey 1974, pp. 215–217, 240–250; Jonsen 1998, pp. 211–215).

The Admissions and Policy Committee consisted of seven members, initially a minister, businessman, homemaker, lawyer, labor leader, and two physicians who were not kidney specialists. The committee reviewed applicants case by case, examining extensive personal and social information about each. It developed a list of factors it considered relevant to its decisions: age, gender, marital status and number of dependents, income, educational background, occupation, "past performance," and "future potential." For several years, the committee used these considerations, which came to be referred to as "social worth criteria," to decide who would live and who would die. In 1962, this process was reported in *Life* magazine by journalist Shana Alexander. Her article set off a national debate over the appropriateness of the committee's selection process. The Seattle approach was criticized for not giving each applicant an equal chance and for being biased toward "middle class values." Among the consequences of the debate, two are especially relevant to the present discussion: doubt arose about the ability of the medical profession to make such decisions in an ethically defensible manner, given that the Seattle method seemed clearly wrong; and it became apparent that it was necessary for American society to address the difficult ethical question of how best to allocate scarce medical resources.

8.2.4.2 End-of-Life Care

In the 1960s and 1970s, new technologies were permitting us to keep alive patients who had seriously compromised functioning of the heart and lungs, patients who would have died in earlier times. Cardiac monitors, respirators, and new resuscitation techniques were becoming part of standard care in neonatal, pediatric, and adult intensive care units. In 1973, neonatologists Raymond S. Duff and A. G. M. Campbell brought new issues to the attention of a wide audience when they reported, in the *New England Journal of Medicine*, that approximately one out of every seven newborn deaths in their nursery resulted from a decision to withdraw or withhold life-saving treatment (Duff and Campbell 1973). In some cases, the fact that the baby was handicapped played a role in the decision to allow death. This article fueled an ongoing national debate over a number of central questions: Do parents have the right to withhold life-preserving treatment from handicapped infants? How aggressively should we attempt to keep alive neonates born so small that they are on the borderline of viability? Is severe brain damage resulting from birth asphyxia grounds for withholding treatment?

Similar issues about quality of life and prolongation of life were arising in adult intensive care units. A widely-recognized case involved Karen Ann Quinlan, who became comatose after consuming alcohol, barbiturates, and Valium. At the emergency department, she was placed on a respirator because of difficulty breathing. After 5 months in the hospital, she remained unconscious and her physical condition was deteriorating; she was becoming emaciated and her body assumed a fetal-like position. At that point, her parents made the agonizing decision to ask the doctors to remove Karen's respirator and allow her to die. However, the medical profession was divided at that time on the question of whether it would be ethical to withdraw life support in a case like this. Doctors also were fearful of legal prosecution for willfully allowing a patient to die. Karen's doctor refused to remove the respirator, after which her parents petitioned the court for an injunction to force the hospital to withdraw life support. After some delay, the case reached the New Jersey Supreme Court, which ruled that Karen had a right to self-determination, that it was appropriate for her parents to exercise that right on her behalf, and that in doing so they could choose to have the respirator removed (*In re Quinlan* 1976). Subsequently, Karen was weaned from the respirator, but she unexpectedly began to breathe on her own. This development raised yet another issue: would it be ethical and legal to withhold artificial nutrition and hydration from an irreversibly unconscious patient?

8.2.5 Dissatisfaction with Medical Education

Medical education was being examined critically, and it had a number of shortcomings. Medical students were expected to memorize huge amounts of information, little of which was retained after examinations. Insufficient attention was being given to independent learning and the development of problem-solving skills. There was generally a lack of clinical relevance in the teaching of basic science during years one and two of medical school, and there was not enough focus on preventive medicine. Medical faculties had grown larger and training programs more complex, leading to less personalized faculty contact with students and residents. Within medical schools, recognition and rewards for faculty favored research and patient care, at the expense of teaching. Medical students and residents were subject to verbal abuse and other forms of mistreatment, and residents were overworked. Medical education had become highly competitive, fostering excessive concern for personal career success rather than attitudes of caring directed toward patients. Amid these various shortcomings, an overarching flaw was that medical education emphasized technology and gave insufficient attention to developing appropriate attitudes toward patients, values, and interpersonal skills. Too many students were graduating without knowing how to treat their patients as persons. The shortcomings in the doctor-patient relationship associated with the increasing estrangement of doctors and patients were not being addressed in medical education (Pellegrino 1978; American Board of Internal Medicine 1983; Jonas 1984; Huth 1984).

8.2.6 Summary

All of these challenges were facing medicine, and there was growing concern that the task of meeting them was beyond the capabilities of the medical profession alone. It was evident that physicians were not adequately taking into account the values of their patients. Insufficient attention was being given to patient autonomy. A new duty to obtain informed consent needed to be clarified. In deciding how best to regulate research, perspectives from those outside of medicine would have to be taken into account. It was apparent that biomedical technology had created, and would continue to create, new ethical issues that were impacting the broader society and were difficult to resolve. Efforts to reform medical education were needed, including a need to address the fostering of humanistic attitudes and interpersonal skills during medical training. The medical profession reached out to others, including the academic disciplines of the humanities, for help in making medicine more humane and in addressing these many challenges.

8.3 The Contrasting Natures of Moral Philosophy and Medicine

The fields of philosophy, religious studies, history, and literature were being beckoned. Looking back, it is ironic that philosophy was included in this call, given the direction that philosophy had been taking during the twentieth century. To a great extent, moral philosophy in the English-speaking world at that time had become removed from "real-world" problems. It had become focused on the analysis of *concepts* relevant to moral discourse. The tone and content of much of moral philosophy had been set by philosophers such as G.E. Moore and A.J. Ayer, and their influence brought about an attack on normative ethics from a number of directions.

Moore critiqued the utilitarianism of Mill, not by challenging the normative claims of utilitarianism, but by arguing that Mill committed a logical fallacy in attempting to define the term 'good' (Moore 1903). In doing so, Moore turned the discussion away from normative principles and toward the analysis of the meanings of words. According to Moore, the concept "good" is indefinable, like the color yellow. Part of Moore's analysis consisted in making a distinction between "natural" and "non-natural" properties. Natural properties can be discerned by the senses; non-natural properties cannot be perceived by the senses. According to Moore, yellow is a natural property that, although indefinable, can be apprehended by means of visual perception. On the other hand, "good"—by which Moore apparently meant "goodness"—cannot be detected by the senses and therefore is a non-natural property. Moore held that Mill's mistake consisted in trying to define what is indefinable. In addition, Moore interpreted Mill as defining "good" as "what is desired," which amounted to defining a non-natural property in terms of a natural property. Moore claimed that non-natural properties cannot be reduced to natural properties, and that Mill had made an additional mistake in this regard. Based on these considerations, Moore labeled Mill's mistake the "naturalistic fallacy." This critique of Mill was enormously influential within moral philosophy during the first half of the twentieth century, in part because the turning of attention toward the meanings of key terms was widely regarded as original and insightful.

A.J. Ayer introduced to the English-speaking world the views of the Vienna Circle, a group of philosophers and scientists whose doctrines are known as logical positivism (Ayer 1936). A theory about the meaning of statements is a central element of logical positivism. It holds that the only statements that have meaning are those that are empirically verifiable and those that are analytic—i.e., are true in virtue of the definitions of their terms. Assuming that normative statements are neither empirically verifiable nor analytic, it follows from logical positivism that normative claims have no meaning. Because they are meaningless, they can be neither true nor false. This doctrine undermines the idea that one can defend the validity of normative claims by giving arguments that purport to support their truth. If ethical claims are neither true nor false, then the question arises as to what is their nature, and Ayer put forward a view about this. His theory, known as emotivism, holds that normative statements express emotions of approval or disapproval.

To say that "X is wrong" is simply to express one's disapproval of X and to attempt to influence others to disapprove of X too. On this view, there is no objectivity in normative ethics, and the idea that there could be a plausible argument in support of a normative statement is mistaken.

With these views as a backdrop, moral philosophy turned more and more toward the analysis of terms and concepts: What is the difference between "right" and "good"? What are the senses of the term "duty"? What kinds of judgments are properly classified as "moral"? Normative ethics gave way to these new philosophical activities, referred to as "metaethics," a term coined by Ayer. A great deal of effort was expended arguing for and against various metaethical assertions. Although some philosophers continued to debate the merits of general normative theories such as utilitarianism, the analysis of concepts came to play a heavy role even in those discussions (e.g., an analysis of the differences between act and rule utilitarianism) and rarely was a normative theory used to try to resolve an actual real-world moral problem. The interest was solely in knowing which theory is "best" according to philosophical standards and not in using a theory to guide actions. Philosophical writing was abstract and general. Attention to specific situations and cases occurred mainly in the context of creating hypothetical counterexamples to refute general principles. Academic philosophy was becoming self-absorbed, further and further removed from any concerns outside of itself. In describing moral philosophy during this period, one commentator stated, "One of the consequences of treating ethics as the analysis of ethical language is … that it leads to the increasing triviality of the subject…" (Warnock 1960, pp. 203–204).

Although it had become largely theoretical, the discipline of philosophy had taught its members valuable reasoning skills. These included an ability to discern the logical validity of arguments and sensitivity to different senses and uses of a term. Philosophers emphasized consistency and clarity. They asked probing questions. They were good at identifying assumptions and exploring where assumptions lead. They could reason rigorously about normative matters. They were attuned to asking about the foundations of claims. They understood that justifiable beliefs require evidence and the support of good arguments. They learned to distinguish empirical claims from those claims that have built into them a disallowance of their own disproof. All of this is part of conceptual analysis. Even so, philosophers were not accustomed to discussing concrete moral problems. When I was first invited to take a position teaching bioethics at a medical school, in 1978, I thought that my job would be to teach these skills to medical students and residents. But I soon realized that my initial expectations required substantial revisions. At that time, schooled in a traditional philosophy department, I did not appreciate the differences between medicine and philosophy.

Physicians have a very different cognitive style, compared to philosophers. Medical science is tangible, in comparison to the abstract theories of philosophy, and the thinking of physicians tends to be concrete, specific, and sequential (Belgum 1983). They are accustomed to reasoning in the face of ambiguities, uncertainties, and probabilities. Moreover, what C.P. Snow described as the "two cultures" does in fact exist within medical schools. Most medical students have a background of heavy study in the natural sciences, and they view medicine as itself a science. Many have come to view ethics as "soft" because it is nonquantifiable and therefore regard it as not susceptible to rigorous thinking. This perception makes communication between the two cultures difficult (Belgum 1983; Brody et al. 1975). Many in the field of medicine did not understand what is involved in medical ethics. It was regarded as focusing on a search for the "right answers," and when there is no clear right answer, a natural conclusion is that ethics is a waste of time.

In addition, clinical ethical issues have features with which philosophers tended to be unaccustomed. First, understanding what the ethical issue *is* in a clinical case requires a certain amount of understanding of the medical facts of the case. Second, relevant facts usually include the psychosocial dimensions of cases—including the patient's family situation, religious beliefs, ethnic and cultural background, and financial factors—and an effort must be made to uncover these. Third, there often is uncertainty about the facts of the case, a common example being uncertainty about the patient's prognosis both with and without treatment. Fourth, a *decision* about how to handle the case is needed, and typically within a relatively short period of time. Fifth, the decision should be sensitive to the moral pluralism of our society and its communities. Although many philosophers reject religious beliefs, for example, as being mere metaphysics, in clinical medicine there is a need to respect the moral pluralism that is found among patients. Sixth, the decision should be within the constraints of law, institutional policies, and professional codes, and therefore should be informed by knowledge of these. Seventh, resolving real cases in which parties disagree requires not only intellectual skills but also interpersonal skills. These include, but are not limited to, the ability to listen well, skill in communicating, and the ability to demonstrate respect, empathy, and support to involved parties. As these considerations suggest, when moral philosophy initially joined with clinical medicine, they were far apart in terms of assumptions and methodologies.

8.4 Emergence of the Teaching of Ethics in Medical Education

The earliest formal teaching programs in ethics and humanities at medical schools began in the late 1960s. The first three were at Pennsylvania State University, the University of Florida, and the State University of New York at Stony Brook (Pellegrino and McElhinney 1981, p. 8). In 1968, the Society for Health and Human Values was founded, attracting medical school faculty from around the country who were interested in promoting the teaching of human values as part of the education of health professionals. An arm of the Society, the Institute on Human Values in Medicine, received a grant from the National Endowment for the Humanities to support the providing of consultations to medical schools on how to initiate ethics teaching programs. With the leadership of Edmund D. Pellegrino, institute members visited 80 medical campuses over a 10-year period, bringing together interested parties and serving as a sounding board for their ideas (Bickel 1987, p. 370).

As ethics and humanities programs came into being at medical schools, the Institute collected and published detailed descriptions of their structure and curricula.

The Institute of Society, Ethics, and the Life Sciences, later to be called The Hastings Center, was founded in 1969 to promote the interdisciplinary study of issues in biomedicine. In 1972, the Institute surveyed all American medical schools to learn the extent of medical ethics teaching (Veatch and Sollitto 1976). Of the 95 schools that responded, only four had a required course in medical ethics and 37 offered an elective course. At an additional 37 schools, ethics was discussed in courses on other subjects and there was no teaching of ethics beyond this. Occasional special programs on medical ethics, such as case conferences or lectures, were reported by 17 schools. A total of 81 schools reported some type of formal teaching of ethics. Among those with no formal teaching, a common response was that ethics was taught by the transmission of values from medical faculty to students at the bedside.

The Institute repeated its survey in 1974 (Veatch and Fenner 1975). Of 112 schools surveyed, 107 responded. The number with required courses in medical ethics increased to 6, and schools with elective courses grew to 47. The number that offered ethics only as part of other courses decreased to 19, and the number offering occasional special programs increased to 56. The total number that offered some type of medical ethics teaching increased to 97. Clearly, a trend toward increased teaching of ethics was under way. As this trend continued, a shift in thinking was taking place; schools were moving from teaching ethics only at the bedside toward the development of formal curricula. At some schools, a shift from thinking in terms of ethics courses toward the development of ethics or humanities programs was taking place. Along with this, there was an increase in the number of faculty with specific responsibility to teach ethics. In the 1974 survey, 31 faculty were identified whose teaching and research was exclusively in ethics, either full or half time, compared to 19 so identified in 1972. Physicians, ethicists, and chaplains each composed about one-third of the 31 (Veatch and Fenner 1975).

Integrating ethics into medical school curricula was not easy, for several reasons. First, the teaching formats commonly used in moral philosophy do not fit well into a typical medical school curriculum. Ethics, as in philosophy more generally, is best taught by Socratic methods that promote dialogue between teacher and student. Small classes are desirable because they give students a greater opportunity to interact with the teacher, whereas large audiences inhibit some students from speaking and make for unwieldy group discussions. For medical schools that have large numbers of students, scheduling small classes presents logistical problems during the preclinical years. Because all students are learning the same material, it is efficient to have large classes for science courses. Assuming that the number of qualified ethics teachers at a school is small, it is unwieldy to schedule small classes for ethics but large classes for all the other courses. Because of this, in part, at large medical schools there was resistance among curriculum administrators to teaching ethics in small groups during the preclinical years.

Second, ethics also found itself competing with the traditional medical school subjects for time in the curriculum. It is characteristic of medical schools that students are expected to learn large quantities of scientific information about a range of subjects. The science departments want as much curricular time as possible so they can teach their material. Medical schools dominated by scientists have a tendency not to give much curricular time (or funding) to a "soft" subject like ethics, particularly when the nature and relevance of ethics teaching is not well understood. Third, medical students tend to focus heavily on the goal of passing their board examinations. They are aware that the board exams tend to contain only a few ethics questions. As a result, in their studies they generally give priority to technical and scientific subjects.

These factors have a great influence on how ethics is taught in medical schools. The initial teaching typically involved the occasional inclusion of ethics topics in ongoing teaching forums, such as grand rounds and case conferences (Brody 1984). An approach that I used was to attend ongoing teaching conferences regularly and comment on ethical issues as they arose. On some days ethical issues would come to the fore and other times they would not. Most philosophers in medical schools avoided the sorts of theoretical discussions that they had been trained to carry out. They also avoided controversial ethical issues and the arguments for each side because that would simply reinforce the belief among students that in ethics there are no right answers. Instead, they focused on practical matters, teaching the skills and knowledge that the students would need in order to take care of patients. Case studies were often used to address curricular goals that typically included the following: how to recognize moral issues and distinguish moral issues from other issues; understanding main ethical principles and how they apply to medicine; and how to resolve ethical problems by using ethical reasoning and thinking clearly and critically. Co-teaching with clinicians was considered to be a good model because it permitted discussions to be informed by relevant medical knowledge.

Slowly, the teaching of ethics in medical schools became more widely accepted. By 1984, the Association of American Medical Colleges (AAMC) had issued a set of recommendations on improving medical education, referred to as the GPEP Report. The document states, "...medical faculties should emphasize the acquisition and development of skills, values, and attitudes by students at least to the same extent that they do their acquisition of knowledge"; and by "values and attitudes," it meant those that "promote caring and concern for the individual and for society" (Association of American Medical Colleges 1984, p. 1). During 1985, the Liaison Committee on Medical Education revised the accreditation standards for M.D. programs to include the ethical, behavioral, and socioeconomic subjects pertinent to medicine (Liaison Committee on Medical Education 1985). The inroads into medical education made by ethics teaching is documented by a survey of medical schools conducted by the AAMC during the 1984–1985 school year. Of 126 schools surveyed, 114 responded. Ninety-five (83%) reported having a required ethics course during the first or second year of medical school, and 38 (33%) had a required course during the third or fourth year (Bickel 1987).

8.4.1 Teaching Ethics in Residency Programs

Ethics teaching in residency programs proceeded more slowly than teaching in the preclinical years of medical school. One reason is that many of the nonphysician ethicists who were coming into medical education lacked the clinical experience needed to teach effectively in the residency years. As time went by, there was an increased degree of integration of ethics teaching into the clinical setting. Philosophers and other ethicists began going on rounds with their physician colleagues, learning about the nature of the clinical setting, the psychosocial dimensions of patient care, and the rudiments of medicine. At my own institution my fellow philosophers, Terrence F. Ackerman and E. Haavi Morreim, and I soon realized that becoming a successful teacher in the clinical setting required establishing our credibility, which in turn required learning enough medicine to understand the facts of a case when it was presented on rounds. We took the view that ethics teaching should not be an "add-on"—something extraneous to the regular training programs—but should be integrated into the ongoing activities of the academic departments with which we worked. We recognized that the task of learning the language of medicine was daunting, and to make it a feasible one, we decided to specialize. Ackerman worked mainly with the departments of medicine and family medicine, Morreim with pediatrics, and I focused on obstetrics and gynecology and neonatology. In doing this, we narrowed considerably the amount of medical knowledge each of us needed to learn in order to do our jobs. I would spend considerable time going on rounds and participating in the typical residency teaching activities—case conferences, grand rounds, and morbidity and mortality conferences.

The teaching of ethics in residency programs received a boost in 1981, when a task force of the American Board of Internal Medicine that was charged to define clinical competence stated that humanistic qualities are essential for certification as an internist. This led to the issuing of a report in 1983 laying out a framework for cultivating the humanistic qualities of trainees. The report stated that the essential humanistic qualities required of candidates seeking certification are integrity, respect, and compassion, and that a major responsibility of those training residents in internal medicine is to stress the importance of the humanistic qualities throughout residency (American Board of Internal Medicine 1983).

The Society for Health and Human Values initiated a project to promote the teaching of humanities and human values in primary care residency programs. In 1981 and 1983, two national conferences were held to promote discussion of this topic. In the proceedings of these conferences, descriptions of teaching programs revealed that there were at least 15 primary care residency programs nationwide in which there was formal teaching of humanities or human values (Boufford and Carson 1984). By 1985 a Residency Interest Group (RIG) had formed within the Society for Health and Human Values. The RIG became the main U.S. professional organization for the teaching of ethics and humanities in residency programs. Its members included persons who either were participating in the formal teaching of ethics or other humanities to residents, were planning to undertake such teaching, or otherwise had an interest in this area.

In 1989 the RIG conducted a national survey to obtain information about the teaching of ethics and humanities in residencies. It began by mailing questionnaires to the 130 members of the group. This enabled the investigators to contact a substantial number of persons who were teaching ethics or humanities in residencies in the U.S. Recognizing that some teachers in this area might not belong to the RIG, the study design involved asking respondents to suggest names of nonmembers who teach these subjects in residency programs. In addition, announcements were placed in the *Bulletin of the Society for Health and Human Values* inviting teachers of ethics and humanities in residencies to participate in the survey. These methods identified 33 additional teachers. The number of respondents was 94, for a response rate of 58%. The results of the study indicated 63 faculty who participated in formal ethics teaching programs for residents, representing 50 graduate medical education institutions. Twenty-seven of the 63 teachers (43%) were physicians. Among nonphysicians, 17 had backgrounds in philosophy, 11 in religious studies or theology, two in literature, and two in law. Other backgrounds mentioned included history, social work, and pastoral care. Among the teaching programs reported, 34 were in internal medicine, 18 in pediatrics, 14 in family medicine, nine in psychiatry, eight in obstetrics and gynecology, five in surgery, two in emergency medicine, one in dermatology, and one in ophthalmology (Strong et al. 1992).

The most frequently reported setting for the teaching of ethics to residents was in hospitals. Thirty-four faculty (54%) reported in-hospital settings exclusively, and 21 (33%) reported teaching in both hospitals and outpatient offices or clinics. A variety of formats were reported, including case management conferences, lectures, ethics rounds, grand rounds, case consultations, lunch conferences, and teaching rounds. In outpatient settings, ethics was taught in case-management conferences, lectures, seminars, and discussions in the office concerning ongoing cases (Strong et al. 1992).

8.5 How Moral Philosophy and Medicine Changed Each Other

Although medical educators were willing to give philosophers a chance to demonstrate what they could contribute, many within the medical profession felt unease about the new ethicists. Their presence was perceived as an incursion into the domain of physicians by outsiders who lacked an understanding of research and practice. There was a fear of having to hear dogmatic pronouncements by ethicists and a fear of being denounced. It became apparent that ethics consultations often did not give "the correct answer" like other consults, and therefore there was some questioning of the value of such consultations. For students, the idea that ethics has little practical value was reinforced by the fact that it was not covered in their course examinations. Many physicians believed that ethicists were self-designated reformers, out to make people be moral. There was a belief that ethics cannot be taught and that attempts to teach it would not make a difference.

Within a few years these feelings fomented into a backlash against ethicists. A paradigm example was an article in the *Journal of the American Medical Association* entitled "Medical Ethics' Assault Upon Medical Values" (Clements and Sider 1983). The theme of the article was that the new emphasis on patient autonomy, put forward by medical ethicists, was an attack on the traditional values of medicine, particularly concern for the well-being of the patient. It lamented that paternalism had become a pejorative term. It should be noted that the article's argument rested on a serious error in the way the authors described autonomy. The concept of autonomy as used by bioethicists was equated with the concept as used by Kant; the two are not the same. In addition to this error, the article appeals to naturalism, a highly controversial view that normative statements can be derived from descriptive statements. The acceptance of this article for publication in a major medical journal indicated a lapse in critical reviewing. Its publication also suggests a tendency among physicians at that time to be sympathetic toward the paternalistic views expressed and a lack of understanding, by the editors, of the concept of autonomy in bioethics.

With the passage of time, however, the backlash faded as the disciplines of medicine and philosophy worked together. A more correct understanding of ethics and the role of ethicists began to spread within the medical profession. Philosophers in the clinical setting were establishing their credibility. Physicians began to understand that ethics is not simply about getting "right answers," but concerns the process of getting to an answer. It encourages one to be explicit about one's principles and the moral arguments supporting one's views (Belgum 1983, p. 11).

Moral philosophy and the other humanities influenced medicine in a number of ways. Doctors have learned some of the basic language of ethics, using the terms autonomy, beneficence, and nonmaleficence in describing their duties. One can also see the effects in medicine's own treatment of ethics. Consider specialty medical professional organizations, for example. Most of these organizations, such as the American Academy of Pediatrics, the American College of Obstetricians and Gynecologists, and the American Academy of Physicians, have come to have bioethics committees that produce position statements on ethical issues that arise within their specialties. That this has come about is a sign that the medical organizations have become more self-reflective about ethical issues. To some degree, philosophical skills are reflected in these position statements, such as making ethically relevant distinctions between types of cases.

Other changes to medicine are related to the various problems that had faced the medical profession. For example, the importance of patient autonomy has become ingrained in physicians, and paternalism is no longer considered to be a generally appropriate approach. With the input of philosophers and lawyers, the questions about how and when to obtain informed consent were addressed, and these topics are now regularly taught to medical students and residents. Concerning the research abuses, a system for overseeing medical research was put in place. This consisted of federal regulations that were based upon the recommendations of the National Commission for the Protection of Human Subjects of Biomedical and Behavioral Research, an interdisciplinary bioethics group established by statute and appointed by the Secretary of the Department of Health, Education, and Welfare. The Commission's recommendations were based, in turn, on an earlier document written by the Commission entitled *The Belmont Report*, whose purpose was to articulate and explain the basic ethical principles that should govern the regulation of human subjects research. The Belmont Report was itself an interdisciplinary project, but it was heavily influenced by the contributions of philosophers H. Tristram Engelhardt, Jr., Tom L. Beauchamp, and Stephen Toulmin (Jonsen 1998, pp. 103–104). Initially, many physicians complained about what they perceived as negative features of the new regulations, such as creating paperwork and imposing rules made by outsiders. Although complaints continue, today the system is generally accepted and is deeply entrenched in academic medicine.

In regard to ethical issues raised by new technologies, an important change in medicine is an increased use of hospital ethics committees and other types of ethics consultations in dealing with difficult cases and end-of-life issues. To be accredited, hospitals are required to have formal mechanisms for addressing ethical issues, and many hospitals satisfy this requirement by maintaining ethics committees. Initially, many physicians resisted using these committees, under the mistaken assumption that their purpose is to make decisions for doctors. With time, physicians came to understand that the committees are merely advisory, and they learned that ethics consultations can be helpful. Concerning medical education, the teaching of ethics has become a standard part of medical school training; the Liaison Committee on Medical Education has made it a requirement that there be formal teaching of ethics and human values during the preclinical as well as clinical years of medical school. Virtually every type of specialty in residency training has a professional organization that formulates curriculum objectives, and it is difficult to find one that does not include ethics among the objectives. Perhaps the most intractable of the various problems has been the too-often impersonal nature of the doctor-patient relationship. Systems for financial reimbursement of doctors have motivated many physicians to schedule more patients and spend less time with each, thereby contributing to the ongoing problem of estrangement.

8.5.1 How Moral Philosophy Changed

Its interaction with medicine resulted in several major changes in moral philosophy. A main one was a shift in direction toward the practical. Philosophers noticed that the new issues in biomedicine raised a number of interesting philosophical questions. At first the emphasis on conceptual analysis remained, but it was directed to topics outside of philosophy itself. An early example was the publication, in philosophy journals, of a number of articles addressing questions about the concept of paternalism: What exactly is paternalism in medicine? Are there different senses of paternalism? When, if ever, is a physician justified in being paternalistic? Philosophy journals began publishing occasional articles on other bioethics topics. There was a steady appearance of articles on the definition of death, another topic ripe for conceptual analysis. Articles appeared that addressed the controversial issue of abortion: What is the moral status of the human fetus? Under what circumstances is abortion morally justifiable? A new philosophy journal appeared, *Philosophy and Public Affairs*, which reflected this practical turn, and many of its articles dealt with issues in biomedicine. Another new journal, the *Hastings Center Report*, proclaimed that its subject matter primarily was ethical issues in medicine and the life sciences. The academic field of bioethics was beginning. A number of other bioethics journals emerged, and they were dominated by discussions of new difficult issues created by biotechnology. Philosophers began taking sides on concrete moral issues!

Within normative ethics, a subset of moral philosophy, changes also began to take place. As moral philosophers in medical centers became more familiar with the clinical setting and the nature of moral problems arising there, they came to realize that there are problems in applying general normative theories like utilitarianism and Kantianism to clinical cases. Advocates of these theories could arrive at any number of disparate resolutions for a given case. This came to be known as the problem of bridging the gap between theories and cases. Clinical ethicists came to realize that it was more fruitful to identify professional duties and "middle-level principles" and to argue for a prioritization of duties and principles in the context of a specific case. This involved a return of attention to the idea that special relationships generate special duties, with a focus on the duties of the health professions. The use of middle-level principles like autonomy and beneficence came to be known as "principlism." The search for arguments that could be used in defending prioritizations led to the rebirth of casuistry (Jonsen and Toulmin 1988). Several approaches to casuistic reasoning were developed (Strong 1999), giving attention to the specifics of the case and following the old maxim "circumstances alter cases." Although these developments were taking place within bioethics, discussions of them were finding their way into traditional philosophy journals.

8.6 And They Named It Bioethics

The new academic discipline of bioethics that emerged has unique characteristics. It embraces many of the features that are typical of philosophy, such as seeking clarification, making distinctions, and analyzing concepts. Like traditional moral philosophy, there are writings in bioethics that are abstract and theoretical. But bioethics also differs in a number of ways from traditional philosophy. By definition, bioethics addresses issues in biology and medicine. It is oriented generally toward the practical—how cases should be handled and what policies would be ethically preferable. This orientation is a result of philosophers and other humanists going into the clinical setting, primarily as medical educators, and dealing with the ethical problems that arise there. This real-life experience profoundly alters one's perspective on how to do ethics. Bioethicists came to regard *usefulness* as one of the main tests for the acceptability of a normative theory, and they turned away from utilitarianism and Kantianism in part because those theories are unable to pass the test.

Another characteristic feature of bioethics is its interdisciplinary nature, with contributors from various backgrounds including philosophy, religious studies, law, and the health professions, among others. In part because of this collaborative environment, bioethicists are generally more willing to accommodate ethical views based on religious beliefs, in comparison to many authors in the traditional field of moral philosophy. Its interdisciplinary nature also influences the style of writing in bioethics, which generally avoids the arcane terminology and pedantic style of more purely philosophical writings. Unlike most work in traditional moral philosophy, it often draws upon empirical data from medicine and the social sciences. Some of the works that count as bioethics actually consist of empirical studies relevant to practical ethical issues. An odd couple worked out its differences and created something that is a blend of the two.

References

American Board of Internal Medicine. 1983. Evaluation of humanistic qualities in the internist. *Annals of Internal Medicine* 99: 720–724.

Association of American Medical Colleges. 1984. *Physicians for the twenty-first century. Report of the panel on the general professional education of the physician and college preparation for medicine*. Washington, DC: Association of American Medical Colleges.

Ayer, A.J. 1936. *Language, logic and truth*. New York: Dover Publications.

Beecher, H.K. 1966. Ethics and clinical research. *The New England Journal of Medicine* 274: 1354–1360.

Belgum, D. 1983. Medical ethics education: A professor of religion investigates. *Journal of Medical Ethics* 9: 8–11.

Bickel, J. 1987. Human values teaching programs in the clinical education of medical students. *Journal of Medical Education* 62: 369–378.

Boufford, J.I., and R.A. Carson (eds.). 1984. *The teaching of humanities and human values in primary care residency training*. McLean: Society for Health and Human Values.

Brody, H. 1984. Michigan State University medical humanities program. In *The teaching of humanities and human values in primary care residency training*, ed. J.I. Boufford and R.A. Carson. McLean: Society for Health and Human Values.

Brody, H., W.B. Weil, B.L. Miller, and J.E. Trosko. 1975. Integrating ethics into the medical curriculum: One school's progress report. *Michigan Medicine* 74: 111–117.

Clements, C.D., and R. Sider. 1983. Medical ethics' assault upon medical values. *Journal of the American Medical Association* 250: 2011–2015.

Collins, S.D. 1940. Frequency and volume of doctors' calls.... *Public Health Reports* 55: 1977–2012.

Duff, R.S., and A.G.M. Campbell. 1973. Moral and ethical dilemmas in the special-care nursery. *The New England Journal of Medicine* 289: 890–894.

Fox, R.C., and J.P. Swazey. 1974. *The courage to fail*. Chicago: University of Chicago Press.

Friedson, E. 1961. *Patients' views of medical practice*. New York: Russell Sage.

Gerber, L.A. 1983. *Married to their careers*. New York: Tavistock Publications.

Huth, E.J. 1984. The humanities, science, and the medical curriculum. *Annals of Internal Medicine* 101: 864–865.

In re Quinlan. (1976). 335 A. 2d 647.

Jonas, H. 1984. The case for change in medical education in the United States. *The Lancet* 2: 452–454.

Jonsen, A.R. 1998. *The birth of bioethics*. New York: Oxford University Press.

Jonsen, A.R., and S. Toulmin. 1988. *The abuse of casuistry*. Berkeley: University of California Press.

Langer, E. 1966. Human experimentation: New York verdict affirms patient's rights. *Science* 151: 663–666.

Liaison Committee on Medical Education. 1985. *Functions and structure of a medical school*. Washington, DC: Association of American Medical Colleges.

Moore, G.E. 1903. *Principia Ethica*. Cambridge: Cambridge University Press.

Peabody, F.W. 1930. *Doctor and patient*. New York: The MacMillan Co.

Pellegrino, E.D. 1978. Medical humanism and technologic anxiety. In *The role of the humanities in medical education*, ed. D. Self, 1–7. Norfolk: Teague & Little.

Pellegrino, E.D., and T.K. McElhinney. 1981. *Teaching ethics, the humanities, and human values in medical schools: A ten-year overview*. Washington, DC: Institute on Human Values in Medicine.

Pratt v. Davis. (1906). 79 N.E. 562.

Rothman, D.J. 1982. Were Tuskegee & Willowbrook "studies in nature"? *The Hastings Center Report* 12: 5–7.

Rothman, D.J. 1991. *Strangers at the bedside: A history of how law and bioethics transformed medical decision making*. New York: Basic Books.

Salgo v. Leland Stanford Jr. University Board of Trustees. (1957). 317 P. 2d 170.

Schloendorff v. Society of New York Hospital. (1914). 105 N.E. 92.

Strong, C. 1999. Critiques of casuistry and why they are mistaken. *Theoretical Medicine and Bioethics* 20: 395–411.

Strong, C., J.E. Connelly, and L. Forrow. 1992. Teachers' perceptions of difficulties in teaching ethics in residencies. *Academic Medicine* 67: 398–402.

Veatch, R.M., and D. Fenner. 1975. The teaching of medical ethics in the United States. *Journal of Medical Ethics* 1: 99–103.

Veatch, R.M., and S. Sollitto. 1976. Medical ethics teaching: Report of a national medical school survey. *Journal of the American Medical Association* 235: 1030–1033.

Warnock, M. 1960. *Ethics since 1900*. London: Oxford University Press.

Chapter 9
The Growth of Bioethics as a Second-Order Discipline

Loretta M. Kopelman

9.1 Introduction: Bioethics' Phenomenal Success in Just 40 Years

Bioethics has become phenomenally popular in a few short decades. In the early 1970s, almost no one used the term "bioethics" and medical ethics was largely a matter of educating health-care students about professional etiquette and behavior.[1] In contrast, today substantive programs, course work and concentrations in bioethics exist in many disciplines including in philosophy, law, medicine, literature, nursing, theology, and the social sciences. (I follow common usage and use the terms "discipline" and "field" interchangeably.[2]) Specialists in bioethics are in demand not only as university professors, but also as also as consultants for news shows, clinics, communities, legislatures, professional organizations, and industry; they are sought out as expert witnesses in courts, and offered memberships on international commissions, on state and federal task forces and on research and institutional ethics committees. Around the world, more and more journals began including bioethical subjects. Over these years even the problems characterizing bioethics grew, crossing many disciplines and encompassing such topics as ethical issues regarding: euthanasia, death and dying, confidentiality, disability, genetics, research policy, patient rights, professional integrity, informed consent, professionalism, abortion, assisted suicide, personhood, health-care resource allocation, reproduction ethics, and environmental ethics.

L.M. Kopelman, Ph.D. (✉)
Department of Bioethics and Interdisciplinary Studies (Emeritus),
Brody School of Medicine, East Carolina University,
East Fifth Street, Greenville, NC 27858-4353, USA

Kennedy Institute of Ethics, Georgetown University, Washington, DC, USA
e-mail: kopelmanlo@ecu.edu

J.R. Garrett et al. (eds.), *The Development of Bioethics in the United States*, Philosophy and Medicine 115, DOI 10.1007/978-94-007-4011-2_9,

Are bioethicists riding an international wave of interest in these bioethical issues or are they creating and directing it? Warren Reich defines bioethics as "the systematic study of the moral dimensions – including moral vision, decisions, conduct and policies – of the life sciences and health care, employing a variety of ethical methodologies in an interdisciplinary setting" (Reich 1995, p. xxi). Albert Jonsen (1998) argues that bioethics is a demi-discipline, part academic discipline and part public discourse. I agree that bioethics is in part a public discourse, but will argue that the part that is a discipline is best understood as a second-order discipline.

From the beginning, questions existed about the interdisciplinary nature of bioethics. What made someone a bioethicist? After discussing the early days of the field, I will show how these tensions arose even in my role as founding president of the American Society of Bioethics and Humanities (ASBH). Some viewed bioethics as an extension of medicine while others saw it as grounded in such disciplines as law or philosophy. Still others denied it could be grounded in any one field and viewed bioethics as public discourse. I will critically examine these views and try to show why bioethics is best understood as a second-order discipline

9.2 The Early Days of an Interdisciplinary Field

In the early 1970s, a series of shocking revelations about health care and human research splashed across the media, implicating prestigious governmental, medical, educational, and legal institutions and undermining their moral authority. I will mention only a few which are now well known. First, risky human research was being done without consent to gain knowledge including: African-American men were kept from treatment of syphilis for decades in a study sponsored by the U.S. Public Health Service at the Tuskegee Institute; elderly Jewish patients were given live cancer cells without their consent in a study conducted by Sloane Kettering Hospital in New York; retarded, institutionalized children at Willowbrook State Hospital were given hepatitis; women were enrolled in placebo controlled randomized clinical trials to test birth control without their consent and, not surprisingly, many became pregnant; prisoners were routinely used in research studies. A shocked public compared these investigations to the medical research "studies" done by the Nazis in World War II and demanded policy changes. Second, the abortion controversy heated up when the US Supreme Court essentially legalized abortion for at least the first half of pregnancy; some saw this as murder of a person, while others were equally adamant it reflected a woman's right to control her body. Third, debates about how to use new technologies emerged when the parents of Karen Ann Quinlan, a young woman who was permanently unconscious, wanted doctors to stop using life-prolonging interventions. Finally, fundamental issues of fairness were debated. Scarce life sustaining resources, including kidneys for transplantation, were being allocated by means of controversial criteria, such as who attended church. Moreover, outcome data showed that most recipients where white men, something that could not be accounted for by medical criteria alone. Women and people of color demanded better treatment.

Those of us who began working systematically on these issues in the 1970s were a small group of mostly young academics who came from many fields. Initially there were so few of us working in bioethics that we knew each other well. A few senior academics, such as Jay Katz and Edmund Pellegrino, encouraged us. We sought out kindred spirits and found them in many fields: law, medicine, philosophy, literature, and so on. For this volume, I have been asked to discuss the growth of bioethics from my perspective.

As a result of my interest, I began teaching bioethics at the University of Rochester, both in the Philosophy Department and in the School of Medicine and Dentistry. My conviction that this was an interdisciplinary field was further strengthened by a meeting at the Hastings Center in June 1972, where people from many disciplines came to discuss these issues. This was also reinforced by a national conference to which H. Tristram Engelhardt, Jr. and Stuart F. Spicker invited me to speak in Galveston on May 9–11, 1974. The conference was entitled "The First Trans-Disciplinary Symposium on Philosophy and Medicine," and its proceedings were later published (Engelhardt and Spicker 1975).

We felt a surge of energy that came from the sense we were part of an important new enterprise dedicated to working on problems in a new way. Because the issues were so complex and important, we believed that they needed to be addressed using experts from many fields. We wanted to break out of our tightly insulated disciplines in the discussions, analyses, and teaching about these problems. One way we did this was through team-teaching in small groups with faculty from other fields. We were aware that new problems were constantly arising and wanted to prepare students to be good problem solvers. That we were succeeding in doing something new and important was demonstrated by the eagerness with which prestigious journals published our papers, and the disproportionate number of us appointed to important panels and commissions.

In 1978, I established the bioethics and humanities program at the Brody School of Medicine at East Carolina University (Kopelman 1995). The following year I hired my colleague and friend John Moskop and we planned a conference on "Ethics and Mental Retardation" which was held on October 1–3, 1981. It was one of the first 2-day national conferences on bioethics. The excitement it generated was evident from the speakers who were willing to come to a new medical school in Greenville, North Carolina. They included: H. Tristram Engelhardt, Jr., Jay Katz, Edmund Pellegrino, David Rothman, Joseph Margolis, Anthony D. Woozley, Cora Diamond, Robert Holmes, Robert M. Adams, William F. May, Larry Churchill, Michael Kindred, Barbara Baum Levenbook, Jeffrie G. Murphy, Larry McCullough, Arthur Kopelman, Stuart Spicker, John Moskop and me. A book of the proceedings was edited and published (Kopelman and Moskop 1984). Today, one would never get most of the leaders of the field to stay for a 2-day conference, especially given the modest honorarium and location of our Greenville conference.

At the conference there was some discussion of what was then a new approach to addressing problems in bioethics offered in writings by Tom Beauchamp and James Childress (1979, 1983, 1989, 1994, 2001, 2008). They recommended using several unranked principles, autonomy, beneficence, non-maleficence and justice, to sort out the issues and frame solutions. (Despite their best efforts in subsequent editions, they are frequently misread as ranking autonomy as the highest principle of the four.) They argued that there was agreement about what principles were important, even though there was sustained disagreements about which moral theories to adopt. Such theories would, for example, rank these principles. Rather than waiting to settle these theoretical disputes, they favored using the principles. Their "four principled" approach, or "principlism" as it has sometimes been called, has been both popular and widely criticized in the intervening decades.

Today, courses in bioethics exist in virtually every medical school, nursing school, university and college around the world, and the interdisciplinary approach has been so successful in bioethics that it is reshaping how other problems are addressed. These courses have also raised questions about who is qualified to teach them. In short, *who is not a bioethicist*?

9.3 An Identity Crisis: SHHV, AAB, and SBC Merge into ASBH

Around the time American Association of Bioethics and Humanities (ASBH) was being founded in 1997, vocal disputes erupted about the nature of bioethics and the role of bioethicists. They were certainly apparent to me as the last president of Society for Health and Human Values (SHHV) in 1996–1997, and as the founding president of the American Society for Bioethics and Humanities (ASBH) in 1997–1998.[3] ASBH was formed from the memberships of SHHV, the American Association of Bioethics (AAB) and the Society for Bioethics Consultation (SBC). The original ASBH board was composed of four members from each group.

ASBH was incorporated in 1998, but board members began working by means of conference calls and e-mail in 1997 before we hired a managing company. Not surprisingly, those were the most difficult months. We worked without pay, a precedent that continues. The board members were all faculty members at universities, and these institutions picked up the costs of phone conferences and travel because funds had not been transferred yet from the three groups from which ASBH was formed. SHHV had 842 members, AAB had 613 members, and SBC had 149 members. By 1998, ASBH had 1500 members enrolled (Jonsen, p. xiiin). Some of us belonged to all three groups, so the merger made sense economically and politically in such a small field. The three physicians who were the presidents of these organizations in 1995–1996, Robert Arnold (SHHV), Steven Miles (AAB) and Stuart Youngner (SBC), worked hard to make the merger a reality.

SHHV was the oldest and largest of the three and was founded primarily by ministers, theologians and physicians in 1965 (Jonsen 1998). By the time I became its last president, it consisted of a plethora of interest groups (program directors, philosophy, nursing, literature, law, history, various medical specialties, etc.) that demanded – and got – money from the board of directors and received dedicated non-peer reviewed time during the national program. The money they received depleted the coffers of the organization, and the time they got during the national conference diminished the time available for peer-reviewed submissions. Many members complained to the board that the national conference seemed fractured and devoid of standards.

Without a doubt, however, these interest groups are extremely important. With a small and relatively stable leadership they can bring up cutting edge issues and give members "homes" in a large organization where they can find like minded scholars, mentors and collaborators. Nonetheless, as the last president of SHHV, I saw how destructive this approach of giving so much money and control to interest groups had been and sought to avoid this as founding president of ASBH. The board agreed. Many members had left the troubled SHHV to join SBC or AAB for these or related reasons.

The first ASBH Board tried to strike a balance. We decided the groups would be given time to meet that would be listed in the program but not money; and that national program would be almost entirely peer reviewed with contributions blinded. During that first year (1997–1998), the interest groups reformed as "affinity groups." Many members of ASBH from the SHHV assumed that they would again have "their" money and dedicated time for their own non-peer review programs at the national meeting, preferably plenary sessions. The new ASBH board spent a considerable amount of time explaining this was a new organization and that no group was entitled to either money or dedicated time in the national program. One small but vocal group was furious that their members and favored subjects got no spot on the national program even when I pointed out that none of them had made any submissions. For the sake of peace, we gave them one hour in the program for that year only.

Worse still, several students familiar with the old SHHV policy and without consulting the ASBH Board, advertised a generous cash award for the best student essay. The students assumed that they would get money from the ASBH Board as they had from the SHHV Board and that they alone would decide who got the prize. In their advertisements they further stirred up the membership by calling ASBH, "the former SHHV." The students were shocked to find they could not simply assume they would get this money and control over who should get the prize. Former AAB and SBC members were irritated at ASBH having been called "the former SHHV." We were able to pull all the students' advertisements except one that, unfortunately, appeared in widely-read *The Hastings Center Report*. In the end, we allocated the money for the student essay award since it had been advertised and because we judged it to be intrinsically worthwhile, but did not give the students the control over selecting the winner. (The ASBH still makes this award.) The affinity groups have gotten no funds from the ASBH Board and typically meet early in the morning or late in the afternoon.

The second organization in the ASBH merger was the SBC. It had been formed by Albert Jonsen and others who were interested in addressing the moral problems and issues arising in the clinical setting. Members sought to become experts in examining or facilitating case consultations and determining what should be done in particular cases. Jonsen and Toulmin's scholarly work in the methodology of casuistry offered a theoretical underpinning for this movement (Toulmin 1981; Jonsen and Toulmin 1988; Jonsen et al. 1982).[4]

The third organization in the merger was the AAB. It was formed out of a meeting called by philosopher Daniel Wickler and others held in the NY Academy of Medicine. Most of its attendees had degrees in medicine, philosophy, or law, and many were representatives of other organizations such as the AMA and SHHV. One goal was to have members who could be called upon to offer expert guidance on complex issues in bioethics. The Hastings Center aspires to similar goals for its fellows, providing contacts to those seeking expertise in some area (such as stem cell research).

In the summer of 1997 in the first year of the ASBH, the board held all but one of its meetings by conference-call. On these phone calls, the board members seemed to me to retain their pre-merger loyalties as representing SHHV, AAB, or SBC. Our first meeting around a table was at a joint meeting of the three soon-to-be-merged organizations in October, 1997. I consulted a friend who was an industrial psychologist about how to make the board more like one group and she recommended we all introduce ourselves. I pointed out that most of us had known each other professionally for many years. But the twist she recommended was that we must introduce ourselves more personally, in terms of influences in our lives, interests, hobbies, family life, etc. This seemed odd and even somewhat invasive to me, I confess. So I asked another psychologist, who gave me the same advice. Somewhat skeptically, I did as they recommended. As luck would have it, the first person to speak was anthropologist Patricia Marshall who promptly leaped up and had us laughing with a hilarious account of who she was and what had influenced her life. Others did the same, and by the time we had gone around the room, the group had gelled. Bioethics had taught me to be open to the recommendations from people in other disciplines.

9.4 An Identity Crisis: Four Views About Bioethics

In the sometimes volatile discussions among the membership about the merger of these three organizations, four views of bioethics emerged that are still apparent. Some conflated bioethics with public discourse, holding that no experts existed in bioethics. Others maintained that there are experts and standards for bioethicists based on one existing discipline (law, medicine and philosophy were often chosen by defenders). Still others argued that bioethics is evolving into a new discipline. In what follows I will argue that bioethics is a second-order, interdisciplinary discipline and the other views are problematic.

9.4.1 Is Bioethics Just Public Discourse?

First, some held that bioethics is just public discourse. On this view bioethics has no special ties to any one discipline with its baggage of acknowledged experts, core texts, and special methods of organizing material, and unique theories or standards of excellence. Defenders often seek diversity, opposing so-called "standards for expertise" as elitist or politically motivated ways to exclude people (Regenberg and Mathews 2005). Other defenders use the term "ethics" to refer to general values, rather than rigorous study of the presuppositions or basis for our ethical values and judgments.

Some link this view with an atheoretical or "situation ethics" approach to solving problems. Those defending atheoretical or situation ethics argue that people making ethical judgments should take in the entire context, viewing an event in its particularity, yet understanding its historical and social settings. Attempts to abstract from the particular features, on this view, will lead to errors since what is relevant in one context is not relevant in another (Fletcher 1966). Simon Blackburn (1996, pp. 80, 352) likens this view to the aesthetic theory of contextualism, which holds that works of art need to be apprehended holistically in their historical or cultural context. On this view, the moral situation needs to be taken as a whole like a work of art. Works of art have features, such as a brilliant color, which are appropriate in one situation but not for another. Similarly, what is relevant in some moral situations is not in others.

Some feminists defending versions of such views (and not all do, of course) are deeply distrustful of so-called experts, rules, or theories (Jaggar 2000; Kopelman 2006).

They focus on individuals' "embodiment" in their social and personal context. They hold that this reveals interdependence, dependence, inequalities, relationships, and communities with unique social and historical features, and it is unrealistic to try to abstract from or transcend them. Moreover, theories miss a great deal of what is important by devoting so much time to abstractions like impartiality, human rights, equality, autonomy, ideal communities, or some notion of universal human conditions.

If bioethics expertise exists, on this view, it should be defined as those having special insight, skill, and knowledge about how individuals are embedded in their social, relational, and personal surroundings, and about how to react properly and sensitively to that circumstance in finding a good basis for action relating to some range of bioethical issues. While this view has been important in focusing attention on how problems arise in specific personal and social contexts and useful in case consultations and counseling, it is problematic.

9.4.1.1 Criticisms

One difficulty with this position is that solving problems in bioethics is a practical activity, not just a reaction to situations. Different features exist in situations and should be identified and ranked or balanced in importance to help us try to be consistent. Consent may "feel" highly relevant in one situation but irrelevant in another. But we need to do the work of determining why. It may be that in one case, the person is competent, but in the other he is not. We can have a consistent policy once we distinguish between competent and incompetent persons. To be consistent and identify the right reasons for acting, people need to take the time to identify relevant similarities and differences among situations and rank important features. Moreover, our experiences and feelings may be unreliable guides for action because of bias and ignorance, or because large or diverse groups have very different experiences that will impede confirmation or generalizations based upon experiences.

A second problem with this position is its denial of expertise and standards. Even "market place" criteria strongly suggest there is more to expertise in bioethics than feelings and insights about particular situations. Successful bioethicists are recognized as experts by means of such "market" criteria as: (1) publication in first tier peer reviewed journals or by well known publishers on bioethics subjects; (2) holding research positions in bioethics and/or gaining promotion and tenure in university appointments; (3) membership on editorial boards or important academic, national or international positions or commissions relating to bioethics subjects; (4) election to important offices in organizations devoted to bioethics; or (5) appointments in sought after and influential positions associated with bioethics. These "market" criteria of success are meant as a rough approximation of who merits recognition as an expert in bioethics.

The use of the market to settle who is an expert has limitations. The above criteria are not necessary, if some people who do not merit success flourish in these ways; moreover, they are not sufficient, if some who merit being viewed as successful are not by these criteria – such as superb bioethics teachers who do not publish, never get promoted or tenured, or win no elections or appointments to prestigious offices. Nonetheless, even though imperfect, these market criteria help solve a rigging problem about how to determine who is an expert. The problem arises because we cannot pick out experts without agreement about standards of expertise, but we cannot set up standards of expertise unless we agree about who are experts. The problem is whether there is a reasonable way to identify either experts or standards of expertise without rigging the outcome to suit unwarranted biases. These market place standards, albeit imperfect, avoid this rigging problem and show successful bioethicists come from a variety of fields such as law, philosophy, medicine, religious studies, social science, literature, dentistry, and nursing.

If this is reasonable, then the view that bioethics is just public discourse without experts seems problematic. Many bioethicists are sought after as experts by discerning people. For example, there are recognized experts in using and critiquing certain codes and guidelines, such as those relating to research ethics, resource allocation, or health or environmental policy. They serve, for example, on institutional review committees or research ethics committees where they use and apply federal research guidelines to judge whether research protocols should be approved. Thus, expertise in bioethics might sometimes justifiably be understood as having special skill and knowledge about how to understand, critique, and comply with some regulations or guidelines that have grown in response to issues in bioethics, and having the knowledge and ability to present arguments that would convince reasonable persons of good will about the meaning, use, and limitations of these guidelines.[5]

Of course, many well-established bioethicists lack expertise in any particular set of issues or regulations and no one has expertise in all of the topics in bioethics. Nonetheless, since a kind of bioethics expertise is recognized in the marketplace, the burden of proof seems to be on those denying experts in bioethics (Kopelman 2006).

9.4.2 *Is Bioethics Grounded in a Single Existing Discipline?*

Second, in contrast to those who view bioethics as public discourse, some hold that it "belongs" to a single discipline and that expertise in bioethics not only exists but is grounded in that field. I will discuss two of the several disciplines who have members making these claims: medicine and philosophy and then criticize them.

9.4.2.1 Medicine's Alleged Claim

Some physicians view medicine as the discipline grounding bioethics. Defenders argue that problems in bioethics generally arise in a medical context, and many of these issues are not new.[6] They can explain the explosion of interest in them *now* as due to medicine's success in developing often costly interventions and technologies that can alter the quality and length of one's life. Moreover, it became increasingly apparent that reasonable and informed people of good will have different views about how to allocate these resources. In addition, they disagree about how to set research policy to balance the interests of individuals and groups in developing better treatments, and deciding how to honor civil liberties and duties to vulnerable people. When little could be done to preserve biological life, for example, we could still agree to "do everything." It is now clear that reasoned differences exist about, for example, how to decide what quality of life merits life-prolonging technologies. It is increasingly apparent that the quality and length of life may depend on one's access to health care and new technologies. It is not surprising, then, that bioethics has grown in popularity as medicine has advanced and raised concerns about access, cost, and control.

Because so many bioethical issues arise in a medical context, some physicians see themselves as uniquely qualified to deal with these problems. Some argue that they should be captains of the ship of bioethics and others (philosophers, nurses, social workers, lawyers, etc.) are their support staff. Other physicians see bioethicists from outside medicine as threats to their traditional roles and duties (Satel 2000). Doctors, on this view, need to reclaim the issues from poorly prepared, albeit well-intentioned, interlopers lacking in experience with the clinical encounter.

9.4.2.2 Philosophy's Alleged Claim

In contrast, some philosophers maintain bioethics is really a part of philosophical ethics (Clouser 1978). A common way to define "ethicist" is as someone who has expertise in ethics and so, on this view, bioethics experts specialize and are competent in that aspect of philosophical ethics associated with bioethics.

This analysis of bioethics expertise focuses on epistemic competency and rational justification. Core curricula and thresholds of competence in bioethics, on this view, are rooted in those of philosophy and ethics, and competency in bioethics is related to competency in philosophy and ethics, such as passing examinations that are prerequisites for the doctoral degree. None of this, however, requires agreement on normative or substantive issues. While most agree this epistemic expertise exists, there is a lively debate among philosophers about whether and to what degree normative decisions lie outside ethics or bioethics (Holmes 1990; Frey 1978). As Lisa Parker has said how we conceive expertise in ethics, "…depends in part on whether expertise is a matter of knowing, thinking, doing, living or being. Different answers to this invite (or depend on) different conceptions of the nature and goals of moral philosophy" (Parker 2005, p. 167).

9.4.2.3 Criticisms

One problem with giving either medicine or philosophy (or law or other contenders) the preeminent place as grounding bioethics, is that they are implausible as the sole basis for the normative and political enterprise that bioethics has become. Its topics increasingly extend beyond the expertise of any one field, incorporating more and more issues related to human flourishing, rights, welfare and dignity (see e.g. UNECSO 2005). For example, doctors may understand how to do kidney transplantation, but are not experts about how states should determine what portion of its budget should go to transplantations. Philosophers may be experts in reflecting on the assumptions underlying just allocation of scarce resources, but many see it as beyond the scope of their expertise to forge agreement about how to make such allocations. Philosophers *qua* philosophers, moreover, are generally uninterested in promoting consensus or advocating for some policy, often preferring to clarify, question, or challenge. Yet advocacy is a primary concern of some bioethicists.[7] Rather, philosophers may seek the most defensible and nuanced arguments – not necessarily those likely to forge a consensus or stir political action.

Another problem is that these approaches minimize the contributions made by people from different fields. This was clear to me from the first conference I attended on what would come to be regarded as a topic in bioethics. It was at Johns Hopkins hospital around 1971, and the then chair of the department, Dr. Robert Cooke, previewed a film he had made about that baby who would soon become widely known as "the Hopkins baby" (Gustafson 1973, pp. 529–530). The infant had been born with trisomy 21 and a duodenal atresia (blockage) that made feeding by mouth impossible. When the parents refused surgery, the clinicians accepted their decision.

It took 10 days for the baby to die of dehydration and starvation. The film, which was shown extensively, begins with a dramatization of these events and ends with an exchange among a theologian, a psychologist and clinicians who disagreed about whether they could have supported the parents' decision. I later learned this story was really a composite of three cases from the 1960s at Hopkins.[8]

At the conference, the clinicians struggled with concepts and controversies that were familiar to those from other disciplines, including what rights children have independent of their parents and what duties professionals and others have to seek court interventions to overrule parental decisions about what is in their child's best interest. It was clear that philosophers and others could make contributions to advancing the discussion of what ought to be done. In part, as a result of this conference, my career path changed from theory of knowledge and philosophy of mind to bioethics.

Thus the difficulty with trying to locate bioethics in any one discipline is that the base is too narrow; it ignores the impact of public discourse and minimizes the contributions of people from other fields. If we look to the market for guidance, it offers the more impartial assessment that many respected bioethicists come from fields other than medicine or philosophical ethics. Similar arguments could be raised against those who regard law, history, sociology, anthropology, or other fields as grounding bioethics. Successful bioethicists come from many fields because they have special contributions. For example, lawyers generally make their contribution to bioethics literature using legal theory and texts, not medical or moral theory.

9.4.3 Does Bioethics Expertise Justify a New Discipline?

Yet another view of bioethics sees it evolving into a new discipline or field. Some argue it is important to establish codes, standards, a core curriculum, degrees, or certification to build public trust that those who call themselves bioethicists are competent (Baker 2005; Wolf and Kahn 2005; Kipnis 2005; Perlman 2005; MacDonald 2003). The version that I will discuss favors developing bioethics as a separate academic field or discipline using what I will call a "functionalist" approach.

Functionalism, as I will use this term herein, is the view that analyzes the meaning of something in terms of its roles, tasks, operations, or purposes. A functionalist analysis of bioethicists' expertise should be understood in terms of (1) their roles in social institutions; (2) the values needed to exercise those roles; and (3) how these values and roles fulfill important purposes in social institutions, including public expectations about how these roles should be fulfilled. I will distinguish two forms. The first describes these codes, roles, values, and purpose but does not judge what they ought to be. Even if bioethicists may agree about these descriptions we know that they disagree about how these codes, roles, values or purposes ought to be understood.[9] Rather than stipulate how to understand them (something that would probably be ignored by those who disagree anyway) defenders want to work for consistency and consensus using what I will call the normative version of functionalism.[10]

In using the *normative version of functionalism*, defenders try to establish how we *ought* to understand bioethicists' roles, tasks, operations or purposes. They seek to do this by beginning with descriptions of what they do. By means of criteria such as consistency and consensus defenders then seek to clarify and refine how these roles, tasks, operations or purposes ought to be understood to ensure competency and build public trust.

This normative version of functionalism has recently been defended by Robert Baker in drafting a professional code of ethics for bioethicists:

> Bioethicists engage in clinical case consultations, serve on research ethics committees (IRBs), provide ethics education to students, professionals, and the public, inform media about complex ethical issues, engage in research and publications, and serve as a community resource. The social authority, trust, and power vested in those assuming the role of bioethicist, or accepting the title of 'bioethicist' makes it imperative that bioethicists' act in a manner worthy of that authority, trust, and power (Baker 2005, p. 38).

Baker recommends a code for bioethicists to make them worthy of this trust, authority and power:

> The public, and the public institutions like the courts, naturally presume that anyone claiming the title 'bioethicist' in an institutional context is publicly asserting some measure of expertise and responsibility about bioethical issues – particularly with respect to their own institution (Baker 2005, p. 36).

Tom Beauchamp is critical of Baker's proposal because it omits a definition of 'bioethics' or 'bioethicists;' he notes that these locutions can be narrowly construed to exclude many people who claim to be bioethicists, or broadly construed to be inclusive. Baker, Beauchamp says, instead offers a standard too weak to be taken seriously as, "anyone who assumes *the role* of bioethicist or *accepts the title* 'bioethicist' is a bioethicist" (Beauchamp 2005, p. 42). Beauchamp also criticizes the lack of specificity in Baker's proposed code: "Basic notions in this code, such as 'conflict of interest,' 'integrity,' 'independence,' 'personal gain,' 'conflicting responsibilities,' 'intellectual interests,' are neither defined nor developed in the form of a body of specific rules" (Beauchamp 2005, p. 43). Many of the points Baker makes in his draft code about these values or notions, argues Beauchamp, are not unique to bioethics and would need to be far more highly specified to be applied to bioethicists alone rather than as general moral considerations. Beauchamp's criticisms seem justified but there may be another way to understand Baker's proposal.

Another way to read Baker's draft of a professional code of ethics for bioethicists would be as offering what I have called a normative functionalist approach; that is, the goal would be to try to overcome discord and offer professional guidance to people who call themselves bioethics; this could, he seems to say, build trust, authority and power for bioethicists by clarifying how they ought to act using consistency and consensus to capture defensible and central values, purposes, roles, and public expectations.

Baker's justification of a professional code or field for bioethicists relies heavily on his discussions the circumstances surrounding Mary Faith Marshall being subpoenaed to testify; it concerned a policy at Medical University of South Carolina

(MUSC) at the time she was the director of its bioethics program. MUSC had cooperated in gathering and reporting information about pregnant patients with addictions to cocaine or other controlled substances that might harm their fetuses. Because of her testimony, the MUSC's president decided not to submit her candidacy for promotion to the Board of Trustees. He backed down when notified of the danger of a possible lawsuit by the American Association of University Professors (AAUP). Marshall had a nursing degree and a doctorate in religious studies. Baker contended that these degrees offered no professional code for "...her responsibilities as a *bioethicist*" (Baker 2005, p. 35). He continued:

> What was at issue in the case was not a relationship with a client, or with a patient, but rather her ethical and legal responsibilities, as a bioethicist, towards the institution employing her, to those whom the institution served, and as someone testifying under oath (Baker 2005, p. 35).

Baker believed that bioethicists have special roles and that a professional code is needed to clarify bioethicists' moral responsibilities in exercising those roles. He also contends that bioethicists fulfill important social purposes, including satisfying public expectations about how these roles should be fulfilled.

Bioethics expertise using a functionalist analysis might be understood as someone having special skill and knowledge about (1) identifying important roles people have in social institutions that address some range of bioethics problems (such as the duties of IRB members); (2) understanding how these roles in institutions serve important social purposes, including public expectations about how to fulfill these roles properly (such as duties to protect research subjects); and (3) clarifying and further understanding bioethicists' responsibilities by first describing bioethicists' roles and then clarifying, refining, specifying, or abstracting the essential features about how they ought to act through such means as consensus or consistency.

9.4.3.1 Criticisms

The overarching problem with a normative functionalist account of expertise in bioethics is that bioethicists disagree about their proper roles in social institutions, about the values or codes needed to exercise the roles, or about what social purposes or public expectations should shape bioethicists' roles. It is possible that agreement will "evolve" as the field matures, as Baker suggests. However, the obstacles to agreement should not be minimized.

First, bioethicists disagree about the proper purpose and scope of defined roles and codes. Some, such as Perlman, agree with Baker that a professional code of ethics for bioethicists would assist those with roles working in industry; it would "provide professional guidance to practitioners in the field [and] afford them protection when threats against professional standards are brought from external forces" (Perlman 2005, pp. 63). Others are sharply critical of these roles and associations, arguing they are inherently presumptuous, self-serving, or corrupting (Cooter 2004; Elliott 2005).

Even those who favor codes to articulate or refine these roles, values, purposes, and expectations, disagree about whether it is possible or important to have one code for the entire field of bioethics. Baker supports one code, while Kenneth Kipnis (2005) favors having a code only for those who do clinical ethics consultations. He disagrees with Baker's method of "trying to extract formulas for cooperation out of discord…. Taking issue with Baker, [Kipnis] would urge that the field narrow its focus initially to hospital ethics consultants." Others favor developing codes, but reject the functionalist methodology and want to begin with clear definitions of what constitutes bioethics or bioethicists (Beauchamp 2005).

Second, bioethicists disagree about the need for its own code of ethics. Baker tries to show the need for a professional code of ethics for bioethicists by discussing the testimony by Mary Faith Marshall that embarrassed her institution. Baker contends that a code for bioethicists was needed because she testified as *a bioethicist*, not as someone with a degree in religious studies or nursing. Others deny that this shows there is a need for a professional code of ethics for bioethicists (Miller 2005; Lantos 2005). Lantos writes,

> …Marshall did not lack substantive guidance about her role in an ongoing institutional controversy. And her choice of actions was supported by a professional organization, the American Association of University Professors. Ultimately, she prevailed. Baker read this case as a sign that the field of bioethics needs a code of ethics. But the case could be read in just the opposite way-that existing codes were adequate, allowing Marshall to ultimately prevail (Lantos 2005, p. 46).

Third, bioethicists disagree about what expectations the public should have regarding bioethicists. Miller disputes Baker's claim that the public has many expectations about bioethicists to act independently of other professions. Miller maintains there is no basis for Baker's analogy that bioethicists are expected to act something "in the manner of an accountant or internal auditor" (Miller 2005, p. 51). Miller questions that Mary Faith Marshall was auditing anything.

Schuklenk and Gallager (2005) write that drafting a professional ethics code for bioethicists ignores that there is a difference between the roles of professionals and bioethicists. Namely, professions such as medicine, law, and nursing have statutory bodies with authority to regulate their affairs and censure members.

> Importantly, there has to be a clear *public interest in granting privileges in return for imposing standards*; we cannot conceive of such interest in bioethics or for that matter any other area of applied ethics. The best we can hope for are the application of academic standards of a discipline on publication in journals of status in peer review… (Schuklenk and Gallager 2005 p. 64).

Still others oppose such codes, anticipating that drafting them will exclude people from the field when what they want is more diversity (Regenberg and Mathews 2005).

Thus, developing a code or set of core competencies for bioethics is controversial. The attempt to forge a new discipline of bioethics is even more divisive. A necessary (but not sufficient) condition of a profession or academic discipline is that members have some unique expertise with its standards defined clearly enough to mark out who does not have this expertise. The current debate about whether bioethics should adopt a code or become a separate discipline is entangled with the debate about whether, or in what way, bioethicists are experts.

Using the current literature, I elsewhere examined and rejected nine different kinds of views about the sort of expertise that might be unique to bioethicists to determine if any one of them might support a new field (Kopelman 2006). In addition to functionalism, philosophical ethics, medical expertise, and atheoretical or situation ethics, discussed above, I also considered analyses of expertise based in "casuistry," "conventionalist relativism," "institutional guidance," "regulatory guidance and compliance," "political advocacy," and "principlism." None succeed in identifying a unique area of expertise for bioethicists that could serve as a basis for a "marketable" new profession or academic field shared by enough successful bioethicists to be justified as non-arbitrary (Kopelman 2006).[11] I argued, in contrast, that expertise in bioethics is rooted in many fields, and bioethics is best understood as a second-order, interdisciplinary discipline.

Even if I am correct in concluding that none of these nine views identify an area of bioethics expertise that could be used to forge a "marketable" new discipline, this would not prove that the interdisciplinary model I will propose in the next section is correct. There might be better defenses of one or more of these nine positions; or some entirely different analysis might be forthcoming about how bioethicists are so uniquely proficient they might sell a new discipline to universities, government, or industry. I hope to show, however, that viewing bioethics in part as public discourse and in part as a second-order, interdisciplinary academic discipline clarifies or explains several important features about this field. I begin by discussing what I mean by a second-order interdisciplinary discipline.

9.4.4 *Bioethics as an Interdisciplinary and Second-Order Discipline*

In the last section I considered claims about whether bioethics might be evolving toward a new discipline. Consider that most new fields splinter off from existing fields, such as psychology did from philosophy, or intervention radiology did from radiology, in order to narrow the subject matter, make it more manageable, and promote greater expertise in the narrower field. In contrast, bioethics goes the opposite direction, with bioethicists expanding the subject matter and seeking additional expertise. This is because the problems addressed are complex, and solutions are envisioned as interdisciplinary.

Consider too an analogy to another field, which, like bioethics, expands the subject and brings in expertise from many fields because solutions are envisioned as interdisciplinary. Associations addressing problems of pain management for patients conceive themselves as interdisciplinary and as welcoming other perspectives. For example, The American Pain Society (APS) describes itself on its website as a "multidisciplinary educational and scientific organization dedicated to serving people in pain."[12] Like the field of bioethics, the goal of APS is not to make the subject narrower and foster greater expertise for fewer persons in a smaller area, but to promote understanding, policy development, or interdisciplinary responses to pain. Nurses, teachers, patients, doctors, social workers, lawyers, social scientists, clergy, and others are encouraged to join this organization because their different professional and academic fields bring important perspectives in trying to solve the various ethical, social, research, teaching, or practice aspects of pain control. For example, policy makers can help address misconceptions about addiction to pain medicines that have been enshrined in laws that have become barriers to better pain management. This interdisciplinary approach is compatible with there being sub-specialties (such as those interested in addressing legal obstacles to prescribing good medications or promoting better care at the end of life).

In what follows I argue that bioethics is best understood as a second-order discipline and offer the following analysis of it: *A second-order discipline* has members who (1) generally do not seek to make their subject matter narrower; (2) envision the problems they address as complex and profiting from the views of people from different professions, disciplines, and academic fields, and seek to broaden the perspective and expertise available to address those problems; and (3) rely in part upon the primary professions, disciplines, or academic fields to set their own educational or other standards of competency for their members.

While bioethics is best understood as a second-order, interdisciplinary discipline, there are some features that roughly unite bioethics as a field (Kopelman 1998). I have argued that this family resemblance clusters around six features.[13] First, bioethicists generally focus a significant portion of their efforts on the momentous moral and social issues that are associated with bioethics (including death and dying, confidentiality, disability, informed consent, abortion, professionalism, euthanasia, assisted suicide, personhood, health-care resource allocation, environmental ethics, and the impact of new technologies including genetic and reproductive technologies). Second, bioethicists generally use interdisciplinary approaches to analyze problems. Third, they generally employ practical reasoning and cases in teaching and writing. Fourth, many use Dewsonian teaching methods focusing upon making students better problem-solvers to give them the skills to address a variety of old and new problems. Fifth, many bioethicists seek not only to clarify the problems, but also to find morally justifiable solutions to the problems driving the field. Finally, bioethicists generally value interdisciplinary and collaborative approaches to solving these problems. While I still maintain that this set of family resemblances exists, they are insufficient to mark a unique area of expertise for the purpose of establishing a new discipline.

9.4.4.1 How This Clarifies Bioethics

Regarding bioethics as a second-order discipline clarifies or explains several of its features. First, it explains why it is difficult for bioethicists from different fields to identify, specify, or set thresholds for a core curriculum, codes, roles, norms, competencies, or duties among bioethicists. Although bioethicists may appreciate the importance of interdisciplinary work, our different academic or professional perspectives set and reinforce different visions about what is important.[14]

Second, it explains why bioethicists may judge that people who never regarded themselves as bioethicists should be so considered when their work, albeit grounded in their own discipline, has great relevance to some issue(s) in bioethics. The committee selecting the 2005 Fellows of the Hastings Center, for example, deliberately selected persons who had made great contributions to addressing problems of bioethics, but who had never (until then) considered themselves to be bioethicists.

Third, expanding, rather than narrowing relevant expertise acknowledges the importance of different perspectives and contributions from many disciplines in trying to clarify or solve problems associated with bioethics. Problems are often so complex that justifiable solutions need to be addressed from moral, social, legal, and empirical vantages. For example, consider policy regarding informed consent for medical treatment. Initially, there was a contentious debate about what consent for treatment or research meant and when it was needed. Important contributions from law, medicine, philosophy, psychology, religious studies, anthropology, and other fields eventually resulted in substantial agreement about how to frame the issues and set policies. As a result of extensive interdisciplinary discussions, within a generation, a largely settled and widely regarded as justifiable policy was established (Faden et al. 1986). This is not to say consensus is good in itself, that everyone should agree, or that there are not sometimes local differences. Rather, the problems that characterize bioethics are complex and difficult, and important views often represent different ranking of values and interests. Solving these problems with interdisciplinary approaches increases the likelihood they will be addressed in a way that helps informed and competent people of good will make the right decisions in the right way for the right reasons.[15]

Fourth, it helps explain why we can both rationally disagree about how to identify basic core curricula and competencies and yet judge that some people have expertise in bioethics and others do not. No one person has expertise in all relevant areas needed to address the broad array of problems characterizing bioethics. *One is an expert with respect to something* and a bioethics expert is someone having expertise *with respect to some area in bioethics*.[16] People are *not* bioethics experts who do not work on some problems associated with bioethics and who lack understanding of the main theories or skills that have grown in response to some bioethics question. For example, a reporter seeking bioethics expertise about surrogate decision-making (in the Terri Schiavo case) sought the views of a lecturer in philosophical ethics who knew nothing about the relevant theories, policies, and arguments regarding surrogate decision-making. Unfortunately this did not stop him from giving an uninformed response. A competent bioethicist speaking on this topic should know something about the literature and policies on surrogate decision-making and he did not. (A competent bioethicist who did not know about this area would know enough not to give an opinion.) Most bioethics experts in North America, and many parts of the world, would be expected to have competency about the questions, theories, and key arguments on this topic. Naturally, no one can be competent in all the complex areas associated with bioethics.

A twofold competency standard exists, then, in having enough expertise to satisfy competency standards from one's own field (as a philosopher, as a lawyer, etc.) and in having enough expertise to satisfy competency standards from your interdisciplinary bioethics colleagues. While there is clearly some agreement, it seems questionable that it is sufficiently robust to grow a new discipline with a set of clear standards of competency for all bioethicists.

Thus, an *Interdisciplinary Analysis of Bioethics Expertise* should be understood as: Given the range of problems associated with bioethics, different areas of academic or professional expertise will be relevant in describing or in clarifying, analyzing, specifying, proposing, or negotiating solutions to particular problems. Those who have the relevant epistemic expertise about a particular bioethics issue have studied the questions to be addressed, are familiar with the main theories and arguments that have grown in response to these questions, and have the skills, knowledge, or ability to present arguments that would convince many reasonable persons of good will.[17]

The relevant skills and knowledge need not be philosophical, and would vary with the problem and how it is framed. Consider the novel *Arrowsmith* by Sinclair Lewis (1925), which raises many moral, social, and policy issues about research ethics, medical ethics, medical education, responses to epidemics, institutional research policies, consent for research, empathy for patients, professionalism, prejudice, duties of physicians, and so on. The perspective that bioethicists might bring to a discussion of some of these points might be as a lawyer, nurse, historian, philosopher, political scientist, literary critic, and so on, with each triggering different relevant theories and expertise. To those who enjoy interdisciplinary work, their different points of view are enriching. Part of the strength of those working in bioethics is the awareness of the overlapping expertise grounded in our different fields. It enables us to see the limitations of our own disciplines and complexities and interrelations from different perspectives.[18]

9.4.4.2 Responding to Anticipated Criticisms

Let me anticipate and respond to some criticisms. First, I have assumed we can avoid the rigging problem by using "market standards" to determine which bioethicists are successful. Some may challenge this assumption and ask why we should not just stipulate what is important – just like a group of highly respected intervention radiologists may get together and stipulate what distinguishes them from other radiologists. But this, I have argued, would not work in bioethics were there is so much controversy about the goals, roles and expectations for bioethicists. Moreover, bioethicists generally seek out multiple perspectives in trying to solve the complex problems addressed in bioethics. What we mean by "bioethics" should, to some degree, square with how the word is used, and currently successful bioethicists are identified as people from many fields with many points of view.

Second, some argue that even if there is not currently one area of expertise for all bioethics, there could be. As discussed, a lively debate already exists about whether bioethics should have a code or become one field. Some find it desirable for bioethics to emulate fields that have broken away from parent disciplines to become unique fields, arguing that this would impact the current practice of granting degrees in bioethics and graduate programs of study in bioethics leading to the Ph.D. (Wolf and Kahn 2005). New fields emerge, and so it is theoretically possible that a new discipline of bioethics might be established; but there would be serious obstacles. Part of what usually marks something as a new academic discipline is that the members initially share a similar education and want to narrow their focus and specialize further. They have mastered a core curriculum, along with those skills and dispositions that prepare them to demonstrate their competency through apprenticeships, boards, or qualifying examinations. Yet, there is likely to be little agreement about what expertise is essential in bioethics because bioethicists do not agree on their roles and come from different professional or academic homes, each with distinct languages, methods, traditions, practices, core curriculum, and competency examinations. Bioethicists, moreover, generally welcome multiple perspectives.

Third, many bioethicists are more at home in bioethicists' organizations and with other bioethicists, albeit educated in other fields, than in their primary disciplines. This can be explained by the current nature of subspecialization once qualifications have been completed. The overlap among bioethicists, I have argued, is genuine; but it is not sufficient to establish a marketable new academic or professional discipline.

9.5 Final Comments

Bioethics will remain an interdisciplinary and second-order discipline as long as most bioethicists want to retain their identity and ties to their parent academic or professional disciplines and recognize the value of many fields in resolving bioethical questions. As long as doctors, nurses, philosophers, lawyers, ministers, those in religious studies, and others who are active in bioethics want to retain their primary academic and professional identifies, we will maintain our different traditions, languages, and honored texts. Bioethics is, moreover, a recognized sub-specialty within each of these fields further grounding bioethics in different disciplines.

If bioethics has a fundamentally interdisciplinary nature with bioethicists rooted in many disciplines, then these different academic or professional perspectives may explain the strong tensions that sometimes exist in discussions about what bioethics is or should be. Given the many different vantages, some see bioethicists as too philosophical and others as too little so; some see them as too much like lobbyists and reformers and others as too little so; some see them as too theoretical and some as not theoretical enough (Elliott 2005; Callahan 2005; Fox and Swazey 2005). Some work to establish a new discipline in bioethics (Wolf and Kahn 2005), while others see the strength of bioethics as grounded in particular existing disciplines (Elliott 2005). Some predict its demise from bioethicists' hubris of trying to answer too broad an array of questions about the human condition (Cooter 2004). Yet others find bioethicists too narrow in their grasp of the context, options, issues, and useful methodologies (Fox and Swazey 2005).

I generally agree with Albert Jonsen that bioethics is partly public discourse and partly a discipline. I regard the proper analysis of the discipline of bioethics, however, to be that of a second-order discipline. Jonsen writes:

> We return to the question, 'is bioethics a discipline?' In the simplest sense, it certainly is; a discipline is a body of material that can be taught, and bioethics is and has been a teachable and taught subject since the mid-1970s. In the strictest sense, it is not a discipline. A discipline is a coherent body of principles and methods appropriate to the analysis of some particular subject matter. Bioethics has no methodology, no master theory … bioethics might be called a 'demi-discipline.' Only half of bioethics counts as an ordinary academic discipline … the other half of bioethics is the public discourse… (1998, pp. 345–346)

I have argued that the part of bioethics that is not public discourse but is academic, is best understood as an interdisciplinary, second-order discipline.

Why has bioethics become an international social phenomenon in such a short time? Perhaps it stems from the importance of its problems and the disagreements over how to solve them among reasonable and informed people of good will. Disputes about these problems often raise deeply held moral values, entrenched cultural expectations, and generate political tensions. Bioethics, with its interdisciplinary approach, gives no single field, religion, or culture the primary responsibility for solving them. Moreover, bioethicists generally rejected both extreme forms of relativism and absolutism, turning to methods of practical reasoning that cross cultural divides. Good solutions in ethics (like good solutions in science or law) do not need to be infallible. Ethics, like law and science, embraces a "self-correcting" method such that solutions to problems have to be reconsidered in light of new information, clarification of the issues, impartiality of application, justifiability of values, or consideration of people's interests, traditions, and law.

I began by asking whether bioethicists are riding an international wave of interest in bioethical issues or are they creating and directing it. Given the broad range of problems addressed in bioethics, I have argued that it is a worldwide social movement that is partly public discourse and partly an interdisciplinary second-order discipline. While bioethicists can be considered experts with respect to particular areas of bioethics, expertise in one area does not transfer to others. No one could be expert in addressing the broad array of issues that currently characterize bioethical problems.

Notes

1. The term *bioethics* goes back to at least 1927; see Jahr (1927). Yet the term "bioethics" became popular, largely by the work of two physicians, Andre Hellegers and Ralph Potter (Jonsen 1998).

2. See Kopelman (2006) for a detailed discussion about why I use the words interchangeably. Portions of this paper, especially on the four views about expertise in bioethics, were adapted from this paper.
3. Finding a title for the organization was a mine field. Some people claimed to do humanities but not ethics, and wanted the title to reflect this. Others insisted that ethics must be in the title. Still others insisted that they did ethics, but not philosophical ethics, and wanted the title to reflect this.
4. Baruch Brody's interest in casuistry was also important to this movement (2003). In addition, many were critical of "principlism," such as Callahan (1999), Jonsen (1991), Gert (1998), Clouser and Gert (1990), and Veatch (1999). A lively controversy exists about casuistry, however, because of its reliance on intuitionism. See Kopelman (2006), McMahan (2000), Nowell-Smith (1957), and Frankena (1973).
5. For an excellent discussion of ethics and expertise, see Rasmussen (2005) and Singer (1972).
6. This position is defended by Satel (2000) and Smith (2000).
7. For example, see UNESCO (2005).
8. For a discussion of the case and film, see Jonsen (1998).
9. See, for example, Baker (2005), Perlman (2005), Elliott (2001), Callahan (1999), Lantos (2005), Miller (2005), Beauchamp (2005), and Schukenk and Gallager (2005).
10. That is, some might ignore the rigging problem and simply *stipulate* the relevant roles, values, practices, and public expectations that are essential to claim expertise in bioethics. The problem with this is that given the lively disputes on these points, there is little reason for those who disagree to take such stipulations seriously.
11. Many well known bioethicists defend many other theories not discussed herein such as utilitarianism (Callahan 1990), egalitarianism (Childress 1970), contractarianism (Daniels 1985), and libertarianism (Engelhardt 1992).
12. The American Pain Society (ASP) website is at: http://www.ampainsoc.org
13. For a discussion of the notion of a family resemblance, see Wittgenstein (1953, 1961).
14. Our fundamentally interdisciplinary character as a second-order discipline also offers a response to those who believe that bioethics as a field is too diffuse (Cooter 2004).
15. Some people without academic or professional credentials have made contributions to the bioethics literature, but if they do so as experts, then their work is judged by the relevant academic or professional discipline.
16. Someone could be a bioethics expert with respect to research ethics, but not environmental ethics, just as someone might be an expert with respect to metaphysics, but not epistemology, or neurosurgery, but not psychiatry.
17. This is also discussed in Kopelman (2006) where I clarify that this analysis has some similarities to Hooker's analysis of moral expertise (1998).
18. Hooker points out that there are concerns with "calling someone a moral expert in Western culture since this suggests the person is judgmental, interfering, condescending, self-important, hypocritical, and perhaps closed minded about morality" (1998, p. 509). Those who are experts need to acknowledge their limitations. Those who are experts in ethics and bioethics similarly need to acknowledge that they have particular areas of expertise and that do not carry over to other areas.

References

American Society for Bioethics and Humanities. 1998. *'Task force on standards for bioethics consultation', core competencies for health care ethics consultation*. Glenview: American Society for Bioethics and Humanities.

Baker, R. 2005. A draft model aggregate code of ethics for bioethics. *The American Journal of Bioethics* 5: 33–41.
Beauchamp, T. 2005. What can a model professional code for bioethics hope to achieve? *The American Journal of Bioethics* 5: 42–43.
Beauchamp, T., and J.F. Childress. 1979, 1983, 1989, 1994, 2001, 2008. *Principles of biomedical ethics*. New York: Oxford University Press.
Blackburn, S. 1996. *The oxford dictionary of philosophy*. New York: Oxford University Press.
Brody, B. 2003. *Taking issue: Pluralism and casuistry in bioethics*. Washington, DC: Georgetown University Press.
Callahan, D. 1990. *What kind of life: The limits of medical progress*. New York: Simon & Schuster.
Callahan, D. 1999. Ethics from the top down: A view from the well. In *Building bioethics*, ed. L.M. Kopelman, 25–35. London: Kluwer Academic Publisher.
Callahan, D. 2005. Bioethics and the culture wars. *Cambridge Quarterly of Healthcare Ethics* 14: 424–431.
Childress, J.F. 1970. Who shall live when not all can live? *Soundings* 53: 339–354.
Clouser, K.D. 1978. Bioethics. In *Encyclopedia of bioethics*, 1st ed, ed. W.T. Reich, 124–125. New York: The Free Press.
Clouser, K.D., and B. Gert. 1990. A critique of principlism. *The Journal of Medicine and Philosophy* 15: 219–236.
Cooter, R. 2004. Historical keywords: Bioethics. *The Lancet* 364: 1749.
Daniels, N. 1985. *Just health care*. Cambridge: Cambridge University Press.
Elliott, C. 2001. Throwing a bone to the watchdog. *The Hastings Center Report* 31: 9–12.
Elliott, C. 2005. The soul of a new machine: Bioethicists in the bureaucracy. *Cambridge Quarterly of Healthcare Ethics* 14: 379–384.
Engelhardt, H.T. 1992. The search for a universal system of ethics: Post-modern disappointments and contemporary possibilities. In *Ethical problems in dialysis and transplantation*, ed. C.M. Kjellstrand and J.B. Dossetor. Dordrecht: Kluwer.
Engelhardt, H.T., and S. Spicker. 1975. *Evaluation and explanation in the biomedical sciences*. Dordrecht: D. Reidel Publishing Company.
Faden, R., T.L. Beauchamp, and N. King. 1986. *A history and theory of informed consent*. New York: Oxford University Press.
Fletcher, J. 1966. *Situation ethics*. Philadelphia: The Westminster Press.
Fox, R.C., and J.P. Swazey. 2005. Examining American bioethics: Its problems and prospects. *Cambridge Quarterly of Healthcare Ethics* 14: 361–373.
Frankena, W.K. 1973. *Ethics*, 2nd ed. Upper Saddle River: Prentice Hall.
Frey, R.G. 1978. Moral experts. *The Personalist* 59: 47–52.
Gert, B. 1998. *Morality: Its nature and justification*. New York: Oxford University Press.
Gustafson, J.M. 1973. Mongolism, parental desires, and the right to life. *Perspectives in Biology and Medicine* 17: 529–530.
Holmes, R.L. 1990. The limited relevance of analytical ethics to the problems of bioethics. *The Journal of Medicine and Philosophy* 15: 143–159.
Hooker, B. 1998. Moral expertise. In *Encyclopedia of philosophy*, ed. Edward Craig, 509. New York: Routledge.
Jaggar, A.M. 2000. Feminist ethics. In *The Blackwell guide to ethical theory*, ed. H. LaFollette, 348–374. Malden: Blackwell Publishing.
Jahr, F. 1927. Bio-Ethik: Eine Umschau über die ethischen Beziehungen des Menschen zu Tier und Pflanze. *Kosmos: Handweiser für Naturfreunde* 24: 2–4.
Jonsen, A.R. 1991. Of balloons and bicycles or the relationship between ethical theory and practical judgments. *The Hastings Center Report* 21: 14–16.
Jonsen, A.R. 1998. *The birth of bioethics*. New York: Oxford University Press.
Jonsen, A.R., and S. Toulmin. 1988. *The abuse of casuistry: A history of moral reasoning*. Berkeley: University of California Press.

Jonsen, A., M. Siegler, and W. Winslade. 1982. *Clinical ethics*. New York: Macmillan Publishing Company.
Kipnis, K. 2005. The elements of code development. *The American Journal of Bioethics* 5: 48–50.
Kopelman, L.M. 1995. Philosophy and medical education. *Academic Medicine* 70: 795–805.
Kopelman, L.M. 1998. Bioethics and humanities: What makes us one field? *The Journal of Medicine and Philosophy* 23: 356–368.
Kopelman, L.M. 2006. Bioethics as a second-order discipline: Who is not a bioethicist? *The Journal of Medicine and Philosophy* 31: 601–628.
Kopelman, L.M., and J.C. Moskop (eds.). 1984. *Ethics and mental retardation*. Dordrecht: D. Reidel Publishing Company.
Lantos, J. 2005. Commentary on "a draft model aggregated code for bioethicists". *The American Journal of Bioethics* 5: 45–46.
Lewis, S. 1925. *Arrowsmith*. New York: Harcourt, Brace.
MacDonald, C. 2003. Draft: Model code of ethics for bioethics. In *The canadian bioethics societies and Ad Hoc Working Group on employment standards for bioethics* [On-line]. http://www.bioethics.ca/resources/PDF%20documents/draftcode.pdf. Accessed Sept 20, 2012.
McMahan, J. 2000. Moral intuitionism. In *The Blackwell guide to ethical theory*, ed. H. LaFollette, 92–110. Malden: Blackwell Publishing.
Miller, F.G. 2005. The case for a code of ethics for bioethicists: Some reasons for skepticism. *The American Journal of Bioethics* 5: 50–52.
Nowell-Smith, P.H. 1957. *Ethics*. Oxford: Basil Blackwell.
Parker, L.S. 2005. Ethical expertise, maternal thinking, and the work of clinical ethicists. In *Ethics expertise: History, contemporary perspectives, and applications*, ed. L.M. Rasmussen, 165–207. Dordrecht: Springer.
Perlman, D. 2005. Bioethics in industry settings: One situation where a code for bioethicists would help. *The American Journal of Bioethics* 5: 62–64.
Rasmussen, L.M. (ed.). 2005. *Ethics expertise: History, contemporary perspectives, and applications*. Dordrecht: Springer.
Regenberg, A.C., and D. Mathews. 2005. Resisting the tide of professionalization: Valuing diversity in bioethics. *The American Journal of Bioethics* 5: 44–45.
Reich, W.T. 1995. Introduction. In *Encyclopedia of bioethics*, vol. 1, 2nd ed, ed. W.T. Reich, xxi. New York: Simon Schuster Macmillan.
Satel, S. 2000. *How political correctness is corrupting medicine*. New York: Basic Books.
Schuklenk, U., and J. Gallager. 2005. Status, careers and influence in bioethics. *The American Journal of Bioethics* 5: 64–66.
Singer, P. 1972. Moral experts. *Analysis* 32: 114–117.
Smith, W.J. 2000. *Culture of death: The assault on medical ethics in America*. San Francisco: Encounter Books.
Toulmin, S. 1981. The tyranny of principles. *The Hastings Center Report* 11: 31–39.
United Nations Educational, Scientific and Cultural Organization (UNESCO). 2005. *Universal declaration on bioethics and human rights* [On-line]. http://portal.unesco.org/shs/en/ev.php-URL_ID=1883&URL_DO=DO_TOPIC&URL_SECTION=201.html. Accessed Sept 20, 2012.
Veatch, R.M. 1999. Contract and critique of principlism: Hypothetical contract or epistemological theory as method of conflict resolution. In *Building bioethics*, ed. L.M. Kopelman, 121–143. Great Britain: Kluwer Academic Publisher.
Wittgenstein, L. 1953, 1961. *Philosophical investigations*. New York: The Macmillan Company.
Wolf, S., and J. Kahn. 2005. Bioethics matures: The field faces the future. *The Hastings Center Report* 35: 22–24.

Part III
The Practice of Bioethics: Professional Dimensions

Chapter 10
The Development of Bioethics: Bringing Physician Ethics into the Moral Consensus

Robert M. Veatch

In order to understand the development of bioethics in the United States, one needs to grasp the sorry state of medical ethics prior to the emergence of what we now call bioethics. After briefly sketching the history of the field beginning in ancient times and moving very quickly to the modern era, I will claim that bioethics emerged in the 6-year period centering around 1970 and that a combination of scientific and cultural factors explains its emergence.

The term "bioethics" was coined, apparently in 1970, as a way of signaling that something very new was happening (Reich 1994). I will argue that what is new involved not only new medical moral issues–transplants, more systematic human research protocols, a new set of end-of-life issues, abortion, contraception, and resource scarcity–but also a new alignment of the participants in the intellectual discussion of the moral dimensions of these issues. The shift from the term *medical ethics* to *bioethics* marks not only the expansion of the subject matter to cover these new issues, but also new domains such as the protection of the environment, animal rights and welfare, and concern about species preservation. More critically, it marks the initiation of a cross-disciplinary conversation (or, more accurately, a resumption of a conversation that had been interrupted about two centuries earlier). This, at least, is the thesis of my book, *Disrupted Dialogue: Medical Ethics and the Collapse of Physician-Humanist Communication (1770–1980)* (Veatch 2005). Before arguing for the thesis that scientific and cultural events in and around 1970 caused the emergence of bioethics at that time, I need to trace the evolution of what is often called "medical ethics."

R.M. Veatch, Ph.D. (✉)
Kennedy Institute of Ethics, Georgetown University, 423 Healy Building, Washington, DC 20057, USA
e-mail: veatchr@georgetown.edu

J.R. Garrett et al. (eds.), *The Development of Bioethics in the United States*, Philosophy and Medicine 115, DOI 10.1007/978-94-007-4011-2_10,

The term "medical ethics" is itself ambiguous. Some people use it to refer exclusively to the systematic reflection *by physicians* on the moral dimensions of their clinical decisions. It is rumored that one distinguished scholar who was not a physician was given the title "professor of bioethics" because physicians on the medical school faculty insisted that only physicians could claim a title using the word "medical."

Surely, however, medical ethics must cover more than the ethics of physician behavior. "Medicine" is a term referring to the domain of health or organic well-being. At minimum the ethical choices made by other health professionals–nurses, pharmacists, dentists, and allied health personnel–appropriately fall under the domain of medicine. They are all professions practiced in medical centers. Hence, it is reasonable to distinguish between the ethics of physicians and the broader field of the ethics of medical professionals. Moreover, medical ethics must be broad enough to include ethical reflection on the choices and responsibilities of medical lay people. If a layperson chooses not to ask her physician about an abortion, surely she has made an ethical decision in the medical realm. Just as education is a domain involving decision-makers other than teachers, so medicine is a domain involving decision-makers who are not medical professionals. I will refer to "physician ethics" when speaking specifically about ethical reflection on physician behavior, and "medical ethics" to cover the much broader sphere of medical ethical reflection.

The term "medical ethics" is ambiguous in a second way. While some have limited the term to reflection about physician behavior when that reflection is carried out by physician groups such as the American Medical Association, others use the term to refer to ethical discourse carried out by those outside the medical profession. The Catholic Church, for example, has long analyzed and developed positions on the medical ethical conduct of both physicians and those who are, from a medical point of view, laypeople. This reflection is typically called "medical morality" or sometimes "medical ethics." Many groups and individuals outside the various medical professions have engaged in medical ethical analysis. Thus, I would object to limiting "medical ethics" to either analysis carried out by medical professionals or analysis about the behavior of medical professionals. I have, therefore, remained comfortable with the titles I have held throughout my professional career, first at the Hastings Center as "Associate for Medical Ethics" and then at the Kennedy Institute of Ethics as "Professor of Medical Ethics."

In spite of the fact that medical ethics should be taken to refer to the domain of medical choices beyond those made by physicians, and should include analysis conducted by groups who are not made up of medical professionals, a valid distinction still can be made between medical ethics and bioethics. Currently, the customary usage employs the term "bioethics" when referring to the systematic exploration of the issues in medicine and biology carried out in interdisciplinary dialogue involving philosophers, theologians, lawyers, social scientists, and health professionals. Bioethics includes ethical issues in biology that are not directly within the medical sphere–high school biology labs, the Environmental Protection Agency, and the sciences of genetics, and animal testing of cosmetics would all be examples. Thus, in distinguishing between medical ethics and bioethics, I would focus not only on the shift to the inclusion of medical laypeople into the conversation and the reflection on medical choices made by lay people, but also on a range of biomedical issues that are not specifically part of clinical medicine. The question of this volume is: why did these perspectives, collectively referred to as bioethics, emerge and exactly when?

10.1 The Sorry State of Medical Ethics Before 1970: The Hippocratic Ethic and Its Flaws

The essence of professionally articulated physician ethics in the Western world is encapsulated in the Hippocratic Oath. It summarizes one version of an ethic for the practice of medicine as written by a physician group (Edelstein 1967, pp. 3–64). However, it has never been the only codification of the ethical norms for physician conduct. Other codes with significant differences have appeared from time to time not only in Western culture (United States Bishops Committee on Doctrine 1994; 'The Oath of Soviet Physicians' Anon 1971), but also in the non-Western world ('Oath of Initiation [Caraka Samhita],' [Hinduism] 1978; 'The 17 Rules of Enjuin' [Japan] Anon 1970). The idiosyncratic nature of the exclusivity of the Hippocratic document is illustrated in the pledge that Hippocratic physicians will not reveal the secrets of what they have learned in medical education to anyone outside their group. The ethical commitments that members of the group will not reveal include opposition to abortion, a prohibition of surgery, a strangely ambiguous confidentiality provision, and an offensively paternalistic commitment to base judgments about what will benefit the patient on the "ability and judgment of the physician." The Oath ends with a plea for moral reward that would count as a Pelagian heresy in other moral systems. If Edelstein is correct in his claim that the Oath represents a medical pledge growing out of Pythagoreanism, then it might be that such controversial and strange moral positions were consistent with and acceptable to both Hippocratic practitioners and patients within that tradition. That physician-articulated ethic is, however, certainly inappropriate for those who stand in other religious and secular traditions including Judaism, Catholic and Protestant Christianity, and the Eastern religious traditions, as well as most schools of secular philosophical thought. If medical ethics follows this set of Hippocratic commitments, it would be understandable that most theological and philosophical world views would not tolerate its medical ethic and would seek significantly different ethical codifications.

By the middle ages, Christian medical practitioners realized the tensions between Hippocratic and Christian ethics and produced "The Oath according to Hippocrates Insofar as a Christian May Swear It" (Jones 1924, pp. 23–26). This radical redrafting removes not only the references to the Greek deities, but also the pledge of secrecy and the prohibition on surgery. It also significantly revises the reward and punishment framework. In fact, in the Middle Ages, ethics for the practice of medicine was not dominated by Hippocratic commitments, at least until the latest portion of this period (Veatch and Mason 1987; Rutten 1996). Much medical ethical thought of this period, just as much other theological and philosophical thought, was dominated by the Thomistic synthesis that relied more on Aristotelian rather than Hippocratic perspectives.

10.2 The Scottish Enlightenment

In the Scottish Enlightenment in the eighteenth century, medical ethics was almost totally devoid of the Hippocratic perspective. Hippocrates was a model for clinical reasoning and his aphorisms were a device for testing medical students on their perfunctory knowledge of Greek, but there was essentially no interest in the ethics of the Greek physician. The great medical ethical leaders of the period–John Gregory, Thomas Percival, Samuel Bard, and Benjamin Rush, for example–were disinterested in the Hippocratic literature as a source for morality (Veatch 2005). They derived their ethical framework from the Scottish philosophy of the day. Drawing on their personal friendships with Thomas Reid, James Beattie, David Hume, Adam Smith, Thomas Gisborne, and Joseph Priestley to ground their ethics, they applied the general ethical theory of the day to the field of medicine. Thus, their medical ethics flowed naturally from their dialogue with the humanists of their period, making medical ethics a tight fit with the philosophical and theological framework of their surrounding culture. For example, the medical students of the University of Edinburgh did not recite the Hippocratic Oath. Rather, they took a slightly modified version of the Sponsio academica that was administered to all students of the University of Edinburgh, an oath that existed even before there was a medical school. That oath required students to pledge loyalty to the king, the Scottish Covenant, and the university (Veatch 2005, pp. 33–35).

This medical ethical orientation spread with the influence of the Edinburgh medical students not only to Scotland and England, but also to the United States, Canada, and the other English-speaking colonies. This close integration of medical ethics into the British empiricism and the Calvinist culture of the day worked well, at least temporarily – especially if one were committed to theology of the Scottish Calvinism or the secular philosophical thought that existed sometimes in company and sometimes in tension with that theological world view.

10.3 The Dark Ages of Medical Ethics

However, soon after this period at the end of the eighteenth century, things began to fall apart. Changes in the culture and in medicine drove physicians apart from the leaders in ethics. Medical students began to be recruited from the middle classes rather than from the elites, who had been prepared with classical education in their homes and in their private schools. Medicine became at the same time more scientific, technical, and specialized. Pressures mounted to remove a humanities requirement from the premedical and medical education of the physician. By the beginning of the nineteenth century, physicians were much less conversant with the classics and with philosophy and ethics. The focus of the curriculum narrowed to a more technical education. In the following decades, when moral crises emerged within the medical profession, organized medicine had to turn to the work of late eighteenth century figures such as Percival, Gregory, and Rush. That, for example, was the approach used from the beginning to the middle of the nineteenth century in the United States when allopathic medicine was confronted with alternative medical practitioners and ethical questions of cooperation arose. Leaders emerged from the state medical societies who, because of their senior status within the profession, were appointed to prepare responses to the events of the day.

This occurred in Boston in 1808 when John Warren, Lemuel Hayward, and John Fleet, local surgeons of considerable professional stature, prepared the "Boston Medical Police," (Warren et al. 1995 [1808], pp. 41–46) a code of ethical conduct for Massachusetts practitioners. These physicians, typically surgeons, had made significant contributions to their profession, but, unfortunately, tended to have little or no experience in medical ethics. They shamelessly borrowed from the writings of the eighteenth century predecessors. The same occurred in Baltimore (Medico-Chirurgical Society of Baltimore 1832, p. 23) and Washington, DC (Medical Association of the District of Columbia 1833) in the 1830s. In Philadelphia, a secret society suggesting a college fraternity was formed in the 1820s. Calling themselves the Kappa Lambda Society of Hippocrates, they wanted to have an Oath reminiscent of the Hippocratic pledge, so they wrote such a document. Unfortunately, none of the authors had much knowledge of the Oath's history, and what was produced had no resemblance to the original Greek document. It drew instead on the non-Hippocratic writing of Percival (Veatch 2005, pp. 110–14).

Finally, by 1847, when the American Medical Association was founded, this same pattern prevailed. Two surgeons with experience in medical editing and professional society leadership took the lead in preparing the Association's code of ethics. Not having any significant preparation in ethics or the medical humanities, they openly borrowed from Percival and Benjamin Rush, thereby importing a medical ethics grounded on a 75-year-old philosophical tradition that was strangely out of place in mid-nineteenth-century America.

During the 150 years from 1800 to 1950, no one contributing to medical ethics as articulated by organized medicine had any significant knowledge of the philosophical or theological traditions that were the foundation for ethics. There was no knowledge of Kant, no serious incorporation of British utilitarianism, no awareness of the work of G. E. Moore or W.D, Ross, no social contract theory, no involvement in the debate over deontology, and no communication with the rights tradition of political philosophy. In theological ethics, there was no sign of the social gospel, no evidence of the Niebuhrs, and no interaction with Catholic influence from Leo XIII or Pius XII.

There were re-edits of the American Medical Association's principles of ethics–in 1871, 1879, 1903, 1912, 1947, and 1957. These new editions saw changes in format, but only minor tinkering in the underlying moral commitments. The re-drafters softened the diatribe against non-allopathic physicians and continued to struggle with the obligation of the physician to society. The physician drafters were more or less on their own, avoiding contact with humanists who might have helped them straighten out the conflicts between Hippocratic focus on the individual patient's welfare and the more social ethical issues that were increasingly imposing themselves onto medicine. On occasion, physicians such as Worthington Hooker, William Osler, or Richard Cabot revealed more openness to philosophical debates over ethical theory, but even in these cases their knowledge was indirect (in the case of Hooker), amateurish (in the case of Cabot), or oriented to the classical works of the Greeks (in the case of Osler). Physician commentators in medical ethics kept repeating their Hippocratic slogans, focusing on the physician's duty to benefit the patient. The only shift was an increasing recognition that physicians also had a responsibility to society, but without non-consequentialist tools to keep social utility in check, the increase in sophistication was slight.

10.4 Signs of the End of an Isolated Professionally-Articulated Medical Ethics: 1950–1970

Any major shift from Hippocratic individualism to a more social ethic for medicine came in for close scrutiny in the years following World War II. The Nuremberg Trials made clear that physicians had not only participated in the Nazi atrocities, but also, in some cases, had taken leadership in them (Lifton 1986; Annas and Grodin 1992; Caplan 1992). The realization that Nazi research was conducted by physicians who had been members in good standing in the profession left Western leaders of organized medicine perplexed in their attempts to insist on an ethic that would remain loyal to the interests of individual patients without declaring all medical research –including that undertaken for the benefit of society rather than the patient – immoral. That the Nazi physicians explicitly ridiculed the Judeo-Christian ethic of compassion for the sick and oppressed and praised the Hippocratic ethic as an alternative that avoided this "weakness" made it even more problematic for organized medicine simply to reiterate the Hippocratic ethic as a protection against abuse of research subjects.

Nevertheless the World Medical Association in 1948 did precisely that; it produced the "Declaration of Geneva" (adopted 1948, published as World Medical Association 1956), a paraphrase of the Hippocratic Oath, as its statement to the world that the profession of medicine would refrain from abuses of patients. That statement reiterated Hippocratic platitudes while avoiding the critical issues of patient rights and the moral justification of using medical skills for purposes other than furthering patient welfare. More generally it avoided providing a codification for physician conduct that was more in conversation with contemporary ethical theory in philosophy, theology, and political theory.

It was left to the governments of the Western victors in World War II to craft a code of conduct for research involving human subjects. In the Nuremberg Trials, international law produced the Nuremberg Code (1946/1978) that, for the first time, looked outside of the medical profession for its concepts, principles, and normative positions. By imposing, as a prior constraint, the foundational requirement of a subject's absolute right of self-determination, the Code resolved the dilemma of justifying the use of subjects for research without letting such use degenerate into abuse by affirming the importance of working for the good of society. This foundation for informed consent in medical research had its precursors in law and political theory, but not in medical codifications for doctors.

It was not until 1964 that the World Medical Association rectified its retreat to the Hippocratic individualism of the Declaration of Geneva by drafting the Declaration of Helsinki (1978 [1964]). It produced a code of ethics articulated by the international professional body of medicine that acknowledged the legitimacy of having physicians undertake actions on patients and healthy volunteers that were not designed primarily to benefit patients. To this day, the rather crude codifications articulated by the World Medical Association and other professional medical groups do not directly resolve the conflict between the exclusive focus on the patient's welfare that is found in the Declaration of Geneva and the condoning of the use of patients for social purposes found in the Declaration of Helsinki or its various revisions.

I take the key mark of the beginning of bioethics to be the restoration of the communication between the medical profession and others disciplines including the humanities, law, and political science. This relocates medical ethical discourse, removing it from the domain of organized medical professional groups and placing it in a dialogue between the medical professions and interested lay groups. Thus, Annas and Rothman are on the right track when they identify the atrocities of World War II and the trials following the war as a marker of the beginning of a new era in medical ethics. However, even though a few signs of an opening of the dialogue between medical professionals and humanists appeared in the 1950s, these preliminary points of contact in the one field of human subjects research are hardly enough to signal the emergence of an era in medical morality so new that it warrants being given a new name. Even in the specific field of human subjects research, only minimal progress had occurred. In December of 1946, the American Medical Association adopted a set of three principles for conducting human experimentation (Advisory Committee on Human Radiation Experiments 1995, p. 135). While these principles endorsed "voluntary consent," they provided few specifics. The American government did have explicit, written directives designed to protect human subjects of radiation experiments. They used much more robust language than the AMA document. They called for "complete and informed consent in writing," but the directives were so novel and controversial that they were apparently kept secret not only from the subjects of the research, but from the investigators as well (Advisory Committee on Human Radiation Experiments 1995, p. 90).

Elsewhere in medical ethics, it was largely business as usual. The lives of terminally and critically ill patients were to be preserved by all available means. That liberal bastion, the Roman Catholic Church, made clear in 1957 that the church's theology did not require preserving life by all means possible and it even acknowledged that the burdens of life-support on other people in society could justify making life-support morally expendable (Pius XII 1958, publishing a 1957 address), but organized physician groups were not to be heard on such radical ideas. Contraception and sterilization were taboo; abortion was not only illegal, but also widely condemned at the moral level. Physicians almost unanimously believed that patients should be treated paternalistically by withholding disturbing information such as a cancer diagnosis (Kelly and Friesen 1950; Branch 1956; Samp and Curreri 1957). The Hippocratic myth that physicians should work for the benefit of their patients was alive and well.

It was in the 1960s that the traditional, paternalistic, physician-dominated ethics for medicine began to be challenged more broadly. Al Jonsen is right that the development in Seattle in 1962 of the artificial kidney–the hemodialysis machine–by Belding Scribner put the subject of allocating a scarce biomedical technology on the map (Jonsen 1998; *see also* Fox and Swazey 1974; Alexander 1962). Thus, the social ethical controversy of allocating a scarce, new medical technology emerged side-by-side with human subjects research as events that made the slogan of benefiting the individual patient morally archaic. Still, organized medicine, with physicians ever-more specialized in developing new research and new technology, was totally beyond its depth in developing an alternative to traditional, professionally-articulated paternalistic and individualistic, consequence-oriented ethics. (By 1978, I was explicitly pressing the difference between organized medicine's ethic and other non-medical grounds for an ethic for medicine; *see* Veatch 1978). Medical professionals desperately needed resources and skills from other disciplines – the humanities, law, and the social sciences – to refurbish an ethic to bring it into line with modern moral dilemmas and modern moral theory.

During this time, medicine was also a victim of its own successes. In the first half of the twentieth century, medicine for the first time really had developed technologies that would preserve lives and cure disease to a significant degree. The discovery of insulin, the development of antibiotics, the invention of the ventilator, the emergence of vaccines, and the beginnings of transplantation of human organs were examples of truly life-saving technologies. One of the effects was an increasing realization that the world's population was growing at an unsustainable rate. A "population crisis" was proclaimed with radical plans to develop mechanisms for limiting fertility in order to avert what was perceived as a coming world crisis (Ehrlich 1968). During these same years of the 1960s, social alarmists discovered that the increasing population was also creating environmental impacts beyond its numbers. We were introduced to the horrors of pesticides and pollution (Carson 1962). While these events were clearly related to medicine, they also pushed beyond medicine's boundaries. We recognized moral issues in biology that went beyond the behaviors of physicians and other clinical health care professionals. As the science of genetics exploded on the scene, people would soon begin to recognize not only the clinical applications of genetic counseling and genetic screening, but also global problems of maintaining the "health" of the gene pool (Hamilton 1972; Etzioni 1973). There was great concern that successful medical intervention to maintain the lives of newborns with serious genetic anomalies could result in the preservation of "bad genes" that previously had worked their way out of the genetic heritage through the early death of those who inherited them or acquired them through mutation. In countless ways, moral issues of biology were on the public agenda. They pressed well beyond clinical medicine. Other issues outside of clinical medicine were quickly added to the list: animal welfare issues, the treatment of specimens in high school biology classes, and the morality of the use of nuclear energy were all examples of why ethics was no longer limited to clinical medicine. It was soon to involve a complex range of biological as well as medical issues. The result was a biomedical ethic or – for short – bioethics.

10.5 The Emergence of Bioethics

My thesis for this essay is that two kinds of socio-cultural events emerged in the 1960s that set the tone for the emergence of bioethics during the 6-year period between the middle of 1967 and the middle of 1973. These events are, first, scientific and, second, social.

10.5.1 The Revolution in Medical Science

The first kind of change that occurred was a change in medical science. It can be referred to as a change in the "modal disease." By "modal disease," I mean the disease that captures the public imagination and motivates medical scientists to press for breakthroughs. Through the first half of the twentieth century, the modal disease, the one that worried everyone, was the acute illness, largely infectious disease: measles, mumps, whooping cough; pneumonia and–most dramatically–polio were the concern of the world. These worries were little different from the plagues of the Middle Ages. It is hard for the younger generation of the twenty-first century to imagine how radically infectious disease shaped daily life a century ago. Infant morality, often resulting from acute infection, could ravage half of the children of a couple. It was not only lack of effective birth control and the need for child labor that led to large families; it was also the rational realization that, if one were typical, a significant percentage of one's children would not survive. If a couple wanted to be confident of surviving children, especially if they wanted male children to survive, they would intentionally have to produce at least four or five.

As a child of my generation, I recall not being permitted during the polio season to go to public places where large numbers of people gathered for fear of contracting the dread polio. Two of my personal friends eventually succumbed. The announcement of the invention of a vaccine for poliomyelitis was a world-changing event.

While acute disease was being conquered, chronic disease was slowly taking its place as the modal disease driving the public concern and research budget. The 1922 discovery of insulin was way ahead of its time. By the 1950s, the ventilator was making it possible to stabilize some patients with what Lewis Thomas (1974) called "half-way" technologies. It would soon produce patients like Karen Quinlan, Nancy Cruzan, and Nancy Jobes. Other chronic diseases gained at least as much attention: cancer, heart disease, and end-stage renal disease. By the end of the 1960s, just as vaccines and antibiotics were making acute illness much less dominant in the minds of the public, we had mastered the art of stabilizing patients suffering from the diagnosis of a chronic disease. It is not that people no longer died of acute illness, or that more research was not needed; it was that the central medical problems to the typical member of the public – at least in the developed world – had shifted from the acute to the chronic.

This had several critical impacts. First, the new biomedical technologies stabilized thousands of patients in a condition that left them living indefinitely, but living in a condition that many increasingly doubted was worth continuing. The ventilator, the hemodialysis machine, the organ transplant, and the various strategies for putting cancers into remission left people asking whether these processes that slowed the terminal phase of people's lives was always worthwhile. Moreover, it often left people not only stabilized, but sufficiently healthy that they could study their own conditions and become knowledgeable about them, sometimes more knowledgeable than the typical primary physician – especially if the chronic disease was rare and the now not-so-sick person had the resources to read and research his or her condition. The end result was a large group of patients who not only had doubts whether it was worth living in the condition they were in, but had sufficient time to understand that clinical choices were constantly being made about how they were treated. These choices radically changed how people would live the last days of their lives. Moreover, these choices were not ones that could be read directly from the medical science related to their condition; they were choices made on the basis of moral and other evaluative premises. If patients knew they were condemned to spend months or years on a ventilator, they often had the chance to think about whether such a life was worth living. If patients were scheduled for painful, nauseating, and often eventually unsuccessful chemotherapy and radiation regimens for their cancers, they often had the time and inclination to ask whether such therapies were worth pursuing.

Thus, scientific progress in medicine changed the modal disease from acute to chronic illness. Chronic illness left patients with good reason to question whether the treatments were always worth it, while sufficient health gave them the time to learn enough about those treatments and to challenge the orders from physicians regarding whether to provide them.

10.5.2 The Socio-cultural Revolution

Questioning of physicians' so-called "orders" pertaining to life-supporting treatments and other interventions was furthered substantially by the social changes of the 1960s. That social revolution began by a questioning of the American participation in the war in Vietnam and the treatment of blacks. The result was a confrontation with authority that began by challenging government officials – civilian and military. War, it was claimed, was too important to be left to the generals. The philosophical insight that was to become crucial for the development of bioethics was the application of the fact/value distinction often attributed to David Hume. Critics of the war claimed that being an expert in military strategy and tactics did not make one an authority on the morality or legality of the war. Normative questions were seen as logically separate from the skills necessary to claim expertise in military matters. The critics claimed that the ordinary civilian – the layperson – was as much an expert in normative aspects of war as military officers or Defense Department officials.

On the civil rights front, there was a rigorous re-discovery of political and legal rights. In particular, the rights of individuals, tracing back to Locke, Hobbes, Rousseau, and the American founding fathers, were extended and set against more consequentialist concerns. The movement for racial equality was also a movement to affirm the rights – and the authority – of low status people and groups outside the power structure and, thus, in important ways, converged with the anti-authoritarian influences of the anti-war movement.

Close on the heels of the anti-war and civil rights movements were similar affirmations of the members of other low status and out-groups. A women's movement pressed for equal status and decision-making authority. A radical student movement demanded the right of learners to be active decision-makers in the governance of the university.

With the advancement of the claims of authority of anti-war activists and the rights of racial minorities, women, and students, could patients' rights be far behind? Thus, the socio-cultural revolution of the 1960s claimed an active role for patients as medical decision-makers just at the time when the scientific revolution in medicine created a new modal disease that gave patients the opportunity to learn about their conditions and the treatment options that were being made for them. That scientific revolution presented treatment options – stabilizing technologies that left patients in uncomfortable and often painful positions – that were controversial and unattractive to many patients. It also left them sufficiently competent to rationally deliberate on the choices being inflicted upon them. Patients who increasingly understood the nature of those treatments began to understand that they had the authority to accept or refuse proposed interventions. Recognizing the rights of lay people to become active decision-makers thus fed off the affirmation of the authority of out-groups, including patient groups. Those two major social changes of the 1960s set the stage for the emergence of bioethics as a new era in the ethics of biomedical decision-making.

10.6 The Birth of Bioethics: 1967–1973

By 1967, the anti-war movement was at its height and the civil rights movement was in full swing. College campuses were in turmoil, and the women's movement was rearranging fundamental social relations. It was in this ferment that a bioethical revolution emerged. It was not merely a questioning of the ethics of research medicine or allocating organs for transplant; it was a basic reordering of the relationship between the health professional and the lay person. The radical change permeated all of medicine. In the fall of 1967, the first medical school with a medical humanities department took in its first class. The department chair, a humanist named Al Vastyan, played a central role in the hiring of other department chairs. In December of that year, Christiaan Bernard performed the first transplant of a human heart. In May of 1968, the Harvard Ad Hoc Committee published its proposal for a new, brain-based definition of death (Harvard Medical School Ad Hoc Committee 1968). In July of that year, Pope Paul VI (1989 [1968]) issued the birth control encyclical that stopped what was looking like an increasing acceptance within the church of oral contraception. During that year, theological ethicist Paul Ramsey, the chair of the religion department at Princeton, took a sabbatical leave to reside at the Department of Obstetrics and Gynecology at Georgetown University's Medical Center. Ramsey was invited by physician Andre Hellegers to help sort out the new moral issues related to emerging birth technologies; his experience led to the 1969 Beecher lectures at Yale and their publication in 1970 as *The Patient as Person* (1970b), perhaps the most influential book in medical ethics of the period. Ramsey's year on the floor of the hospital also led to the publication, in the same year, of his exploration of the ethics of birth technologies, *Fabricated Man* (1970a). In 1969, an ad hoc committee began to evaluate the Tuskegee syphilis study, a long-term, controversial project that purposely left men with syphilis untreated. Also in 1969, a new kind of social research institution was founded: the Institute of Society, Ethics and the Life Sciences – an interdisciplinary research institution founded by the cooperation of a physician and a philosopher working outside the academic setting of either a medical school or a philosophy department. Its original research focus included issues of clinical medicine – death and dying and the control of human behavior – as well as issues of population control and genetics, some of which were concerned with the gene pool and related topics of biology. That same year, a legal opinion was issued in the first of a series of cases that revolutionized the foundation of informed consent (Berkey v. Anderson). In 1970, in addition to the publication of Ramsey's two volumes, three important books on birth control appeared (Noonan 1970; Callahan 1970; Grisez 1970). In December of that year, the U.S. Department of Health, Education, and Welfare (1971) issued its Yellow Book – guidelines on the protection of human subjects of biomedical research, formulated not from a medical professional framework, but from a governmental policy perspective. In 1971, the nation's first academic research institute devoted to biomedical ethics was created. Known as the Joseph and Rose Kennedy Institute for the Study of Human Reproduction and Bioethics (and later to be called simply the Kennedy Institute of Ethics), its founder and its prime benefactor–Andre Hellegers and Sargent Shriver–were apparently co-creators of the new term "bioethics." Independently, Van Rensselaer Potter (1970) coined the same term, perhaps a few months earlier. However, in Potter's case, he was referring to the cluster of ethical issues related to ecological and environmental issues, as well as the medical implications of the biological revolution.

By 1972, several more legal cases forced a rethinking of the foundation of the informed consent doctrine, leading to a general shift away from the now-outdated belief in the professional standard by which the level of information that a patient must be told was determined by professional consensus (e.g., *Canterbury v. Spence* 1972; *Cobbs v. Grant* 1972). It was replaced by a "reasonable person standard" relying on the standard of what reasonable laypeople similarly situated would want to be told. This was a change entirely consistent with the newly established power of lay people to have the information appropriate to their new role as primary medical decision-makers.

By January of 1973, the United States Supreme Court issued the critical ruling growing out of the women's movement that legalized abortion (*Roe v. Wade* 1973). That year, the American Hospital Association (1978 [1973]), working with an interdisciplinary committee involving only a minority of physicians, developed the Patient's Bill of Rights that, for the first time, gave primary attention to the rights of patients rather than the moral duties of physicians. This rights-oriented ethic supplanted the traditional, paternalistic, consequence-oriented ethic of organized medicine. Later that year, the Congress held hearings in response to the Tuskegee syphilis study exposé (U.S. Senate Committee on Labor and Public Welfare, Subcommittee on Health 1973). The hearings led to the law creating the National Commission on the Protection of Human Subjects of Biomedical and Behavioral Research, once again transferring leadership on ethical issues in medicine from the profession to the public, and setting the basis for a moral theory in medicine that replaced consequentialist Hippocratism with a more deontological ethic of rights and duties including, for the first time, a moral principle of justice. Theologians, religious ethicists, and eventually philosophers would emerge as dominant players in the process. By law, the National Commission would have to have a majority that was not physicians.

In the short space of 6 years, almost every major aspect of medical ethics had been rearranged: human subjects research, death and dying, organ transplant, abortion, and informed consent. More fundamentally, ethics in biology and medicine was wrested from the grasp of the medical professional organizations and placed in the hands of public, interdisciplinary authority. The moral theory underlying bioethics was no longer the antiquated and isolated individualistic consequentialism of the Hippocratic tradition; it was now the moral theory of philosophers and lawyers and theologians: a theory based on human rights, a Kantian conception of autonomy, and an anti-utilitarian notion of justice. Bioethics was born.

References

Advisory Committee on Human Radiation Experiments. 1995. *Final report*. Washington, DC: U.S. Government Printing Office.

Alexander, S. 1962. They decide who lives, who dies. *Life* 53: 102–125.

American Hospital, Association. 1978 [1973]. A patient's bill of rights. In *Encyclopedia of bioethics*, vol. 4, ed. W.T. Reich, 1782–1783. New York: The Free Press.

Annas, G.J., and M.A. Grodin (eds.). 1992. *The doctors and the Nuremberg Code: Human rights in human experimentation*. New York: Oxford University Press.

Anon. The 17 rules of Enjuin: For disciples in our school. 1970. In *Western medical pioneers in feudal Japan*, ed. J.Z. Bowers, 8–10. Baltimore: The Johns Hopkins University Press.

Anon. The oath of Soviet physicians. 1971. Trans. D. Zenonas. *Journal of the American Medical Association* 217: 834.2.

Berkey v. Anderson: 1969, 1 Cal. App. 3d 790. 82 Cal. Rptr. 67.

Branch, C.H. 1956. Psychiatric aspects of malignant disease. *CA: Bulletin of Cancer Progress* 6: 102–104.

Callahan, D. 1970. *Abortion: Law, choice and morality*. New York: Macmillan.

Canterbury v. Spence: 1972, D.C. Cir, 464 F. 2d 772.

Caplan, A.L. (ed.). 1992. *When medicine went mad: Bioethics and the Holocaust*. Totowa: Humana Press.

Carson, R. 1962. *Silent spring*. Boston: Houghton Mifflin.

Cobbs v. Grant: 1972, Cal., 502 P.2d 1.

Edelstein, L. 1967. The Hippocratic oath: Text, translation and interpretation. In *Ancient medicine: Selected papers of Ludwig Edelstein*, ed. T. Owsei and C.L. Temkin, 3ff. Baltimore: The Johns Hopkins Press.

Ehrlich, P.R. 1968. *The population bomb*. New York: Ballantine Books.

Etzioni, A. 1973. *Genetic fix*. New York: Macmillan.

Fox, R.C., and J.P. Swazey. 1974. *The courage to fail: A social view of organ transplants and dialysis*. Chicago: University of Chicago Press.

Grisez, G.G. 1970. *Abortion: The myths, the realities, and the arguments*. New York: Corpus Books.

Hamilton, M.P. (ed.). 1972. *The new genetics and the future of man*. Grand Rapids: Eerdmans.

Harvard Medical School Ad Hoc Committee. 1968. A definition of irreversible coma: Report of the Ad Hoc Committee of the Harvard Medical School to examine the definition of brain death. *Journal of the American Medical Association* 205: 337–340.

Jones, W.H.S. 1924. *The doctor's oath: An essay in the history of medicine*. Cambridge: The University Press.

Jonsen, A.R. 1998. *The birth of bioethics*. New York: Oxford University Press.

Kelly, W.D., and S.F. Friesen. 1950. Do cancer patients want to be told? *Surgery* 27: 822–826.

Lifton, R.J. 1986. *Nazi doctors*. New York: Basic Books.

Medical Association of the District of Columbia. 1833. *Regulations and system of ethics of the Medical Association of Washington*. Washington, DC: Barron.

Medico-Chirurgical Society of Baltimore. 1832. *A system of medical ethics*. Baltimore: James Lucas and E.K. Deaver.

Noonan, J.T. 1970. *The morality of abortion: Legal and historical perspectives*. Cambridge, MA: Harvard University Press.

Nuremberg Code, 1946. 1946/1978. In *Encyclopedia of bioethics*, vol. 4, ed. W.T. Reich, 1764–1765. New York: The Free Press.

Oath of Initiation (Caraka Samhita). 1978. In *Encyclopedia of bioethics*, vol. 4, ed. W.T. Reich, 1732–1733. New York: The Free Press.

Paul VI, Pope. 1989 [1968]. Encyclical letter on the regulation of births (July 25, 1968). In *Medical ethics: Sources of Catholic teachings*, ed. K.D. O'Rourke and P. Boyle, 85–91. St. Louis: The Catholic Health Association of the United States.

Pius XII, Pope. 1958. The prolongation of life: An address of Pope Pius XII to an international congress of anesthesiologists. *The Pope Speaks* 4: 393–398.

Potter, V.R. 1970. Bioethics, the science of survival. *Perspectives in Biology and Medicine* 14: 127–153.

Ramsey, P. 1970a. *Fabricated man*. New Haven: Yale University Press.

Ramsey, P. 1970b. *The patient as person*. New Haven: Yale University Press.

Reich, W.T. 1994. The word 'bioethics': Its birth and the legacies of those who shaped its meaning. *Kennedy Institute of Ethics Journal* 4: 319–335.

Roe v. Wade: 1973, 410 U.S. 113, 93 S.Ct. 705.

Rutten, T. 1996. Receptions of the Hippocratic *Oath* in the Renaissance: The prohibition of abortion as a case study in reception. *Journal of the History of Medicine and Allied Sciences* 51: 456–483.

Samp, R.J., and A.R. Curreri. 1957. A questionnaire survey on public cancer education obtained from cancer patients and their families. *Cancer* 10: 382–384.

Thomas, L. 1974. *The lives of a cell: Notes of a biology watcher*. New York: Viking.

United States Bishops Committee on Doctrine. 1994. Ethical and religious directives for Catholic health care services. *Origins* 24: 449–462.

U.S. Department of Health, Education, and Welfare. 1971. *The institutional guide to DHEW policy on protection of human subjects*. Washington, DC: U.S. Government Printing Office.

U.S. Senate. Senate Committee on Labor and Public Welfare. Subcommittee on Health. 1973. *Quality of health care–human experimentation, 1973: Hearings before the Subcommittee on Health of the Committee on Labor and Public Welfare United States Senate, Ninety-Third Congress: First session on S. 974, S. 878, S.J. Res. 71: Part 1*. Washington, DC: U.S. Government Printing Office.

Veatch, R.M. 1978. The Hippocratic ethic: Consequentialism, individualism and paternalism. In *No rush to judgment: Essays on medical ethics*, ed. D.H. Smith and L.M. Bernstein, 238–265. Bloomington: The Poynter Center, Indiana University.

Veatch, R.M. 2005. *Disrupted dialogue: Medical ethics and the collapse of physician/humanist communication (1770–1980)*. New York: Oxford University Press.

Veatch, R.M., and C.G. Mason. 1987. Hippocratic vs. Judeo-Christian medical ethics: Principles in conflict. *The Journal of Religious Ethics* 15: 86–105.

Warren, J., L. Hayward, and J. Fleet. 1995 [1808]. Boston medical police, Boston Medical Association. In *The codification of medical morality: Historical and philosophical studies of the formalization of Western medical morality in the eighteenth and nineteenth centuries. Volume Two: Anglo-American medical ethics and medical jurisprudence in the nineteenth century*, ed. R. Baker, 41–46. Dordrecht: Kluwer Academic Publishers.

World Medical Association. 1956. Declaration of Geneva. *World Medical Journal* 3(Suppl. 10–12): 1769–1771.

World Medical Association. 1978 [1964]. Declaration of Helsinki–1964. In *Encyclopedia of bioethics*, vol. 4, ed. W.T. Reich, 1769–1771. New York: The Free Press.

Chapter 11
Professionalism vs. Medical Ethics in the Current Era: A Battle of Giants?

Edmund L. Erde

Preface

The author of Ecclesiastes 1:9 declares that there is nothing new (under the sun). Likewise, in his *Four Quartets*, T.S. Eliot (n.d.) declares that the past, present, and future contain each other, and his end is in his beginning.[1] Perhaps these intuitions apply as well to philosophy, at least after Plato. If they apply that to the struggles we have identifying and using the norms related to ethics and morality, we may, to adopt an echo of a phrase, always have our problems with us. This may even apply to norms and virtues appropriate to medicine and its practitioners. We may simply have to rotate the conceptual tools and social practices that have been with us, and give each a turn at bat from time to time. But, enough about metahistory.

11.1 Introduction

Although this essay is for a history book, it is not essentially history (or metahistory). While it draws on and/or (unintentionally) invents faulty and simplified accounts of the past, it is not about the past. Rather, its temporal orientation involves both a present trend and a timeless concern. As of this writing, the trend began (again[2]) between 15 and 30 years ago, though its birth-date is very difficult to establish. It began when prominent doctors started to summon their colleagues to professionalism. Shortly, there were calls to make it a central part of the medical school curriculum. Naturally, there were then attempts to define it. Eventually, there were efforts to measure it (Hafferty and Levinson 2008).

E.L. Erde, Ph.D. (✉)
School of Medicine, University of Medicine and Dentistry of New Jersey (Retired),
409 Loral Dr., Cherry Hill, NJ 08003, USA
e-mail: ele5@verizon.net

J.R. Garrett et al. (eds.), *The Development of Bioethics in the United States*,
Philosophy and Medicine 115, DOI 10.1007/978-94-007-4011-2_11,

These calls are a current (perhaps trendy) answer to a timeless question: "How shall values, rules and principles bear on medical practice and practitioners?" To represent both the present and the timeless, I contrast two complex replies. One, I label *secular academic medical ethics* (which I give various nicknames[3]). The other, everyone calls *professionalism*.

In large measure, Thomas Kuhn's (1962) highly influential notions of *paradigms* and *paradigm shifts* shape my approach. Kuhn developed these while doing the history of sciences. I use them here in connection with moral matters.

Paradigms are abstract constructs that create a perspective on a field of phenomena and inquiry. They contain various components (Malone 1993). Some (a) identify kinds of entities relevant to the field, (b) generalize about the entities and their interactions, and (c) define what counts as a problem in the field. Other components (d) specify models of explanations of interactions or (e) endorse styles of problem solving. They establish (f) exemplars of accomplishments (books, papers, experiments, etc.) and (g) values within the field. They also (h) provide a lexicon, which shapes communication among colleagues. Although I will only rarely mention these, they have guided me.

Several dangers lurk for one who tries to specify or define a field of study by comparing its rival paradigms and choosing which rival to adopt. One danger is failing to allow sufficient overlap among the rivals' topics and methods to allow comparisons. Another is failing to identify the tensions and disagreements between them to foster contrasts. Finally, one can ask biased questions, thus favoring one paradigm over the other. I offer the following question to both frame the field and avoid the dangers: "Which norms should govern the practice and practitioner of medicine—and why those?" This question allows for substantial overlap between *professionalism* and *secular academic medical ethics*. It also allows for significant gaps and tensions between them (for they are credible rivals, after all).

To show the differences, I draw on Edmund Pellegrino's sketch of the history of medical ethics, and sometimes go beyond it. Pellegrino (1993) traced ethics' course through four periods. Each disclosed failings of its own and produced a longing for a successor.

The first period ran from the time of Hippocrates to the mid-twentieth century. It was a pre-paradigm grab bag of precepts and opinions lacking conceptual organization.[4] But it did not help solve problems that were new in the twentieth century. The second period brought "a paradigm." It was *secular academic ethics*. Its first "edition" was *principlism*, and it offered convincing resolutions to some problems. However, it also produced an abundance of other vexing problems, which led to the third period—*antiprinciplism*. It stressed practical wisdom and case-based decision making, rather than mastery of theory. Thus, principlism and antiprinciplism became rivals *within* the secular academic medical ethics paradigm. Their rivalry led to what Pellegrino called the fourth period: *crisis*.

Thus, secular academic ethics *contains* rival views. But does it have a rival at a "higher" level? I turn to show that it does—and that the rival is professionalism. Next, I characterize secular academic medical ethics as a paradigm. After an interlude, I characterize professionalism as a paradigm. After that, I compare and contrast the two. But please note, dear reader: these comparisons and contrasts are impressionistic. I do not claim or suppose that those in either camp agree with their camp-mates on all major points. Nor do they disagree on all major points with those in the opposing camp. I will remind the reader of this from time to time.

11.2 Rivalry?

In 2003, Jack Coulehan and Peter Williams noted that, over the previous decade, talk of professionalism had become "a hot topic" in medical education (2003, p. 7).[5] There was much evidence for their claim. For, in that decade, the sophistication and quantity of the literature about professionalism increased vastly. In addition, beginning in 1999, The Accreditation Council for Graduate Medical Education (ACGME) went beyond adding to the literature. It made professionalism a core competency of residency program evaluation (see ACGME 2002). Also, formal courses on it entered the medical school curricula, and the National Board of Medical Examiners began to develop an instrument for evaluating it in medical students (2009).

However, those who would move professionalism into the curriculum would not be inserting it into a vacuum. Over the previous quarter of a century, medical ethics had gained a foothold in medical schools. Had professionalism's advocates been asking for more of that under a different name, their request would have been confusing. In invoking the term *professionalism*, they must have meant something different from *ethics*. Thus, they must have thought that the established approach—secular academic medical ethics—differed from what they wanted. To advocate for the change, they must have thought that medical ethics had failed somehow and that some other stuff—professionalism—would be better. They were calling for a major course correction—a "revisioning"—to expel medical ethics and replace it with professionalism. So, what was medical ethics when calls for professionalism began?

11.3 Paradigm-I: Secular Academic Medical Ethics

Recall Pellegrino's characterization of the first period of medical ethics as a collection of isolated precepts and rules, with no organization. Individual norms lacked foundation and thus were no more than opinion. The collection offered no system; it suggested no way to solve new problems. Still, ethics in that period contained much of value.[6] It, Pellegrino noted, worked well for 2,500 years.

In the 1960s, people ceased to defer to the ethics from the first period. Pellegrino noted two causes of this. First, social movements for civil rights, women's rights and consumers' rights built on America's long-lived reverence for law (Glendon 1991). Ideas about rights in these contexts became a model for patients' rights. Second, medicine made vast gains in technical power; thus, it became specialty-oriented. This depersonalized care, which had the effect of alienating doctors and patients from each other.

Two other social forces that undermined first-period ethics are worthy of being added to those that Pellegrino listed. First, attempts to control costs, such as from "managed care," brought changes to the organization of medical practice (see below). Second, in response to German and American abuses of human beings in research from the 1930s on, the law (treaties, case law, statutes) came to emphasize subjects' right to informed consent. Later, informed consent became prominent in medical practice as well.[7] It had not been part of first-period ethics, but the times forced it front and center. Thus, first-period ethics would no longer do.

Thus, the second period, *principlism*, began. It was revolutionary. Scholars drew on the methods of the humanities and the social sciences to clarify and resolve conceptual and normative dilemmas. They also changed the norms so their content would be secular. In the first period of ethics, medicine's morality was rooted in various religious traditions. That would not do in our contemporary pluralistic society. For, each religious tradition could consider alternative positions as mere opinion. Thus, we needed a secular system of morality for a highly diverse society.[8]

This remade medical ethics as *academic* medical ethics—a new discipline.[9] It promised to be revolutionary. Albert R. Jonsen and Andre E. Hellegers (1976) saw society as being in great need of such a paradigm, because it did not even understand what the phrase *medical ethics* should cover. As evidence, they noted that newspapers used the phrase when reporting on doctors performing euthanasia, owning stock in pharmaceutical companies, and coercing poor women into sterilization in order to protect public coffers. Jonsen and Hellegers thought that only a new discipline/paradigm could narrow the meaning properly.

They celebrated the emerging paradigm as "a species of the genus 'ethics'" (1976, p. 19). The methods of moral philosophy could clarify key moral ideas, rationalize and justify norms, and deduce and apply rules. This would provide decision-makers with a deep grasp of key notions and ways to approach dilemmas—i.e., it would provide a paradigm.

As it emerged, secular ethics came to do what many paradigms do. It accepted one text as classic. This was Tom L. Beauchamp and James S. Childress's *The Principles of Biomedical Ethics* (1979). The book highlighted four principles: *autonomy, beneficence, nonmaleficence,* and *justice.*[10] Like all paradigms, its expanding use disclosed both philosophical *and* practical strains. For example, Beauchamp and Childress admitted that they could not crown any one of the four principles sovereign over the others. They recommended that the four be weighed on a case-by-case basis. However, the weighing metaphor is disquieting. Often, we do not know how to weigh any one value, much less how to compare the weight of competing values. Weighing returns us to pre-paradigm thinking—all is just matters of opinion. Thus, the paradigm generated problems for itself from within. It also encountered new "external" problems that it could not convincingly address.

Another strain arose, because in spite of Beauchamp and Childress's reluctance to make any principle preeminent, others were willing to do so. In powerful works, Veatch (1981) and Engelhardt (1986) promoted patient *autonomy* as preeminent.[11] Often, arguments seemed based on a tacit model from business. For example, a valid contract takes the parties to be free and able to negotiate the terms of their commerce.

The rules give them equal power to deal. If medicine conforms to this model, practitioners should offer to diagnose a patient's problem, explain the risks and benefits of each test in detail beforehand, and then comply with whatever the patient chooses. Treatment should work analogously. The patient, as reasoning and free, should make his choices and assume the associated risks. Thus, as applied to medicine, this ethic was about specifying the shape of the rights of patients and the correlative duties of doctors.[12]

Medical paternalism was the main moral problem. Respect for autonomy was its answer. Autonomy helped organize the contents of the new paradigm, placing *negative rights* at the center. These rights leave people free to enjoy (or suffer with) their choices. They are sometimes cast as "the right to be let [or left] alone." They amount to this: one cannot enter someone's personal sphere and act on that person without the person's permission. That permission must be based on honest disclosure about the purpose for entering that personal sphere and the likely consequences of the interaction. This is the right of informed consent.

Thus, autonomy came to define many problems within the ethics paradigm and offer solutions to these problems. However, it also generated moral strains. For example, it significantly demoted the importance of *patients' best interests* (Veatch 1995). This disquieted practitioners' sense of responsibility,[13] and raised their concern about malpractice suits.

Autonomy's proponents had to say how one should manage patients who lack rationality, e.g., children, the mentally ill, the demented, etc. In partial response, they forged the principle of *substituted judgment* to apply to patients who had lost their ability to reason. The principle mandates that practitioners follow the values that their patient had before the loss. Someone who knew a patient's values well would ponder how he would have understood the meaning or quality of life of his current life. That person would act as proxy—i.e., choose as though she were the patient.

This sometimes strained the paradigm, too. It empowered proxies to make bad choices. A proxy might reject productive care and let a patient die "unnecessarily." Or a proxy might order treatment for a hopeless condition as though it could be cured, thus imposing avoidable suffering on the patient and wasting resources. The proxy's power trumped Medicine's common sense about the patient's best interests.

So the notion *best interests* became a shadow and a beggar. It begged important questions and often made no sense (Veatch 1995). Could a doctor tell what was best for patients? In a multicultural world values can vary greatly and might be incommensurable. Thus, specifying an interest as "best" (or not) lacks foundation and may be audacious. Hence, ethics does not, but professionalism does, incorporate the tenet that "the doctor knows best." Similarly, the idea of *harm* loses some substance in the ethics paradigm. The culture of medicine trains its members to think of harms as objective and absolute. It casts death as a clear harm, but pain as something less than harm. When, slowly and with difficulty, medicine finally recognized pain as a harm, it did so because of pain's impact on eating, sleeping, and exercising—not because of its intrinsic offensiveness. These views of death and pain were part of

Medicine's cultural common sense. However, the public found them odd (or worse). Patients wanted to value death and pain in their own way and to act accordingly.

So, *respect for patient autonomy* produced both answers and strains in the form of puzzles and dissatisfying results. Autonomy proved to be a dangerous two-headed snake on a stick. I have shown some dangers from one head—negative rights. I turn now to show dangers from positive rights.

Positive rights pose a significant set of puzzles. Rights are abstract social structures. They entitle persons to make claims on their own behalf, which others have a duty to accept. Positive rights entitle a person to demand goods or services and oblige others to comply with those demands. In healthcare, patients have some positive rights, and doctors or institutions must conform to them appropriately. For example, a federal law, EMTALA, grants positive rights to those who present to Emergency Departments for care. It establishes the person's right to be examined, and, if found to be in medical crisis, to be stabilized (US Government 2010).

Note how vexing questions arise about positive rights. For example, if autonomy enables patients to refuse a test or treatment, why does it not also enable them to secure one they want? Doctors might refuse to comply on grounds of professional autonomy.[14] Can this justify rejecting a patient's medical choice? If professional autonomy cannot justify forcing a treatment on a patient, how can it justify overriding a patient's demand for a treatment? Shouldn't a patient be able to decide which risks to run for particular gains? Shouldn't she be able to get what she thinks is good for her even if her doctor disagrees? How can doctors justify refusing to do an elective cesarean section? How can they justify limiting the number of embryos to implant in a woman or the number of children to help her produce?[15] How can surgeons set age or health limits on those who can have bariatric surgery? Insurance companies might be able to limit whom they will cover for which treatment, but how can doctors defy patients' rights to run risks? Governments may forbid doctors to prescribe antibiotics for viral infections on grounds of protecting public health. But what justifies a doctor refusing to provide antibiotics on their own authority?

Cardiopulmonary resuscitation (CPR) raises acute issues about positive rights. CPR's success rate is low. It can greatly harm a patient. However, doctors rarely are empowered to order that CPR be withheld without valid consent from the patient or her proxy. When a patient is gravely ill, family members often say that they want "everything done." Hospitals generally accept (or pretend to accept) that as the exercise of a positive right. Thus, doctors are supposed to do CPR, even when they think it is medically foolish.

Another strain on the ethics paradigm came from clinicians' resenting it. It was an outsider's paradigm, mandating actions contrary to ***M***edicine's traditional values. And it morphed into quasi-law, by drawing heavily from law. Meanwhile, practitioners' lexophobia[16] enhanced the resentment. Thus, academic ethics vastly reduced doctors' professional autonomy (discretion). For some of professionalism's proponents, regaining professional autonomy was a major goal (Cruess et al. 2002)

In sum, the conflicts within principlism burdened it enough to lead to Pellegrino's third period: antiprinciplism. In it, Jonsen and Hellegers' vision for secular academic medical ethics blurred. Its content had become legalistic, and it failed to achieve the depth that advocates had promised. Both negative and positive rights gave answers that were sometimes contrary to Medicine's common sense and to conscientious practitioners' *sense of professional autonomy*. The desire to regain it was a strain that took ethics into Pellegrino's third period—*antiprinciplism*—and then led to the fourth period: *crisis*.

11.4 A Note on Philosophical Method

Thus, professionalism's "temperature" as a topic rose. Nevertheless, its meaning got scant light. There were many attempts to define it. Some are insightful. Still, each fails (Erde 2008). Thus, instead of defining it, some authors just list its elements. The lists have not given us the accuracy and clarity we want (Erde 2008). Gaining accuracy would require a narrative about the separate items on such a list. The result would be too unwieldy and diminish clarity. It would likely omit much of relevance and include false leads. Hence, we must settle for less accuracy and clarity than we want.

To clarify *professionalism* to the extent possible, I take yet a different route. I identify and explore the major problems that medicine's leaders think professionalism will solve or reduce. However, before discussing them, I recall an analogy from Plato.

In the *Republic*, Plato's characters wanted to understand *justice* as a virtue of individuals. Socrates rejected that as too difficult. He proposed that they discuss the just community instead. An individual is too small a subject for us to discern his makeup in great detail. A larger locus, the community, would be easier to study. This, they thought, would answer their question, "What is justice in man?" It would work because Socrates cast *justice* as the same trait, whether in a community or an individual.

Think of *professionalism* like that. Think of it as two-sided. First, it is the essential ideology of medicine [*M*edicine] as a tradition and an institution—as something like a macrocosm with its own integrity.[17] Second, it is also a character trait—a virtue—of individual practitioners. As such, as Aristotle said, it is to be understood "relative to its proper work" (*Nicomachean Ethics, 1139a17-18*). That is what proponents of professionalism sought to instill—the ability to make proper choices as doctors. They wanted doctors to have a "second nature" that is the congealed product of inculcated ideals. On this model, practitioners learn the ideals from a macrocosm. *M*edicine is the macrocosm. It is the bearer of the ethos of professionalism. Because the ethos was rightly developed, it can train its newcomers to make proper choices. Thus, professionalism is a variant of virtue theory from the ancient Greek paradigm. ***It is the disposition to sustain life and maximize health. And that includes being oriented to improve the skills and knowledge that one must have to achieve these goals.*** Thus, professionalism is a set of habits, chosen by reason and refined through practice.

However, professionalism's advocates overlooked some of its complexity. They did not take the strata of values into account. Values come as (1) ideals, (2) practical interests and (3) predilections, including temperament and expectations of etiquette. It is important to consider the role of each of these.

Ideals state professionalism's ethic. They are *chosen*—adopted, after reflection. They serve as reasons that inform such notions as *praiseworthy*, *required*, *optional*, and *forbidden* or *condemnable*. They arise from *Medicine's mission* in combination with at least four other sources: (a) society's dictates, (b) patients' demands/needs, (c) tradition/history,[18] and (d) an inclination to help those who are sick or suffering.

Practical interests contain a sense of the worth of resources such as money, time, reputation and health. They also include social arrangements that foster these practical goods. One such arrangement is being a monopoly. Another is being entrusted to self-police rather than being highly accountable to outside authorities.

Predilections or preferences are prejudices, tastes, and dispositions that produce reactions. Members hold them independent of reflection. For example, students and faculty (unwisely) come to esteem those who can work on very little sleep.

Medicine communicates explicit and implicit norms at all three levels. Together they shape professionalism as a perspective, an ethos. The ethos stigmatizes doctors who testify against defendants in medical malpractice cases. It leads them to expect free lunches at pharmaceutical seminars. It gets residents to scold students for taking too long dealing with patients. It traps doctors into becoming workaholics. This is *M*edicine as a subculture.

11.5 Professionalism

In Sect. 11.3 above, I noted Jonsen and Hellegers' complaint that society applies the term *medical ethics* to too varied a set of issues. Interestingly, some of the topics they might ostracize are part of professionalism. To present a detailed, values-focused account of it, I offer two guides. The first is an argument claiming that doctors have a duty to accept as patients even persons who have a disease that can endanger a doctor's health and business. The second guide is a set of ten problems that medicine's leaders seem to think professionalism would resolve or lessen.

11.5.1 First Guide—The Duty to Accept AIDS Victims into a Practice

In the 1980s, persons with AIDS had great difficulty finding doctors who would accept them as patients. Doctors did not want to risk becoming infected from a patient. Some feared that patients who did not have AIDS would leave their practice to avoid exposure to it from those who did. The doctors' stance was consistent with what the American Medical Association (AMA) claimed for doctors—the right to decline any prospective patient.

Edmund D. Pellegrino denounced doctors' refusal to accept these people as patients (1987). He grounded their obligation to take such risks in six facts:

1. doctors have a calling—they are called to heal,
2. patients are vulnerable,
3. patients lack, but doctors have, the knowledge and skill to heal,
4. doctors hold their technical knowledge in trust for those in need,
5. taking their oath at graduation commits doctors to provide care and comfort, and
6. self-effacement is a core disposition of professionalism.

Item (6) means that improving the welfare of the sick takes precedence over the interests of the professional. Pellegrino contended that doctors are like firefighters, police officers, or soldiers. Accepting their roles obliges them to run risks. This is not true of merchants, shoemakers or plumbers.

Pellegrino thus concluded that doctors have a moral duty to accept new patients that bring danger. This makes no sense under secular academic medical ethics' respect for autonomy. That paradigm would see Pellegrino as claiming that one has a duty to do more than one has a duty to do. Though this claim is paradoxical or contradictory in secular academic medical ethics, for Pellegrino it is constitutive of professionalism.

Facts (1)–(6) are medical professionalism's foundation. These define the ideal doctor. She is someone who frames her work, *and constructs herself*, to fight disease, reduce pain, and prevent death. She learns so that she can do the work well. She is dedicated to it and, in effect, has promised to focus entirely on helping those in need, even to the extent of self-deprivation. She has a calling to do the art well.

Pellegrino's account, then, was more than a sketch of facts and mandates for action. It defined *doctorhood*. It specified what constitutes being a doctor—a "true" professional. To be in accord with this ideal, a doctor always acts as *the* champion-of-afflicted-persons'-health, always fosters their "best interests."

AIDS victims' lack of access to doctors has now faded as an issue. Still, it was instructive, in that Pellegrino responded to it by asserting that doctors have duties to people even if there is no doctor-patient relationship between them. Further, he proffered six facts to use as the foundation for professionalism *generally*. They bear on every problem that prompted medicine's leaders to promote it.

11.5.2 Second Guide—Concerns of Conscientious Leaders

In the 1980s, leaders within medicine mentioned various problems that they thought professionalism could solve. Were those problems of equal concern to medical ethics, in the same ways? To contrast the two paradigms, I here discuss what each paradigm implies for ten problems. Clearly, each problem is of interest to professionalism, because its advocates alluded to each when explaining why they raised it as medicine's flag. Yet medical ethics might not be interested in all ten. Even when both paradigms are interested in a particular problem, the intensity of their interest may differ.

One idea that many of these problems share is that the doctor serves as a *fiduciary* for the patient. This means that a doctor's primary moral virtue is to be trustworthy as the unfailing champion of his patient's life and health. Thus, he must be worthy of blind trust in his dedication to patients' medical well-being. He does what is ***best*** for the patient, even if it costs the doctor to do so. The first nine problems overlap in varying degrees. They are mostly about practitioners. The last is about medical education.

Again, please recall my declaration at the end of the first section—i.e., the comparisons and contrasts are impressionistic. No one from one paradigm need agree with all others from that camp. Neither must anyone disagree with the other camp on all major points.

Problem (1): Doctors faced with a **conflict of interests** when providing care fail to do right by their patients. These conflicts come in many forms (Spece et al. 1996). The classic type occurs when a doctor must choose between providing best care and maximizing her own economic wellbeing, most commonly her income (Erde 1996).

The structure of economic conflicts varies with what the doctor can bill for or who pays for care and how. For example, "fee for service" is one form of charging for care. It charges the patient for each individual service she received and for the costs of the materials used. Therefore, fee for service generated temptations for a doctor to do medically unnecessary tests or treatments solely to increase his bill. These conflicts had an added dimension if a doctor owned a lab or radiological center. Wanting the business to thrive could tempt the doctor/owner to order more tests.

This is very important because ***tests pose significant dangers*** of several sorts. They can injure a patient. They can produce false findings that cause the patient unwarranted worry. They can expose patients to injuries from steps involved in following up on abnormal results. They impose costs to patients and the nation without providing value. When a doctor orders medically unnecessary tests (for any reason), she acts as the exact opposite of a fiduciary. Because the misuse of tests is intentional, the doctor's action is *treachery*.

Conflicts of interest may involve something less than treachery. Imagine a doctor committed to being champion of her patients' health. She still needs a policy for when a patient's condition does not clearly indicate the need for a test, treatment or referral. There are two ways of being cautious in such cases. One is **active pursuit**. In this view, a doctor would never allow a disease to elude diagnosis. If a patient has signs or symptoms, the doctor orders any plausible test or referral in hunting down the answer. Alternatively, cautious practice may be **passive observing**—refraining from testing, treating or consulting until the need for that action is clear.

The culture of Medicine supported the first approach. It constitutes its members to pursue answers and to correct problems—even potential problems. In addition, doctors often have a subliminal profit motive, putting a conflict of interest just below full awareness. The inclination to diagnose and the subliminal conflict of interest can combine to produce overuse of tests, without treachery. Still, non-treacherous testing puts patients in danger and adds to the cost of care as much as treacherous tests do. Thus, passive observing is the wiser policy, even though it defies the values of the culture.

HMOs employed several strategies to have doctors adopt passive observing. They would not pay for certain tests unless they had granted permission for their use prior to the patients being tested. Also, they monitored doctors' charts to assess whether interventions were overused. Finally, they paid bonuses to doctors who ordered tests under a particular frequency. All of this was both to improve care and lower costs. However, those bonuses did not always improve care. They made conflicts of interest a glaring problem in a new way. For, patients now suffered because doctors did not test when it was appropriate to do so. Whether treacherously or not, doctors tested less, in order to get the bonuses.

How would the two paradigms process all this? Each would find it worthy of attention, but in different ways. Secular ethics would require the doctor to provide the patient with full, clear information about tests' risks and benefits, and help the patient make an autonomous choice. The doctor would then act accordingly, even if it were not the choice most likely to maximize the patient's health. By contrast, professionalism would condemn taking a course that is not health maximizing. Its virtue of self-effacement would have practitioners disregard bonuses and do what was best for the patient even if that meant losing income. Anything less violates self-effacement and trustworthiness.

Professionalism sends a message about self-effacement even when the relationship is not of the classic doctor-patient type. For example, in testifying before the United States Congress, Dr. Linda Peeno (1996) discussed her work as a medical director for a health insurance company. She confessed to having refused permission for a patient to get potentially life-saving care to save the insurance company money (1996). The patient died. Dr. Peeno's action came to haunt her, and she condemned it as unprofessional. This assessment seems correct. Dr. Peeno used her medical credentials to get her job, but transgressed the explicit contractual commitments that her company had to the patient. It was treacherous and violated the mission of medicine.

Suppose a doctor is going to operate on an AIDS patient. The doctor knows of two approaches to the operation. One is generally better for the patient. The other lowers the surgeon's risk of sticking herself with a contaminated needle. The situation raises a few questions. For example, does the doctor have a conflict of interest? Is her health or safety a resource, quite like her income?

Professionalism as a virtue may be too weak to help doctors resist conflicts of interests. More importantly, its stance on conflicts of interests seems impossible to live by. One simply cannot be obliged to provide *the best* care. No one may know what is best. The best diagnostician, medication or equipment may be too far away to access. The wait for the best doctor or materials can be too long to be useful. There may be no best treatment because the problem is routine and anyone can do a reasonably good job of treating it—nothing is best because all are equally good. *Best* care may require the doctor to move into the patient's home and never let her out of sight. Or it may require a doctor to donate her own blood or kidney for the patient. Would it oblige a doctor to refrain from retiring because a patient will trust no one else and will suffer as a result? Professionalism's rhetoric suggests that doctors have the duty to sacrifice themselves in such cases. Its advocates seem blind to these implications. It promises too much. Nothing in secular ethics suggests that excess.

Problem (2) involves **substandard quality of care**. Secular academic medical ethics does not concern itself with the quality of care. It seems to presume that quality will be good enough. It would be interested in the topic only if different treatments posed different risks and benefits. Then knowing about quality would be of interest for the informed consent process.

By contrast, professionalism would have great interest in deficient care and its causes. One cause may be betrayal from the classic conflict of interest just discussed. Another is doctors tolerating an environment or system that fosters inferior care—too large a caseload, too full a schedule, etc. A doctor may tolerate defective equipment (e.g., suboptimal lighting), not wanting to alienate administrators by complaining about it. A doctor may not know enough about the condition, the case or the treatment she is to use. Or, she may mistake the meaning of test results. Her attention may lapse. A key member of her team may be missing. The chart may be defective or the handwriting difficult to read. Staff may check equipment poorly. The doctor may have been swayed by pharmaceutical marketing. Professionalism would have professionals eliminate or reduce these. Academic medical ethics would not. It would leave correcting these causes to others (perhaps the advocates of professionalism).

Problem (3): Doctors **communicate poorly, are not readily available, and have poor social skills.** Complaints about these aspects of behavior are of low concern to academic medical ethics. It only requires doctors to be clear and truth*ful* for purposes of obtaining a valid informed consent. It does not require empathy, kindness, or good manners. Indeed, it practically ignores them. Nor does it contemplate the doctor-patient relationship as a healing force. However, it might care about some behavioral themes. It would condemn misleading a patient via false hope. It mandates that doctors (and everyone else, for that matter) keep their promises.

Presentation of self is part of this problem. Professionalism's concern with it is multifaceted. If, as Pellegrino contended, professionalism begins with being called to medicine to help the sick and suffering, doctoring is based on an emotional investment. One will care deeply about the quality of one's work and its outcome for patients. Thus, doctors must avoid driving their patients to deceive or to evade them. For example, patients should not fear to admit to smoking, abusing drugs or skipping exercise, etc. Professionalism would not want a patient to give a misleading history, which she may do to avoid a scolding. That misleading history may lead the doctor to misdiagnose her and order the wrong treatment.

Secular academic medical ethics may differ. It could take the view that individuals are supposed to develop some fortitude as part of their autonomy; they must take the responsibility to live with the consequences of their own nature and choices.

Under this paradigm, doctors need not be solicitous. If a patient deceives or avoids care, that is her problem. However, nothing in the ethics paradigm supports scolding a patient for poor behavior.

Problem (4): Doctors **inadequately advocate** for the sick. This failure looms as a large danger under HMOs. One way it arises is familiar from the story of Dr. Peeno, above. Recall that for insurers to pay for some tests or treatments, patients need the insurer's permission in advance. A company sometimes refuses permission, as Dr. Peeno did for hers. When this happens, doctors advocate for patients by appealing the refusal.[19]

Advocacy was not yet a theme when secular academic medical ethics began. At that time, insurance companies did not refuse to pay for tests or referrals. Thus, there was less need for a doctor to advocate for his patients. It may have been needed if the patient were applying for workers' compensation or life insurance. However, as the context changed, advocacy became necessary to secure good care. Vigorous advocacy would be a duty under professionalism. Under secular ethics, whether there is this duty depends on whether the doctor somehow promised to advocate.

Doctors sometimes elect to advocate for sick or suffering persons, or for those running particular risks, even if there is no doctor-patient relationship. [20] Through much of the 1980s, several doctors argued against boxing because it can cause brain damage (Lundberg 1983). If one could get doctors to boycott matches, in some states the match would be illegal (Sammons 1989). This would protect all would-be boxers—patients and non-patients alike. Campaigns against tobacco or against pregnant women consuming alcohol are also advocacy for non-patients. Joining such campaigns conforms to ideals of professionalism. By contrast, secular academic ethics should oppose any public health measure that infringes on autonomy; it would tolerate only educating the public.

Problem (5): The profession should resolve **doctors' confusion** about their duties as it arises from changes in medicine and society. Consider three arenas in which confusion has been significant: (a) revolutionary changes arise from technologies and techniques; (b) major changes arise from new commercial aspects of insuring patients or paying for practice; and (c) payers or governments use health care and medical science for social control. Many of these are topics that the ethics paradigm addressed as problems to solve for individuals' use in the conduct of their lives. Professionalism, however, addressed them as divisive perplexities to resolve so that members would have a unified answer to appeal to.

Arena (a): New technologies and techniques confuse practitioners, e.g., regarding end-of-life care. This could happen from harvesting still-beating hearts from "donors." This could provoke confusion over the meaning of "death" and establishing when and whether someone died. One could also be confused when asking family members for permission to harvest organs. Professionalism's commitment to health should dispose doctors to urge organ donation. Secular ethics' commitment to autonomy requires informing family members about the choices open to them, and it obliges doctors to refrain from urging any particular choice.

Other confusions in this arena arose in the late 1960s. Patients began to *refuse* life-saving treatments. Acceding to their refusals seemed contrary to the commitments to life and health. The ethics paradigm's stress on the right to be left alone clarified the issue in favor of patients' self-determination, and Medicine slowly accepted *withholding* such treatments. Later, negative rights supported what sometimes feels like homicide: acceding to a patient's demand to remove life-prolonging treatments.[21]

Still, some puzzles about *withdrawing* life-prolonging treatment endure. For example, the "terminal wean" involves removing respiratory support from a patient who cannot breathe on her own. These patients may experience "air hunger." To prevent this dreadful aspect of dying, professionals can administer morphine before removing the ventilator. But the morphine can cause breathing to stop before the patient dies "naturally." Thus, it can accelerate dying. Therefore, using it goes beyond respecting negative rights, which would only require removing offending equipment. If administering morphine may cause a patient's death, is using it a form of homicide? The doctor need not intend to kill in such scenarios, but she may still be killing. So, does using morphine in this way offend professionalism's commitment to life and health? Is it something doctors should not do? Could it be both ethical and unprofessional?

Arena (b) involves confusion about commercial aspects of practice. Incentives to change practice could confuse doctors. Professionalism took a clear stance in favor of patients' wellbeing in the face of conflicts of interest. What stance does it take in the face of a patient in great need but who lacks insurance or whose insurance does not apply? There is evidence that it leads doctors to be dishonest in order to promote patients' health.[22]

Academic ethics, however, says little about such confusions. All that it requires is honesty. Thus, it requires disclosing the existence of conflicts of interest to patients so they can factor its influence into their decision. Moreover, its logic would condemn misrepresenting a record if that helps a patient get health care or coverage.

Economic confusions also arose from changes in the types of cases with which patients present. In the twentieth century, chronic conditions became a large part of daily practice and raised the cost of care greatly. Doctors could easily be confused about their duties to profoundly demented patients or patients in a persistent vegetative state. What is a doctor's duty when conditions and treatments are novel, prolonged, expensive, and show little to no prospect of benefiting the patient?

Arena (c) involves confusion from social uses of medicine. Many third parties (employers, life-insurance companies, and health clubs) require a report on the health of a person. Doctors report on persons they have never seen before. What if they find a serious problem but the contract forbids them to inform the patient? Would an explicit agreement among the doctor, those asking for the assessment, and the person to be assessed settle the doctor's confusion about dangerous findings? Under secular ethics, no breach of trust occurs from complying with the agreement, but complying could offend professionalism. The call to help the sick also urges warning, regardless of the agreement.

Confusion is latent in many types of screening. A transplant team wants to know whether a patient is likely to stay sober and protect the transplanted organ. It asks a psychiatrist to assess this. To answer well, the doctor wants information that she can best obtain if the patient trusts her. But is she trustworthy in this context? After all, her findings may lead her to recommend against an individual receiving an organ. That does not foster the patient's health. It runs counter to professionalism.

Doctors also assess whether an individual is competent to be executed or able to sustain torture. Giving the go-ahead defies professionalism's mandates. Such cases could well confuse a doctor who works for a government *and* has adopted professionalism.

This problem of confusion is more important to professionalism than I first realized. It is symbolic of practitioners' need for help from Medicine as an institution. Confusion often calls for the *community's* decision to settle what is confusing. We get the community's answer from a process. Members think about the confusing topic. They discuss it, write position papers about it, and vote on motions about it. The vote can be a crystallized answer on behalf of the group. Thus, getting the community to resolve it relieves individual doctors of guesswork, using intuition, or relapsing into the ways of the first period of medical ethics. Thus, it gives an answer as doctrine—the institution's doctrine—rather than as individual conscience (but see Savage 2008). This marks a great difference between professionalism and ethics; ethics would have each individual decide autonomously.

Problem (6): Doctors are **dishonest in business dealings**. Doctors defrauding insurers gets wide publicity at times. Typically, they do this to enhance their personal wealth.

Fraud was not a topic for secular academic medical ethics. Its moral status is too easy. Professionalism is concerned about fraud and the damage scandals can do to medicine's reputation and image [problem (9), below]. Not all fraud targets patients. Some may leave the patient's autonomy intact. It may target an insurance company to further a patient's "best interests." Professionalism seems to approve of that sort of "Robin Hood" behavior (Novack et al. 1989).

Problem (7): Doctors **fail to respect boundaries**. When someone is the doctor in a doctor-patient relationship, all of her interactions with the patient should be role-related (or conform to custom/politeness).[23] Boundaries outline the shape of the moral territory within which actions are *appropriate* **as** *doctors' actions*. They also indicate what doctors must refrain from doing vis-à-vis their patients, so as not to undermine their doctoring.

The paradigms have different degrees of interest in boundaries and draw them in different places. For the most part, academic medical ethics will find them worthy of attention when a violation is a step toward exploitation. Professionalism, however, has greater interest in them and understands them much more broadly.

The classic example of a troubling boundary crossing is a doctor having sexual relations with their patient. Medical ethics did not pay attention to this. Howard Brody (1992) explained why, for example, medical ethics textbooks do not include the topic. Such books aim to provoke thought on complex and morally controversial issues. Sexual battery is not morally controversial. Condemning it requires no ethical analysis. It is easy to denounce a doctor's exploitation of a heavily sedated patient or someone undergoing a physical examination. Boundary violation of this sort are rape or akin to it. Both paradigms would find them easy to condemn.

Even if Brody is correct that medical ethics has no interest in sexual boundary issues, he may be wrong to accept that the issues are too easy. Behavior that is less blatant than rape can have many nuances that are worthy of moral analysis. Analyses of manipulation and seduction could teach us a great deal. For example, does *autonomy* imply that individuals (patients) have a *responsibility* to control doctors' attempts to intimidate or flatter them?

To trade in stereotypes: presumably, the doctors are males (in mid-life crisis) and the patients are young women in need of protection from sexual exploitation. Secular ethics might defend the freedom to be intimate as follows.

> Years ago, young American women were vulnerable. Today, this is generally not true. Many have fine careers. They are unlikely to be impressed by a person just because he is a doctor. Many young women do not live with their parents. Such a person would have no need to explain to her parents why she changed doctors. Objections to crossing sexual boundaries are usually framed in terms of doctors having power and patients not. Today, however, no power differential is evident in many doctor-patient relationships. *Perhaps*, then, one need not have concerns about patients' weakened autonomy. The parties could be consenting adults.

Secular ethics, then, does not find being a patient so debilitating as to destroy autonomy.

Proponents of professionalism could take a contrary stand at two levels. First, they could find sexual boundary issues complex enough to be worthy of analysis. As proof of this, they could, second, contend that a sexual relationship between doctor and patient cannot be consensual because of the power discrepancy. Doctors' notable powers derive from many sources, including technical knowledge, social status, and legal authority.[24] Some power is from knowing a patient's history. This includes their sexual and emotional histories as well as their history with sexually transmitted diseases (STDs). All this could cue a doctor about a patient's libido or self-control, and ways of successfully manipulating the patient.

Professionalism would care about doctors ***or*** patients falling into emotional traps. In a healing relationship, either party can have complex and confusing reactions. Patients may feel loyalty, gratitude and relief. Doctors may feel loyalty, validation, and empowerment.[25] Either doctor or patient could confuse all such feelings with romantic interest.

Professionalism would take an interest in how crossing sexual boundaries may seriously compromise care. Patients may be less honest with their doctor/lover/partner to keep the image of being a desirable mate. They may refrain from undertaking treatments because they want to end a sexual relationship and do not want to start a course of treatment with a soon-to-be ex-lover. Alternatively, a doctor's desire to end a sexual relationship may coincide with their starting or continuing a prolonged course of treatment. That desire may influence whether he evades starting the treatment or how well he does the work. Moreover, if the doctor ends the affair, how is a patient (especially a very sick one) to take "being dumped" by someone trained to be discerning and knows her so well? Would she avoid medical help in the future? Would it affect how she relates to other doctors?

Sexual relationships may reduce a doctor's ability to be objective. He might miss signs of disease or increase his vigilance, leading to inappropriate tests and treatments. Dual relationships could also weaken a doctor's power to refuse an inappropriate request. In a sexual relationship, doctors—anyone, really—can lose power that they need for work. Doctors need the power to resist inappropriate requests, e.g., for opioids, for a procedure they do not do well, or that they not initiate the process for psychiatric commitment.

Another kind of boundary issue involves one party soliciting small favors and gifts from the other.[26] A doctor should be able to tell friends or relatives that their request is "over the line." Being able to say so is an important tool, especially when the other relationship is very significant. On the other hand, a doctor's requesting small favors from patients may be exploitive. Patients doing even unsolicited favors can reduce the doctor's power, objectivity and independence. Doing favors for patients can lead them to have unrealistic expectations and feel abandoned if the favors cease or fail.

There are boundaries other than those related to sexual contact. Patients may want certain medical treatments even if there is no disease or threat to their health. A young woman may want her healthy uterus removed. Doctors resist this, viewing it as a request to tamper with nature when nature is fine. Those rare patients who hanker to have healthy limbs removed have difficulty getting a doctor to do it or a hospital to allow it. The refusals appeal to medicine's ideals (Henig 2005). Some persons want surgical sex change. Some want their deaths hastened. Secular ethics would condemn a doctor's complying with these desires only if they violate the patient's autonomy. Professionalism, however, would condemn many. For, it takes ***N***ature, "nature taking its course," or "the fullness of being," to be normative, and fulfilling these requests would stray from the natural.

Those who oppose such actions often accuse doctors who do them of "playing god." The phrase has two uses. First, it characterizes an act as forbidden to all persons. Alternatively, it characterizes the act as forbidden to those in the role *doctor*. The latter use implies that the act offends the nature of ***M***edicine. Leon Kass took this stance on *doctors* euthanizing autonomous patients who ask for it (1989). The AMA and other authorities oppose doctors participating in capital punishment, which sometimes causes havoc when a state schedules an execution (O'Reilly 2009).[27]

Problem (8): The profession **self-polices poorly.** It is not clear how secular academic medical ethics would process or care about this. Its concepts do not entail that a physician "police" other physicians. However, the profession has made self-policing integral to itself. Why?

Recall the facts that Pellegrino appealed to in establishing doctors' obligations to accept AIDS victims. I believe we can modify three and justify the duty to self-police. Pellegrino's original observed that the sick are vulnerable, they cannot heal themselves, and doctors commit to helping the sick. My modifications are these.

The sick are vulnerable to imposters and incompetent doctors, they cannot sort incompetent healers and imposters from competent ones, and doctors' commitment to help the sick includes protecting them from imposters and incompetent persons. Self-policing is that protection. Unfortunately, the record on fulfilling this duty is poor (Public Citizen 2007).

Problem (9): Doctors damage or diminish the **public image of medicine**.[28] The terms *professional* and *professionalism* connote a high social status. This is probably left over from when the terms offered useful contrasts to being a laborer, an artisan or a merchant. The group's concern with status endures—professionalism takes medicine's public image seriously. That is a natural predilection of members of a group. A good image would serve many of its interests, including securing its monopoly and freedom to self-regulate. A good image could also facilitate its accomplishing its mission, if the doctor-patient relationship is itself therapeutic.

On the other paradigm, medicine's public image does not interest secular academic medical ethics. Image has no bearing on patients' rights of self-determination. Thus, secular medical ethics is indifferent to the façade. In fact, some have advocated demolishing it. Allen Buchanan (1996) has argued that, given physicians' shortcomings, patients would be better off if they do not trust their doctors. Secular ethics might even find concern with image merely haughty.

Problem (10) involves **medical education's** weakness vis-à-vis promoting professionalism. There are two types of flaws. First, medical schools do not teach professionalism enough. Second, through "the hidden curriculum," medical education damages students' morale (Hafferty 1998; Brainard and Brislen 2007).[29] Evidence shows that medical education leaves its students cynical and less empathic. Secular academic ethics has been silent about this. Some authors who focus on professionalism consider this important and very difficult to remedy (Coulehan and Williams 2003).

11.6 Professionalism as an Evolving Paradigm

By this point, a reader might infer that professionalism is preferable to medical ethics. Does it seem that professionalism takes the higher moral ground, and has the virtue that virtues are supposed to have? It is in the mold of the virtues, implying that practicing well is "its own reward." For, perhaps one way to view all ten problems as a set shows a strength of professionalism. A proponent of professionalism might complain as follows:

> You ethicists omit so many problems of everyday practice. For most of us, euthanasia comes up rarely. For others of us abortion is also rare. Truth-telling and informed consent do not even scratch the surface of what doctors go through. Some of the truth and informed consent issues are in the boundaries problem, anyway. So you exclude much of what we care about. And now you are on to discuss stem cells, cloning, and who owns tissue lines.

> So many issues you attend to are not in the lives of most practitioners. And you are too idealistic about much of what does overlap—e.g., defensive medicine.

However, I do not believe that professionalism wins. Consider some of its serious flaws. Pellegrino's first period of medical ethics was "really" a sort of professionalism. Its lack of organization led to the need for secular academic medical ethics in the first place. Further, paternalism permeated that period. Though it means well, paternalism both belittles patients and deprives them of valued choices. It seems that its good intentions and rhetoric about the best care make professionalism incapable of eliminating paternalism.

Professionalism brought another nagging problem. It forbade patients to refuse life-saving care. This sprang either from paternalism or from a commitment to valuing life and health as absolutes. Either way, resisting patients' right to refuse life-saving care runs deep in professionalism. Only a legal revolution could alter this, and the revolution is not complete. Hospitals remain difficult places in which patients can exercise their rights.

Professionalism also sends a troubling message about accountability to authorities outside of medicine. Medicine's culture teaches what many cultures teach: a "we-they" mentality. The bonding that this fosters means that doctors do little self-policing even when a colleague is a danger to patients. Further, the culture stigmatizes those who are willing to testify for a plaintiff against another doctor in malpractice cases. "Peace in the house" of medicine often trumps what seem to be fundamental duties (Stewart 1979).

Hence, the two paradigms can differ a good deal over the importance of a question or issue. Medical ethics views abortion as a fundamental question. Professionalism, if it took the Hippocratic Oath as a serious exemplar, might find the question easy and incline its membership to oppose abortion. For it, a more pressing question would be how professionals should relate to others who hold a different position on abortion.

From paternalism to peace in the house, professionalism has its dark side and dangers (Erde 2008). These are evident in organized medicine's advocacy on behalf of doctors' rights to refuse to treat certain patients. The dark side is also evident in the lack of self-effacement on which Pellegrino focused. Self-effacement is a key element of being a fiduciary.

However, demanding that doctors be a fiduciary in general and self-effacing in particular may be asking too much. Do we think, for example that, when someone takes up practicing medicine, s/he accepts a duty analogous to that of ships' captains to "go down with the ship?"[30] Does the role *doctor* require radical emulation of Mother Teresa or Albert Schweitzer? Perhaps the model that professionalism promotes is too like the ideal Roman Catholic priest.[31] It casts professionals as having a calling. Though joining the profession entails obligations and costs, the language of being a fiduciary may exaggerate the sacrifices to which patients are entitled. We can expect *some* income deprivation (not poverty), *some* sexual self-restraint (not chastity), and *some* physical and legal courage. But at the extreme, some of these commitments are too difficult for practitioners to live by.

Moreover, professionalism's take on the role *doctor* seems out of date. It may no longer be possible to have a strictly fiduciary relationship with patients. Such relationships require the agent acting on behalf of someone to do what is *best* for that recipient. This may have made sense in a solo, low-tech practice, in which doctors did not face issues from health insurers or partners in a large group, and served a homogenous population. Doctors in that context did not face the prospect of buying very expensive equipment that rapidly becomes outdated. They did not face a great deal of oversight from government and other regulatory bodies. At the same time, most illnesses were acute and short-lived or quick to kill. All that made fiduciary duties clear and made fulfilling them possible.

But that is not today's medicine. Now, role-confusion is intrinsic and not resolvable by appeals to self-effacement. Some practice designs make some kinds of self-effacement impossible. In large group practices, business managers have a great deal of power and the discretion of individual doctors is greatly limited. Moreover, some other kinds of self-effacement endanger patients and doctors themselves: working without getting sufficient sleep is a "selfless" badge of honor in medicine's subculture. It is also very dangerous.

We can derive many of professionalism's concerns from Pellegrino's six facts or from the ten issues that I discussed. For example, it is easy to derive the mandate to help impaired or burnt-out colleagues from the notion of a doctors' calling, the vulnerability of the sick, a concern to avoid sullying medicine's image, and the responsibility to self-police. Likewise, we can derive professionalism's concern about doctors' relationships with pharmaceutical companies. The companies' marketing strategies often cross boundaries, provide incentives that create conflicts of interests and thus can undermine the quality of care. The public's learning about how incentives work has also damaged the image of medicine and led to demands for accountability (Pear 2005). Still, the six facts and ten issues do not entail some imperatives in any obvious way. For example, it is difficult to derive the stigma that attaches to doctors who testify against a negligent colleague.

11.7 The Metamorphosis of Professionalism's Paradigm

Professionalism has changed its content over time. For example, it has ceased to condemn advertising (Tomycz 2006). It has also weakened its resolve to promote life and health and now accepts some limits on life-saving medical treatments.

Aside from changing individual tenets like these, professionalism has also changed (or tried to change) in more radical ways. One constructive attempt was "Medical Professionalism in the New Millennium: A Physician Charter" (Medical Professionalism Project 2002). Its authors included a substantial number of prominent doctors. The Charter appeared simultaneously in Britain's *The Lancet* and in America's *Annals of Internal Medicine*.

The Charter is unusual. It combines tenets from both paradigms. However, it has not received universal acceptance or praise. Kerry Breen saw nothing new in it and found it weak in several ways (2002). One weakness is that it is defensive and merely an expression of frustration. Second, it is self-serving in that it defends medicine against unwanted changes that society may impose on it. The charter *poses* as a social contract through its rhetoric, but it is not a genuine social contract. For only representatives from medicine formulated it. No one represented society in its formulation.

Jay Johansen (2002) found the Charter vague. He questioned the meaning of its urging doctors to "provide health care that is based on the wise and cost-effective management of limited clinical resources." The Charter says, "[p]atients' decisions about their care must be paramount, as long as [they] ... do not lead to demands for inappropriate care." Johansen charged that this is merely playing "word games." As formulated, he claims, it destroys patients' rights. He found its self-important air offensive, and mocked the introduction in the *Annals*. There, the editor called its publication "a watershed" and hoped "[e]veryone ... [will] ... ponder its meaning" (Sox 2002, p. 243).

Frederic W. Hafferty and Dana Levinson took the opposite position (2008). They found the Charter new and praiseworthy, a great improvement over the early essays about professionalism. Those essays addressed the personal motivations of individual practitioners and focused on conflicts of interests, and took professionalism to be the doctrine/virtue by which a doctor would steel herself to be self-effacing.

Hafferty and Levinson saw these early positions as worn out. The Charter's advance was to focus on society rather than individuals. The authors of the Charter took the view that achieving professionalism's ideals requires restructuring society, not psyches. Hafferty and Levinson agree. Indoctrinating individuals' character will not succeed; professionalism as character building *did not work* (2008, p. 607).

This position, however, renounces or destroys professionalism. As an "ism," professionalism is a doctrine. It informs and shapes those who are "called to serve the sick and suffering." It tells them how to conduct themselves to fulfill their calling. Like *justice* in Plato's *Republic*, professionalism is part of a microcosm and a macrocosm. It directs and shapes the group *and* individual members. But Hafferty and Levinson reject this take on professionalism.

They contended that the original take did not work. As an alternative, they endorsed shifting the effort to improve medicine, from an emphasis on remaking individuals to remaking work environments. In doing this, however, they are no longer discussing *professionalism* as a character trait or virtue of individuals or of *M*edicine. They urge us to drop concern with individuals or groups in order to get the quality we want. Achieving the goals in the Charter is beyond what practitioners or the institution of *M*edicine can do. To succeed in securing those goals, medicine must engage society at large, and have society reorganize itself.

Thus, although they are discussing ways to secure some of Medicine's great goals, their proposal is a radical reformation of the means of doing so. They add much that is not currently part of professionalism, namely reorganizing society. They drop much that currently is part of professionalism, namely self-management. Thus, though they praise the Charter, they do it much more damage than do Breen and Johansen. They unwittingly abandon *professionalism* as such.

Abandoning it and virtue ethics may be wise. Recent work in philosophy and psychology undermines the ideas of *self* and *character* (Metzinger 2009; Hofstadter 2007). We are all highly fragmented. Our composition makes those concepts inapplicable. Moreover, recent work in ethics suggests that there is no steady moral character. Individuals take their cues about right and wrong from the social structures in which they find themselves (Doris 2002)—this seems to be Hafferty and Levinson's position. It implies that we must forsake the vocabulary of *moral character*, of which *professionalism* is an instance. But that is just what secular academic ethics did. It organized medical ethics in terms of rights and duties. It is not much to add systems of accountability in order to control the bad man in all of us (Holmes 1897).

11.8 Conclusion

Frequently, Plato represented the human psyche as having three elements in strife with one another. The *Republic* specifies one combination as *wisdom*, *spirit* and *the urge for gain* (580ff). Plato considered each crucial for promoting the happiest life. None should be destroyed. Rather, they must be coordinated. That is the role of *justice*. Justice, Plato concluded, is the only way to happiness. It is its own reward.

Are proponents of professionalism as savvy as Plato? Do they recognize, as Plato did, that practical interests, spirit, and desire are part of the human make-up? Professionalism's advocates need not deny—and have not explicitly denied—that urge for gain is part of the human condition. They could admit such desires and forbid acting on them. But this makes their ethos too demanding. Few applicants to medical school could live up to it. Graduating only persons who fit that ideal would render our supply of doctors inadequate.

Plato recognized an analogous difficulty about a just city. Perhaps that is why he declared that it need not be built (*Republic* 592b). Rather, he would have us think of it as an ideal for each person to use to construct a plan by which to live. And so the *Republic* is an exercise in *hypothetical* soul-making. Plato could hope that, if his readers read and used his work well, they would design a proper blueprint for themselves.

Does that bring us full circle? We have seen that professionalism provides a blueprint of the character of the doctor. But the structure has not been built. The plans did not work, perhaps in part because they ignore too much of human nature or because *character* is a bogus concept. However, even if some few persons have character, there may well be too few compared to the numbers of doctors we need. Thus, we should not expect each doctor to realize professionalism's virtues.

At one level, then, professionalism's rhetoric is nostalgic utopianism. To break its grip, we must realize that patients will always need protection from doctors. Although patients may be autonomous, their rights are often ignored or overridden as professionals do their work.

At the start, I took the big question to be, "Which norms should govern the practice and practitioner of medicine?" I showed how two paradigms conflict about how best to answer that question. Do the conflicting paradigms and our observations of doctors' performance imply that antiprincipIism will reign and we will remain in crisis? Mayhap we just have to muddle through. I do not mean that we go back to the ethic of Pellegrino's first period. That was a mere catalog of values and rules. I think we can have more system than that, though it will not be perfect.

Does that mean we should be demoralized or cynical? There will be gaps in our paradigm and gaps between normative answers and the conduct of our doctors. Shall we merely lament them?[32]

Professionalism has ancient roots. Off and on over centuries, those roots came above ground or retreated. Conscientious leaders have invoked professionalism both from insight and out of nostalgia (Erde 2008). Reverting to old values and models is common across the millennia. But such reverting is frequently "too little, too late," and it too leaves us adrift.

Perhaps the trend of invoking professionalism exemplifies a thesis of Albert William Levy's (1974). Levy showed that when philosophers characterize an ideal person or ideal life, they are often recapitulating what had been society's ideal a generation earlier. These ideals just rationalize what is current as common sense. Perhaps that is what is taking place with advocates for professionalism. Professionalism expands our vision of the ethical needs we have regarding doctors. However, it is too idealistic and it fails due to its vagueness and its dark side. Secular academic medical ethics mostly opposes paternalism. If it could come to recognize some of professionalism's insights and concerns but address them through social structures, this might be a better answer.

11.9 Postscript (Scholarly and Unscholarly)

I get my subtitle from a metaphor of Plato's. In the *Sophist* (246a), discussants allude to classic debates over the nature of reality. They note that some "giants" hold that the only reality is the unstable material stuff of experience. Knowledge is a matter of perception and truth is relative to perspective. The other "giants" are rational philosophers. They hold that there are abstract, unchanging ideal objects, which we know a priori, regardless of any perspective.

Over the last 2,500 years, though mostly in the last century, many great philosophers came to renounce the philosophers' camp as Plato described it. We can consider modernism, from its emergence in the 1600s, as the successor to the views of the first tribe of giants. Its tools are reductionism, linguistic analysis, pluralism, and respect for individual autonomy and human differences. Eventually modernism gave rise to pragmatism. Under that, we accept a view as true only insofar as it works for us in explaining and predicting.

Alasdair MacIntyre has argued that modernity's approach is a dead end (1981, and many publications thereafter). It has come to naught and even done substantial damage. MacIntyre holds there cannot be a settled and productive society without social practices and traditions that shape the virtues in individuals. He finds academic secular medical ethics bankrupt. It is rooted in the ethical writings of Immanuel Kant and John Stuart Mill. As MacIntyre sees it, both made the fundamental mistakes of modernism, in different ways. They tried to develop universal methods that exclude historical and contextual perspectives. MacIntyre considers this impossible and destructive.

MacIntyre would reject secular academic medical ethics. It fails badly enough to not be a paradigm. A MacIntyrean should contend that we could not even muddle through the dilemmas and problems without conceptual and evaluative supports from a tradition and social practice. Thus, we cannot work in as thin a context as the paradigm of secular academic medical ethics.

I believe that MacIntyre is wrong. I think academic secular pluralism has enough within it to assist us in making progress on many normative problems related to medicine and health care. And I think that professionalism's many flaws show that it has failed. Each paradigm has goods to offer, but neither will have the last word.[33]

Notes

1. See the first three lines of Burnt Norton, the first of The Four Quartets, and the first four lines of East Coker, the second of The Four Quartets. Available online at: http://www.coldbacon.com/poems/fq.html.
2. According to Laurence B. McCullough, in the mid-eighteenth century John Gregory began a movement in professionalism very like the current one (1998). Some writers mark the current trend from Arnold Relman's (1980) warning about commercialization. Jack Coulehan and Peter Williams (2003) place it in the early 1990s.
3. I sometimes use *ethics* or *medical ethics* without *academic* and *secular*. Unless the context indicates otherwise, I mean the whole phrase.
4. Kuhn found such disorganization characteristic of "pre-paradigm" states. Pellegrino did not use Kuhn's terminology. I use both Kuhn's and Pellegrino's terms.
5. Counting back a decade from 2003, when Coulehan and Williams noted that professionalism had become a "hot topic," one reaches 1993—the year that Pellegrino published his history. Coincidence? Or did the crisis (the fourth period) lead to calls for professionalism?
6. Its most famous precept, "above all, do no harm," is certainly invoked a great deal.
7. The course that informed consent took to its present role was not as smooth as this paragraph suggests. In a fine summary of informed consent's history in the US, Rebecca Skloot (2010, Chapter 17) notes that the first legal use of it in America involved a clinical case (Salgo v. Leland Stanford Jr. University Board of Trustees (Civ. No. 17045, First Dist., Div. One, 1957)). In the mid-1950s, American researchers often did not know about the requirement. Many did not know of the Nuremberg code or its emphasis on obtaining valid informed consent from subjects. Some who did know thought such constraints did not apply to them as well-meaning good guys.
8. It is important to distinguish between (a) rejecting the moral/political *relevance* of input from religious traditions and (b) rejecting the intelligibility of particular religious positions related to medical ethical matters. H. T. Engelhardt, Jr. (1986) fits (a). K.D. Clouser (1973) fits (b).
9. My courses at University of Texas Medical Branch in the late 1970s fit this account, with classes focused on *personhood*, *rights*, *paternalism*, etc.
10. Pellegrino reported that critics called the set of four "a mantra" (1993, p. 1160).
11. The philosophical literature apart from medical ethics also gave a great power to autonomy (Nozick 1974; Dworkin 1984).
12. Negotiating fees was not part of this model, though now some economists stress it.
13. Paul Appelbaum and Thomas Gutheil (1979) indicate the depth of professionals' resentment of patient autonomy. They characterize the results of importing it into treatment of the mentally ill as allowing patients to rot with their rights on.

14. It is important to distinguish between *personal autonomy* and *professional autonomy*. The former involves choosing values, rules and goals in leading one's life. It defines authenticity and one's *sense of self*. The latter involves discretion in applying one's expertise to cases; one appreciates the novelty of the case in deciding what to try in that situation. It is about judgment but not strongly connected with one's sense of self. In a given case, a doctor could reasonably expect colleagues to agree that there are several defensible choices. However, one need not expect agreement of choices related to one's personal autonomy.
15. Dr. Jeffrey Steinberg ran the clinic that implanted eight embryos in Nadya Shuleman ("the octomom"). Their births gave her a total of 14 children. In the uproar afterwards, Steinberg asked, "Who am I to say that six is the limit? There are people who like to have big families" (Watkins and Neergaard 2009). Without formulating it in these terms, he may have thought that she had a positive right to have them implanted. This would be a position within academic ethics. A rival position would be to refuse to assist Ms. Shuleman, with that refusal based on professional judgment. Vague as this is, it became the grounds for the California licensing board invoked to revoke Dr. Steinberg's license (Anonymous 2011).
16. Years ago I discovered this phobia—fear of lawsuits. It infects many roles.
17. For example, in its advance directives statute, New Jersey avows four state interests that shaped it (NJADHC). One is a desire to preserve medicine's integrity.
18. Traditions and history supply models, metaphors, and analogies. One powerful analogy for doctoring is the priesthood and saintliness. See below.
19. This sort of advocacy can be taxing, time consuming, and ultimately expensive for a doctor. It may also bring retribution or sanction from the HMO.
20. Some of professionalism's proponents complain of doctors' diminished *altruism*. I take *altruism* to mean something close to *generosity* or *self-sacrifice*, when there is no clear duty to act in those ways. Thus, it is not altruistic to resolve a conflict of interest in a patient's favor because one has a duty to do that. Advocating for people who are not one's patients—e.g., the uninsured—is a good example of altruism.
21. Still, using negative rights in this way does not settle all questions about end-of-life care. A question of positive rights remains. Why is there no right to assisted suicide or euthanasia (in most jurisdictions)? Part of the answer is that professionalism's background assumptions are not secular. It takes each individual human life as of absolute, intrinsic value in the cosmic order. It subscribes to "the fullness of being" and finds *N*ature to be normative and interfering with nature to be suspect (Erde 1989). Thus it sees ending a life as illegitimate, but letting "nature take its course" as legitimate. For more on this, see the discussion of boundaries below.
22. See the discussion of fraud below.
23. Martinez (2005) makes a convincing case for some exceptions.
24. Brody's analysis, *The Healer's Power*, illuminates the powers very well (1992).
25. Freud made much of "transference"—patients' inevitable infatuation with whoever is their doctor—and "counter-transference," a doctor's infatuation with a patient.
26. Reportedly, when George Harrison of the Beatles was dying, his oncologist asked Mr. Harrison to sign a guitar for him. The Harrison's estate sued for violations of privacy. The doctor's actions were handled as a boundary violation and the doctor lost his job (Bahrampour 2004).
27. The American Medical Association grants doctors a right to a personal opinion about capital punishment. But it sees their participating in executions as contrary to the nature of the profession, and forbids it. One doctor who did participate declared that the prisoner is not clearly *a patient* (Crabbe 2007).
28. Concern with image was one of the reasons that, from its inception, the American Medical Association opposed advertising. Tomycz considers it a lesser reason (2006).
29. The logic of each paradigm would structure educational method differently. Secular academic ethics would care about respecting students' autonomy. Thus, it should teach *Socratically*. The teacher has students rigorously explore ideas so they can decide about right and wrong. Professionalism, by contrast, aims at conformity, not individual conscience. It should favor indoctrinating newcomers. They must learn the rules of the tribe. Students can only learn them from instruction.

30. Do all captains still accept it? Consider the controversy surrounding the captain of the Costa Concordia, a cruise ship that sunk off the coast of Italy. Captain Francesco Schettino, was placed on house arrest, for among other things "abandoning a ship before its passengers were safe…" (Pianigiani and Cowell 2012).
31. Not only are the codes for Roman Catholic clergy and medical professionals closely analogous, the general doctrines are similar as well. One similarity regards *equality*. Both Roman Catholic theology and the doctrine of Medicine tie the equality of persons to the intrinsic value of human life and oppose anyone's judging a life as not worth living. Another similarity regards death. Both cast it as an evil that cannot be used as a means to good ends. Secular ethics differs on these points. It casts *equality* in terms of rights, protection of rights, and self-determination. Nor does it consider death an intrinsic evil.
32. Ernest Hemingway offered us such a niche in his poem "The Earnest Liberal's Lament" (1922). There, he notes failures of monks and pet cats to be chaste, failures of girls to refrain from biting (in lovemaking?), and he ends this tiny poem with a rhetorical question, a plea, or expression of exasperation. He asks what he can do to rectify such failings in the world.
33. Thanks to Drs. Carry Burns, Marvin Herring and Matt Holtman, Ms. Jacqueline Giacobbe and Fatmata Kabia for comments that helped me improve this essay.

References

Accreditation Council for Graduate Medical Education (ACGME). 2002. Competency requirements [On-line]. Available at http://www.acgme.org/acWebsite/irc/irc_compIntro.asp. Accessed 24 Sep 2012.

Anonymous. 2011. 'Octomom' doctor's license to be revoked, state medical board rules. *L. A. Times (local)* June 1 [On-line]. Available at http://latimesblogs.latimes.com/lanow/2011/06/a-matter-octomom-doctor-michael-kamrava.html. Accessed 24 Sep 2012.

Appelbaum, P.S., and T.G. Gutheil. 1979. "Rotting with their rights on": Constitutional theory and clinical reality in drug refusal by psychiatric patients. *The Bulletin of the American Academy of Psychiatry and the Law* 7(3): 306–315.

Bahrapmour, T. 2004. Ex-Beatle's oncologist to lose post at Staten Island Hospital. *New York Times* January 9 [On-line]. Available at http://query.nytimes.com/gst/fullpage.html?res=9A02E2DE1F31F93AA35752C0A9629C8B63&sec=&spon=&pagewanted=print. Accessed 24 Sep 2012.

Beauchamp, T.L., and J.F. Childress. 1979. *Principles of biomedical ethics*. New York: Oxford University Press.

Brainard, A.H., and H.C. Brislen. 2007. Viewpoint: Learning professionalism—A view from the trenches. *Academic Medicine* 82: 1010–1014.

Breen, K. 2002. Letter to the editor. *The Lancet* 359: 2042 [On-line]. Available at http://www.thelancet.com/journals/lancet/article/PIIS0140-6736(02)08809-8/fulltext. Accessed 24 Sep 2012.

Brody, H. 1992. *The healer's power*. New Haven: Yale University Press.

Buchanan, A. 1996. Is there a medical profession in the house? In *Conflicts of interest in clinical practice and research*, ed. R.G. Spece, D.S. Shimm, and A. Buchanan, 105–136. New York: Oxford University Press.

Clouser, K.D. 1973. The sanctity of life: An analysis of a concept. *Annals of Internal Medicine* 78: 119–125.

Coulehan, J., and P.C. Williams. 2003. Conflicting professional values in medical education. *Cambridge Quarterly of Health Care Ethics* 12: 7–20.

Crabbe, N. 2007. Doctor: Execution not medical work. *Gainesville Sun*, February 20 [On-line]. Available at http://www.fadp.org/news/gs-20070220/. Accessed 16 Feb 2009.

Cruess, S.R., S. Johnston, and R.L. Cruess. 2002. Professionalism for medicine: Opportunities and obligations. *The Medical Journal of Australia* 177: 208–211.

Doris, J.M. 2002. *Lack of character: Personality and moral behavior*. Cambridge: Cambridge University Press.
Dworkin, R. 1984. Rights as trumps. In *Theories of rights*, ed. J. Waldron, 153–167. Oxford: Oxford University Press.
Eliot, T.S. n.d. The four quartets [On-line]. Available at http://www.coldbacon.com/poems/fq.html. Accessed 24 Sep 2012.
Engelhardt Jr., H.T. 1986. *Foundations of bioethics*. New York: Oxford University Press.
Erde, E.L. 1989. Studies in the explanation of issues in biomedical ethics: (II) On 'on play[ing] god' etc. *The Journal of Medicine and Philosophy* 16: 593–615.
Erde, E.L. 1996. Conflicts of interests in medicine: A philosophical and ethical morphology. In *Conflicts of interest in clinical practice and research*, ed. R.G. Spece, D.S. Shimm, and A. Buchanan, 12–41. New York: Oxford University Press.
Erde, E.L. 2008. Professionalism's facets: Ambiguity, ambivalence and nostalgia. *The Journal of Medicine and Philosophy* 33: 6–26.
Glendon, M.A. 1991. *Rights talk: The impoverishment of political discourse*. New York: The Free Press.
Hafferty, F.W. 1998. Beyond curriculum reform: Confronting medicine's hidden curriculum. *Academic Medicine* 73: 403–407.
Hafferty, F.W., and D. Levinson. 2008. Moving beyond nostalgia and motives. *Perspectives in Biology and Medicine* 51: 599–615.
Hemingway, E. 1922. The earnest liberals' lament. In several collections (for example, Gerogiannis, N. (ed.). 1979. *88 poems*. New York: Harcourt Brace Jovanovich, Inc. Available at http://books.google.com/books?id=Ic4nHr_6JXMC&pg=PA52&lpg=PA52&dq=ernest+hemingway+poems+monks+bite&source=bl&ots=J0wu8AqoIt&sig=zmdKilsTf5pICccg4DH7I18EDh8&hl=en&ei=-W2VSYqfPI-EtgeO05GhCw&sa=X&oi=book_result&resnum=1&ct=result). Accessed 24 Sep 2012.
Henig, R.M. 2005. At war with their bodies, they seek to sever limbs. *New York Times*, March 22 [On-line]. Available at http://www.nytimes.com/2005/03/22/health/psychology/22ampu.html?_r=1&scp=1&sq=seek+to+sever+limbs&st=nyt. Accessed 24 Sep 2012.
Hofstadter, D.R. 2007. *I am a strange loop*. New York: Basic Books.
Holmes, O.W. 1897. The path of the law. *10 Harvard Law Review* 457. Available at http://www.constitution.org/lrev/owh/path_law.htm. Accessed 24 Sep 2012.
Johansen, J. 2002. The charter on medical professionalism: The new Hippocratic Oath? "Pregnant pause"—A blog. Available at http://www.pregnantpause.org/ethics/medprof.htm. Accessed 24 Sep 2012.
Jonsen, A.R., and A.E. Hellegers. 1976. Conceptual foundations for an ethics of medical care. In *Ethics and health policy*, ed. R.M. Veatch and R. Branson, 17–34. Cambridge, MA: Ballinger Publishing Company.
Kass, L. 1989. Neither love nor money: Why doctors must not kill. *The Public Interest* 94: 43–44.
Kuhn, T. 1962. *The structure of scientific revolutions*. Chicago: University of Chicago Press.
Levy, A.W. 1974. *Philosophy as social expression*. Chicago: University of Chicago Press.
Lundberg, G.D. 1983. Boxing should be banned in civilized countries. *Journal of the American Medical Association* 249: 250.
MacIntyre, A. 1981. *After virtue*. Notre Dame: University of Notre Dame Press.
Malone, M.E. 1993. Kuhn reconstructed: Incommensurability without relativism. *Studies in History and Philosophy of Science* 24: 69–93.
Martinez, R. 2005. A model for boundary dilemmas: Ethical decision making in the professional-patient relationship. *Ethical and Human Sciences and Services* 2: 43–61.
McCullough, L.B. 1998. *John Gregory and the invention of professional medical ethics and the profession of medicine*. Dordrecht: Kluwer Academic Publishers.
Medical Professionalism Project: ABIM Foundation. 2002. Medical professionalism in the new millennium: A physician charter. *Annals of Internal Medicine* 136: 243–246.
Metzinger, T. 2009. *The ego tunnel: The science of the mind and the myth of the self*. New York: Basic Books.

National Board of Medical Examiners (NBME). 2009. Assessment of professional behaviors [On-line]. Available at http://professionalbehaviors.nbme.org/index.html. Accessed 24 Sep 2012.

NJADHC. New Jersey Advance Directive for Health Care Act 26: H2-53 paragraph d. Available on line at: http://lis.njleg.state.nj.us/cgi-bin/om_isapi.dll?clientID=211591960&Depth=4&TD=WRAP&advquery=%22Advance%20Directive%20for%20health%20care%22&headingswithhits=on&infobase=statutes.nfo&rank=&record={A927}&softpage=Doc_Frame_Pg42&wordsaroundhits=2&x=50&y=16&zz=. Accessed 24 Sep 2012.

Novack, D.H., B.J. Detering, R. Arnold, L. Forrow, M. Ladinsky, and J.C. Pezzullo. 1989. Physicians' attitudes toward using deception to resolve difficult ethical problems. *Journal of the American Medical Association* 261: 2980–2985.

Nozick, R. 1974. *Anarchy, state and utopia*. New York: Basic Books.

O'Reilly, K.B. 2009. Doctor quits prison job over execution. *American Medical News* [On-line]. Available at http://www.amaassn.org/amednews/2009/02/09/prsc0209.htm. Accessed 24 Sep 2012.

Pear, R. 2005. Panel seeks better disciplining of doctors. *New York Times*, January 24 [On-line]. Available at http://query.nytimes.com/gst/fullpage.html?res=9F07E6DA1339F936A35752C0A9639C8B63. Accessed 24 Sep 2012.

Peeno, L. 1996. Testimony before the US House of Representatives. May 30. Available at http://www.thenationalcoalition.org/drpeenotestimony.html. Accessed 24 Sep 2012.

Pellegrino, E.D. 1987. Altruism, self-interest, and medical ethics. *Journal of the American Medical Association* 258: 1939–1940.

Pellegrino, E.D. 1993. The metamorphosis of medical ethics: A 30 year perspective. *Journal of the American Medical Association* 269: 1158–1162.

Pianigiani G., and A. Cowell. 2012. Divers suspend underwater search of stricken Italian liner. *New York Times*, January 31 [On-line]. Available at http://www.nytimes.com/2012/02/01/world/europe/divers-suspend-underwater-search-of-costa-concordia-in-italy.html?scp=1&sq=costa%20concordia%20captain%20arrested&st=cse. Accessed 24 Sep 2012.

Public Citizen. 2007. Public citizen issues annual ranking of state medical boards [On-line]. Available at http://www.citizen.org/pressroom/release.cfm?ID=2450. Accessed 24 Sep 2012.

Relman, A.S. 1980. The new medical-industrial complex. *The New England Journal of Medicine* 303: 963–970.

Sammons, J.T. 1989. Why doctors should oppose boxing: An interdisciplinary history perspective. *Journal of the American Medical Association* 261: 1484–1485.

Savage, D.G. 2008. President Bush unveils conscience rule for health-care industry: Last-minute declaration lets providers refuse to participate in care. *Chicago Tribune*, December 19 [On-line]. Available at http://articles.chicagotribune.com/2008-12-19/news/0812180751_1_right-to-refuse-rule-includes-abortion-health-care-health-care. Accessed 24 Sep 2012.

Skloot, R. 2010. *The immortal life of Henrietta Lacks*. New York: Broadway Books.

Sox, H.C. 2002. Editor's introduction to "medical professionalism in the new millennium: A physician charter". *Annals of Internal Medicine* 136: 243.

Spece, R.G., D.S. Shimm, and A.E. Buchanan (eds.). 1996. *Conflicts of interest in clinical practice and research*. Oxford: Oxford University Press.

Stewart, J.B. 1979. *Blind eye: The terrifying story of a doctor who got away with murder*, 1999. New York: Touchstone.

Tomycz, N.D. 2006. A profession selling out: Lamenting the paradigm shift in physician advertising. *Journal of Medical Ethics* 32: 26–28.

US Government. 2010. Overview EMTLA [On-line]. Available at https://www.cms.gov/EMTALA/. Accessed 24 Sep 2012.

Veatch, R.M. 1981. *Theory of medical ethics*. New York: Basic Books.

Veatch, R.M. 1995. Abandoning informed consent. *The Hastings Center Report* 25(2): 5–12.

Watkins, T., and N. Neergaard. 2009. Octuplets spur debate over ethics. By the Associated Press, Saturday January 31 [On-line]. Available at http://www.msnbc.msn.com/id/28939439/. Accessed 24 Sep 2012.

Chapter 12
The Role of an Ideology of Anti-Paternalism in the Development of American Bioethics

Laurence B. McCullough

One embraces ideology when one holds a position or takes a stance for which argument is not given or for which a flawed argument is given but one is unaware of or, worse, elects to ignore those flaws. The first form of ideology is egregious in any intellectually and practically serious field of inquiry, including bioethics, because it patently violates the standards of argument-based ethics (McCullough et al. 2004). The second form of ideology is at once more subtle and powerful, for it shapes perceptions and therefore the discourses of any intellectually and practically serious form of inquiry and does so in such a thoroughgoing way that these perceptions and the discourses that they shape are taken for granted. They become part of the accepted self-understanding of fields of inquiry, making challenges to them especially unwelcome to those wedded to them.

In this chapter, I will undertake the unwelcome but essential task of showing that this second kind of ideology—in the form of an ideology of anti-paternalism—has become entrenched in contemporary American bioethics and that this contemporary self-understanding is a legacy of the anti-paternalism that profoundly, and mistakenly, shaped American bioethics in its formative years four decades ago. American bioethics in its formative period made both empirical and historical claims about medical paternalism that do not withstand sustained critical scrutiny; bioethics was founded on errors about medical paternalism. Bioethics' anti-paternalism became and has remained an ideology for which flawed argument has been, and continues to be, given or taken for granted. Having exposed this ideology of anti-paternalism, I call for bioethics to leave it behind in favor of sustained inquiry into more fundamental conceptual and practical moral challenges.

L.B. McCullough, Ph.D. (✉)
Center for Medical Ethics and Health Policy, Baylor College of Medicine,
One Baylor Plaza, MS 420, Houston, TX 77030-3411, USA
e-mail: Laurence.McCullough@bcm.edu

J.R. Garrett et al. (eds.), *The Development of Bioethics in the United States*,
Philosophy and Medicine 115, DOI 10.1007/978-94-007-4011-2_12,

12.1 Paternalism and Anti-paternalism

The *Encyclopedia of Bioethics* in its first edition appeared in 1978 (Reich 1978). The *Encyclopedia* provided the first comprehensive, scholarly, non-polemical overview of the then-new field of bioethics. The *Encyclopedia* also consolidated the field and became the point of departure for much subsequent scholarship and teaching in bioethics. The *Encyclopedia* therefore is one of the most reliable, if not the most reliable, sources for the state of the field of bioethics in what we can call its formative, classical period (the 1960s and 1970s).

In the entry, "Paternalism," in the first edition of the *Encyclopedia*, Tom L. Beauchamp (1978) provided the following definition:

> "Paternalism" is used in the ethical literature, and in this article, to refer to practices that restrict the liberty of individuals, without their consent, where justification of such actions is either the prevention of some harm they will do to themselves or the production of some benefit for them that they would not otherwise secure. The main ethical issue centers on whether paternalistic justifications are morally acceptable. The *paternalistic principle*, as it will be referred to here, says that limiting a person's liberty is justified if *through his own actions* he would produce serious harm to *himself* or would fail to secure an important benefit (Beauchamp 1978, p. 1194, emphasis original).

On the basis of this remarkably clear account, the defining components of a paternalistic action can be identified. An action counts as an instance of paternalism if and only if: (1) the action interferes with the liberty of an individual, his or her freedom to make and implement choices or decisions; and (2) the action is reliably judged to benefit that individual, either by preventing harm to himself or herself that the individual is about to bring about as a result of his decisions or absence of decisions or by bringing about some good or goods for that individual that would otherwise be foregone as a result of not interfering with that individual's decisions or absence of decisions. Both components must be satisfied for an action to count as paternalism. If an action does not interfere with an individual's liberty, that action does not count as paternalism. If an action is undertaken to interfere with an individual's liberty, not to benefit that individual but to benefit *others* by protecting them from harm or securing some good for them as a predictable consequence of interfering with the individual's liberty, that action does not count as paternalism. It may count as interference with an individual's freedom of decision making, but not an interference that is properly conceptualized as paternalism.

Beauchamp went on to provide an equally clear and useful account of anti-paternalism:

> Those who oppose paternalism argue either that the [paternalistic] principle is not a valid moral principle under any conditions or that it can be used only under very restricted circumstances (Beauchamp 1978, p. 1194).

In the second edition of the *Encyclopedia* (Reich 1995), Beauchamp (1995) provided a succinct account of paternalism, emphasizing and refining its two components:

> Paternalism is the intentional limitation of the autonomy of one person by another, where the person who limits autonomy justifies the action exclusively by the goal of helping the person

> whose autonomy is limited.... Following this definition, an act of paternalism overrides the value of respect for autonomy on some grounds of beneficence. Paternalism seizes decision-making authority by preventing persons from making or implementing their own decisions (Beauchamp 1995, pp. 1914–1915).

Beauchamp acknowledged a competing definition that does not require that the individual interfered with is autonomous. Because anti-paternalism in bioethics has focused on interfering with autonomous individuals for their own good—most prominently, by refusing to be truthful with them about their diagnosis, its clinical management, or its prognosis—this second definition is not relevant here. The entry in the current edition of the *Encyclopedia* (Post 2004) reprints the second-edition entry without alteration.

12.2 Medical Paternalism as the *Bête Noire* of Bioethics

From its inception bioethics regarded medical paternalism as its *bête noir*, which required slaying and which bioethics understands itself indeed to have slain. It was commonplace for textbook collections from the classical period of bioethics to have sections on paternalism and on truth-telling. In one of the first and most widely used textbooks from the classical period of bioethics the message was that physicians were paternalists who, for patients' own good, were not honest with them (Gorovitz et al. 1976). Allen Buchanan (1978) opens his important paper on medical paternalism with the claim that "[t]here is evidence to show that among physicians in this country the paternalist model is the dominant way of conceiving the physician-patient relationship" (Buchanan 1978, p. 370). In the first edition of their landmark work, *Principles of Biomedical Ethics*, which is the most commonly taught textbook in the field, Beauchamp and Childress write:

> Physicians often think in terms of patients' needs and interests. This can lead them to a paternalistic stance rather than to an emphasis on patient autonomy (Beauchamp and Childress 1979, p. 204).

Albert Jonsen, Mark Siegler, and William Winslade, in *Clinical Ethics*, which is widely used in health professions education, state:

> Medical practice has traditionally been strongly paternalistic: Physicians have often concealed diagnoses from patients "for their own good" (Jonsen et al. 1982, p. 52).

In the first edition of the *Encyclopedia* Beauchamp (1978) documented medical paternalism. He wrote:

> There are so many individual examples of controversial paternalistic justifications in biomedical and behavioral contexts that only a few selected examples can be treated here (Beauchamp 1978, p. 1197).

The examples are patients' refusals of surgery, research on human subjects, truth-telling, behavior control, biological control, and informed consent. The examples in the second entry on paternalism (Beauchamp 1995) are overriding refusals of and requests for treatment, incomplete disclosure of information to patients, the therapeutic privilege, health policy designed to prevent excessive risk, government restrictions, "the model of paternalistic authority," preventing suicide, and involuntary commitment. Both entries accurately summarized the state of anti-paternalism in the field in its first three decades.

Medical paternalism was a major problem and bioethicists set out to correct it, once and for all. Just as the paternalism of the state should not be accepted, so too the paternalism of physicians should not be accepted. Instead, the autonomy of patients should be recognized and take primacy.

This self-understanding of bioethics as a corrective to medical paternalism, bioethics' anti-paternalism, has become pervasive. In the "Paternalism" entry in the *Stanford Encyclopedia of Bioethics*, for example, Gerald Dworkin writes: "Doctors do not tell their patients the truth about their medical condition" (Dworkin 2005).

In a recent review in *The New Republic* of the book, *Patient, Heal Thyself,* by Robert Veatch, the distinguished surgeon-writer Sherwin Nuland introduces his reader to bioethics as an antidote to the "old paternalistic approach that continues to guide the attitudes of some physicians even now" (Nuland 2009, p. 48). Nuland calls this approach "old," because he adopts the received view in much of bioethics that the seeds of medical paternalism can be found in the admonition in the Hippocratic Oath to "use regimens for the benefit of the ill in accordance with my ability and my judgment, but from [what is] to their harm or injustice I will keep [them]" (von Staden 1996, p. 406). Nuland notes that the paternalistic attitude of physicians

> …was already encouraged [in ancient Greece], as it still is, by the simple fact that the healer has always been possessed of a body of knowledge and skills unavailable to the patient (Nuland 2009, p. 49).

Nuland accepts the self-understanding of bioethics as a corrective to such authoritarian paternalistic attitudes and the actions they supported.

> As the centuries rolled by, that specialized corpus of abilities grew slowly larger, its acceleration magnified many-fold when real science entered medical thinking near the middle of the nineteenth century. At the same time, but not obviously related to the advent of biotechnology and molecular medicine, in the 1960s there occurred the social upheaval that manifested itself within the ancient art of healing as "the patient self-determination movement." As the bonds of unquestioned authoritarianism in all arenas were thrown off or loosened, patients began to demand a greater degree of participation in deciding how they should be treated. And all the while clinical decision-making was becoming increasingly complex, and its interplay with considerations of ethics and morality came to the forefront as never before, especially as certain high-profile cases made their appearance before the courts (Nuland 2009, p. 49).

The antidote, of course, is the principle of autonomy, which "holds that, except in cases of incapacity, the patient is a rational person with rights, opinions, and aims, who is the final arbiter of his or her own best interests" (Nuland 2009, p. 49).

Put aside that Nuland accepts the myth of what Robert Baker (1993) has described as the "Hippocratic footnote" reading of the history of Western medical ethics, which has been thoroughly debunked (Baker 1993; Nutton 2009). Put aside that the remedies that Hippocratic physicians offered to the sick were already available to them but were kept secret or hidden from them in order to protect the physician's economic self-interest. Put aside that scientific thinking began to shape medicine and professional medical ethics already in the eighteenth century (McCullough 1998). Put aside, too, that the ethical concept of the sick person as a patient was not introduced until the invention of professional medical ethics in the eighteenth century, which concept replaced the centuries-old concept and discourse of *aegrotus* or the sick individual (McCullough 1998).

Notwithstanding these serious scholarly and therefore conceptual problems, Nuland gets the self-understanding of bioethics as the anti-paternalistic corrective to centuries of medical paternalism pitch-perfect: Physicians for centuries interfered with the autonomy of patients by being silent when they should have been forthcoming about the patient's diagnosis, its prognosis, and its clinical management and therefore systematically excluded patients from the decision-making process about their own medical care. Recognition and routine implementation of patient self-determination is the remedy for this paternalism. Nuland unwittingly expresses bioethics' ideology of anti-paternalism beautifully.

12.3 Origins of Anti-paternalism in Bioethics: A Critical Assessment

Let us now examine the origins of this self-understanding of bioethics as the anti-paternalistic correction of authoritarian medical paternalism of (alleged) ancient lineage. The central front in the struggle, self-styled as a revolution by some, was medical paternalism in the form of not telling the truth to patients that had the immediate and obviously unacceptable consequence of not involving them in decisions about their medical care, an egregious violation of their autonomy that cried out for permanent correction. In the classical literature of bioethics, two sources of evidence for medical paternalism were frequently invoked: an empirical study of physicians' practices regarding truth-telling and major texts in the history of medical ethics. I will critically assess each source of evidence in turn.

12.3.1 Empirical Evidence for Medical Paternalism: The Oken Study

Donald Oken's (1961) study of what physicians told patients with cancer about their diagnosis, prognosis, and treatment planning looms large in the early bioethics literature (Oken 1961). According to the *ISI Web of Knowledge* database (accessed January 21, 2011), it has been cited 329 times. In one of the major texts from the classical period of bioethics, Veatch reflected the (almost) iconic status of the Oken study:

> Physicians are strongly committed, in the ideal at least, to protecting patients from potential harm, both physical and mental. According to one important study [Oken 1961], 88 percent of physicians responding reported that they follow a usual policy of not telling patients that they have cancer. Some of these physicians may have decided not to tell for irresponsible reasons: to tell someone he is about to die is not a pleasant task and can be time-consuming. But physicians for the most part are morally dedicated to the principle of

> judiciously withholding information that they feel would do serious harm to the patient. To transmit it would violate their medico-moral responsibility to do what they consider to be in the best interests of their patient. The Hippocratic Oath has the physician pledge to follow that regimen which, "according to my ability and judgment, I consider for the benefit of my patients, and abstain from whatever is deleterious and mischievous." Bernard Meyer, a physician who has written on the ethics of what should be told to the terminal patient, says: "Ours is a profession which traditionally has been guided by a precept that transcends the virtue of uttering truth for truth's sake; that is, 'So far as possible, do not harm'" (Veatch 1976, p. 206).

Buchanan also refers to the Oken study in support of the claim that physicians have adopted an ethos of paternalism. He, too, cites the higher number, "[a]lmost 90%," who usually do not tell the truth to patients with cancer.

> The physicians' justifications for withholding or falsifying diagnostic information were uniformly paternalistic. They assume that if they told the patient he had cancer they would be depriving him of all hope and that the loss of hope would result in suicidal depression or at least in a serious worsening of the patient's condition (Buchanan 1978, p. 373).

A close reading of the Oken study reveals a very different picture. He did not represent his sample as representative; nor, in fact, was it. In the discourse of contemporary quantitative methodology in descriptive bioethics, Oken's study should be characterized as surveying a convenience sample of 219 physicians in a single institution. His return rate of 95% was excellent but only makes the results representative of his sample, not of physicians generally. Veatch was careful at first to write "of physicians responding" but moved to a greater level of generalization when he incautiously wrote "physicians for the most part."

Veatch then reported that 88% of Oken's sample has "a usual policy of not telling." Buchanan rounds this number up to "[a]lmost 90%." These are accurate but not complete reports of Oken's results. Oken went on to point out that 32% make exceptions often or occasionally and 47% very rarely. Only 8% never tell their patients. Thus, in this non-representative, convenience sample of physicians from a single institution there is no evidence of widespread paternalism as claimed by Veatch and Buchanan; only 1 out of 12 respondents could be reliably described as systematically engaging in medical paternalism regarding truth-telling to patients with cancer in the form of routinely not disclosing their diagnoses to them.

At first glance, the first component of the concept of paternalism, restricting the liberty/interfering with the autonomy of patients with the capacity to make their own medical decisions, is satisfied by the self-reported behavior of withholding information. Not providing patients with information about their diagnosis and prognosis and alternative treatment plans obviously restricts their choices and therefore their liberty or autonomy. A close reading of the Oken text, however, does not lend credence to an interpretation that the practice of his 8%, much less the other 80% who at least sometimes withheld diagnostic information from patients with cancer, satisfies this first component, because Oken also reported that "[a]greement was essentially unanimous that some family member must be informed if the patient is not made aware of the diagnosis" (Oken 1961, p. 1123). This statement was *not* reported by Veatch, and my search of references to Oken's study in the early bioethics literature finds no instance of citing this crucial statement by Oken. This statement is very important, because if physicians were fulfilling the first component of the ethical concept of paternalism, they would have to have taken the view that the patient with cancer should not be told his or her diagnosis *by the physician or anyone else*. Instead, Oken's results indicated that the patient with cancer should not be told his or her diagnosis *by the patient's physician*. Patients with such diagnoses should, however, be told by their family members. Thus, Oken's data cannot be cited as an empirical basis for satisfaction of the first component of the ethical concept of paternalism.

What of the second component, acting for the good of the patient? Oken went on to challenge the assumption that cancer patients cannot bear the news, e.g., because of depression that might be made worse by giving them what we now call "bad news." Instead, Oken suggests that not telling cancer patients their diagnoses may be *self-protective behavior of physicians*. As Oken put it, "There is a strong tendency to avoid looking at the subject of cancer and the facts related to it" (Oken 1961, p. 1127).

Oken provided an important elaboration of this insight into his findings:

> Medicine is a difficult and exacting profession making heavy psychological as well as physical demands. Our personalities, feelings, and attitudes play a major role in determining the manner in which we communicate with and treat patients. They can constitute a tool of incalculable value: the art of medicine. But they can also interfere. No other area in which we work makes heavier claims than the treatment of cancer patients, with the suffering, pain, and death which are its frequent attendants. Pressed by these demands, we turn away in order to blunt their awful impact (Oken 1961, p. 1128).

Note what Oken wrote: the attitudes of physicians can affect the "manner in which" physicians communicate bad news to patients. The medical paternalist would focus on *whether* patients with cancer should be informed. Moreover, the medical paternalist would focus on attitudes concerning the potential harm of honest communication of bad news *to patients,* but Oken's concern is the potential psychological and even physical harm *to physicians* of having to give patients bad news. Physicians, Oken concluded, "turn away" from giving bad news because they are trying to blunt the biopsychosocial impact on themselves of having to do so, especially the promotion of "pessimism" *among physicians* about patients with cancer (Oken 1961, p. 1126). The second component of paternalism was, simply, absent from Oken's interpretation of his own results.

Despite the absence of empirical support in the most frequently cited empirical study of truth-telling in the bioethics literature, the ideology of anti-paternalism has been remarkably durable. Consider, for example, the following from a recent book by Veatch:

> Moral conflicts involving the principle of veracity [show respect by telling the truth] follow the same pattern as the other principles grouped under the heading of respect for persons. Once again, we have a conflict between doing what is best for the patient, in terms of benefit and harm, and fulfilling some general obligation, in this case the obligation to tell the truth (Veatch 2000, p. 74).

Veatch then cites Oken and adds:

> The reason [most physicians reported that they usually do not tell cancer patients their diagnoses] is easy to understand if one understands the Hippocratic principle and the depth of the commitment of 1960s physicians to that principle. They were afraid that if they told

> the patient, the patient would become psychologically upset, and the Hippocratic Oath says not to do things that will upset the patient (Veatch 2000, p. 74).

The Hippocratic Oath says no such thing. The Oath states that the physician should avoid iatrogenic harm to patients, i.e., iatrogenic morbidity and mortality, because bad outcomes were (and still are) bad for attracting market share (and the market for medical services in ancient Greece was intensely competitive and economically perilous for physicians), and because causing such harm is a kind of madness that sets in when physicians forget and then attempt to exceed the limits of medicine, as explained in *The Art* (Hippocrates 1923).

Oken also says no such thing. Thus, even if, for the sake of argument only, we were to grant that Oken's sample was representative of physicians in the United States in the early 1960s, his results documented not paternalism, but self-protective attitudes and behaviors of physicians that resulted in patients with cancer being given bad news by someone other than the physician, because of the onerous nature of their directly discharging this duty themselves. Unwittingly, as we shall see just below, these attitudes reflect major discussions in the history of medical ethics of truth-telling and decision making with patients and the durability of the ethics of truth-telling offered in these texts.

12.3.2 The Ethics of Truth-Telling and Decision Making with Patients in Major Texts in the History of Medical Ethics

In addition to the Oken study, major texts in the history of medical ethics were cited in the classical literature of bioethics, to support the claim that paternalism was a medical ethical norm. Jay Katz's landmark work, *The Silent World of Doctor and Patient*, announced and reinforced in its very title the ideology of anti-paternalism in bioethics (Katz 1984 [2002]). This enormously influential book was distinctive especially for its taking the history of medical ethics seriously, a then-uncommon approach to bioethics, with the very important exception of the *Encyclopedia of Bioethics* (Reich 1978, 1995), which has a large section on the history of medical ethics. Katz invoked this history to support his central claim that physicians engaged in a medical paternalism of silence with their patients. His conclusion about his reading of the history of medical ethics was striking for its force and breadth.

> The history of the physician-patient relationship from ancient times to the present bears testimony to physicians' caring dedication to their patients' physical welfare. The same history, by its account of the silence that has pervaded this relationship, also bears testimony to physicians' inattention to their patients' right and need to make their own decisions. Little appreciation of disclosure and consent can be discerned in this history, except negatively, in the emphasis on patients' incapacities to apprehend the mysteries of medicine and, therefore, to share the burdens of decision with their doctors (Katz 1984 [2002], p. 28)

In support of this interpretation of the historical ethical norms of medicine, Katz cited, among others, two of the major figures in the history of modern medical ethics, the physician-ethicists John Gregory (1724–1773) of Scotland and Thomas Percival (1740–1804) of England.

Katz represents Gregory as not supporting "patient involvement in decision making" (Katz 1984 [2002], p. 13). Katz goes on to say that Gregory's "lectures were filled with observations of traditional Hippocratic vintage" (Katz 1984 [2002], p. 14) and that Gregory "is silent on the question: When should such difficulties [bad news] be shared with patients?" (Katz 1984 [2002], p. 14).

Gregory nowhere mentions the Hippocratic Oath or any other ethical texts attributed to Hippocrates. Far from being silent on the question of giving patients bad news, Gregory explicitly addressed truth-telling to patients:

> A physician is often at a loss in speaking to his patients of their real situation when it is dangerous. A deviation from truth is sometimes in this case both justifiable and necessary. It often happens that a person is extremely ill; but yet may recover, if he be not informed of his danger. It sometimes happens, on the other hand, that a man is seized with a dangerous illness, who has made no settlement of his affairs, and yet perhaps the future happiness of his family may depend on his making such a settlement. In this and other similar cases, it may be proper for a physician, in the most prudent and gentle manner, to give a hint to the patient of his real danger, and even solicit him to set about his necessary duty. But, in every case, it behooves a physician never to conceal the real situation of the patient from the relations. Indeed justice demands this; as it gives them an opportunity of calling for further assistance, if they should think it necessary. To a man of a compassionate and feeling heart, this is one of the most disagreeable duties in the profession: but it is indispensible. The manner of doing it, requires equal prudence and humanity. What should reconcile him the more easily to this painful office, is the reflection that, if the patient should recover, it will prove a joyful disappointment to his friends; and, if he die, it makes the shock more gentle (Gregory 1772 [1998], pp. 34–35).

To be sure, Gregory claimed that a "deviation from truth is sometimes in this case both justifiable and necessary" and Katz appears to have read this to mean justifiable and necessary for the good of the patient. However, in the latter half of this passage, Gregory set out his justification: telling seriously ill patients their diagnosis and prognosis is "*one of the most disagreeable duties in the profession*." The context for Gregory's analysis of the obligation to be truthful with the seriously ill must be understood correctly. The loss of patients by physicians—of female adult patients to the complications of pregnancy and childbirth, of younger children to infections, and of parents, siblings, and older children to injury and war—was a constant in eighteenth-century life. Gregory's experience was typical: his beloved wife died delivering their sixth child and three of their children preceded her in death. It should therefore come as no surprise that Gregory expressed concern that the constant experience of patients' deaths would harden the heart of physicians, deadening them to the Humean sympathy that formed the foundations of his medical ethics (McCullough 1998). Truth-telling to the seriously ill was, we would say, biopsychosocially very burdensome on physicians. It still was in the 1960s, as Oken's interpretation of his results indicates. To Oken's interpretation Gregory would, in my judgment, say: "Just so."

Note that Gregory went on immediately to state that the duty to tell the truth to seriously ill patients was also "indispensable." It would be better, Gregory added, if the burden of discharging this indispensable duty could be passed on to others, namely family members. The reader should be aware that Gregory is addressing truth-telling mainly to patients in the private practice of medicine, in which well-to-do patients were seen in their homes and did not go to infirmaries (hospitals), which were for the working sick poor (Risse 1986), a population addressed below.

The second component of the concept of paternalism is thus not satisfied when we go to Gregory's texts. What of the first component, interfering with the liberty or autonomy of the patient, expressed in rights? In the discourse of eighteenth-century medical ethics, this topic was addressed under the rubric of the "government" of the patient. Gregory wrote:

> The government of a physician over his patient should undoubtedly be great. But an absolute government very few patients will submit to. A prudent physician should, therefore, prescribe such laws, as, though not the best, are yet the best that will be observed; of different evils he should chuse the least, and, at no rate, lose the confidence of his patient, so as to be deceived by him as to his true situation. This indulgence, however, which I am pleading for, must be managed with judgment and discretion; as it is very necessary that a physician should support a proper dignity and authority with his patients, for their sakes as well as his own (Gregory 1772 [1998], pp. 22–23).

Contemporary physicians will, I am confident, recognize the enduring truth of Gregory's clinically grounded, pragmatic account of the limits of physicians' abilities to control, i.e., govern, their patients. He added to this pragmatic justification the following:

> Every man has a right to speak where his life or his health is concerned, and every man may suggest what he thinks may tend to save the life of his friend. It becomes them to interpose with politeness, and a deference to the judgment of the physician; it becomes him to hear what they have to say with attention, and to examine it with candour; If he really approves, he should frankly own it, and act accordingly; if he disapproves, he should declare his disapprobation in such a manner, as shews it proceeds from conviction, and not from pique or obstinacy. If a patient is determined to try an improper or dangerous medicine, a physician should refuse his sanction, but he has no right to complain of his advice not being followed (Gregory 1772 [1998], pp. 33–34).

To my knowledge, this is the first occurrence of the discourse of patients' rights in the history of medical ethics, long predating what we should now properly characterize as its recrudescence, not invention, in bioethics two centuries later. More to the point, Gregory's antipathy to complete government of the patient and his explicit endorsement of the rights of patients, as we would now put it, to participate in decisions about their medical care and to make and implement what the physician judges to be clinically "improper or dangerous" decisions, make it unequivocally clear that there is no textual support in Gregory's medical ethics for the first component of the concept of paternalism.

In short, one cannot find in Gregory's medical ethics either component of the concept of paternalism. Indeed, he can reliably be read as *rejecting* both components of paternalism.

How does Percival's medical ethics fare? After acknowledging that Percival calls for others than the physician to give bad news to patients, Katz writes:

> Percival seemed oblivious to the hopeless charade he had perpetrated [by calling for others to inform patients of grave illness]. How could physicians, after having asked "others" to inform patients of impending death, expect "to be ministers of hope [and to] revive expiring

life," without hopelessly confusing their patients at the same time? In any case, the retreat from conversation was total (Katz 1984 [2002], p. 18).

Percival, a captive of Hippocratic paternalism, was, according to Katz, "not concerned... with liberty, with patients' participation in decision making" (Katz 1984 [2002], p. 17).

Let us consider Percival's text. Percival addressed himself to the physician's responsibilities of truth-telling to patients in the infirmary, a free hospital for the worthy, working sick poor:

> It is one of the circumstances which softens the lot of the poor, that they are exempt from the solicitudes attendant on the disposal of property. Yet there are exceptions to this observation. And it may be necessary that an hospital patient, on the bed of sickness and death should be reminded, by some friendly monitor, of the importance of a *last will* and *testament* to his wife, children, or relatives, who otherwise might be deprived of his effects, of his expected prize money, or of some future residuary legacy. This kind office will be best performed by the house surgeon, whose frequent attendance on the sick diminishes their reserve, and entitles him to their familiar confidence. And he will doubtless regard the performance of it as a duty. For whatever is right to be done, and cannot by another be so well done, has the full force of moral and personal obligation (Percival 1803 [1975], Part I, Section 7, emphasis original).

Like Gregory, Percival took the view that patients in the infirmary should be given bad news, but not by physicians. Family members would not be routinely available to fulfill this obligation. Someone as ready to hand in the infirmary as was the family in the home was needed. Percival's remedy is that the physician should enlist the "house surgeon" to discharge this duty. In current medical argot, the house surgeon was then the lowest form of life on the medical totem pole and the physician the highest, reflecting the longstanding rivalry between physicians and surgeons. In the emerging hierarchy of hospital power the physician had gained the new authority to assign this task effectively to a subordinate. Patients with serious illnesses having thus been informed, they could then participate in conversations about their clinical management with their physicians.

A similar line of reasoning and a clear debt to Gregory's medical ethics appears in Percival's account of the physician's obligation to be truthful with patients in private practice.

> A physician should not be forward to make gloomy prognostications, because they savor of empiricism, by magnifying the importance of his services in the treatment or cure of the disease. But he should not fail, on proper occasions, to give to the friends of the patient, timely notice of danger, when it really occurs, and even to the patient himself, if absolutely necessary. This office, however, is so peculiarly alarming, when executed by him, that it ought to be declined, whenever it can be assigned to any other person of sufficient judgment and delicacy. For the physician should be the minister of hope and comfort to the sick; that by such cordials to the drooping spirit, he may smooth the bed of death, revive expiring life, and counteract the depressing influence of those maladies which rob the philosopher of fortitude, and the Christian of consolation (Percival 1803 [1975], Part II, Section 3).

Making "gloomy prognostications" meant that the physician would exaggerate the seriousness of the patient's condition, to increase the gratitude of the patient when treatment went well (as expected) and to protect the physician from retaliation should treatment go badly. This is not about medical paternalism but about inappropriate influence of the physician's self-interest.

More to the point, note that the physician, when "absolutely necessary" in terms of the gravity of the patient's condition and a severely time-limited need to act, should directly communicate bad news to patients. Contra Katz, conversation of physicians with such patients should occur. Their liberty is not ignored, much less violated, meaning that the first component of paternalism is not satisfied. There is indeed a duty to be truthful with seriously ill patients in private practice, but fulfilling this duty is "alarming," a choice of words even stronger than Gregory's "most disagreeable." The second component of the ethical concept of paternalism cannot be found in *Medical Ethics*. To be sure, the closing sentence of this passage might invite a different interpretation, but only if read in isolation, which is not permissible in sound scholarship. So strong had the ideology of anti-paternalism become in the 1980s that Katz could not appreciate the plain language and concepts of Gregory's and Percival's medical ethics.

Kathleen Powderly (2000) has undertaken remarkable scholarly research on the practice of a nineteenth-century gynecologist, Alexander Skene, of Brooklyn, now a borough of New York City. Powderly demonstrates that Dr. Skene developed and implemented what we would now recognize as a forerunner of the informed consent process and did so with female patients, belying the chauvinist image of the paternalistic male physician disrespecting female patients. Katz did not have the advantage of Powderly's scholarship, but we do. It completely undermines the historical claims that undergird bioethics' ideology of anti-paternalism.

12.4 Conclusion

The anti-paternalism of what can now be called the classical literature of American bioethics was marked by the claim that medical paternalism was well documented empirically and was the accepted norm in the Western history of medical ethics. I have shown that neither claim has textual support in sources such as Oken, Gregory, and Percival that were cited in support of this claim. The development of American bioethics, we can therefore now say with confidence, was predicated on two crucial claims, one empirical and one historical, and both errors about medical paternalism (McCullough 2011).

As a consequence of these deeply flawed claims, the anti-paternalism that defined early American bioethics—and still defines it to a considerable degree—should be classified as an ideology of the second type described at the beginning of this chapter. The elements of bioethics' ideology of paternalism can now be stated. Physicians were usually silent or were not honest with their patients about serious diagnoses, their prognoses, and their clinical management, and the reason for what was taken to be egregious interference with the autonomy of patients was the psychological and clinical benefit that patients derived from the silence of their physicians. Physicians did not involve patients in decisions about the patients' care, again for their own good and in an unwarranted assertion of physicians' scientific and clinical authority in violation of the decisional rights and therefore the liberty and autonomy of patients. Both components of medical paternalism were thought to be satisfied, resulting in the emphasis—from the origins of bioethics to the present—on medical paternalism as an ethically unacceptable practice and the ethical principle of respect for patient autonomy and practices based on it, such as informed consent, as the fully justified and effective remedy. Physicians were once systematic paternalists but were no longer justified in their medical paternalism. Indeed, they never were justified in their medical paternalism (an egregious assertion that the ethical concept of patient autonomy was available to physicians and physician-ethicists before the concept was invented, at least in prototype, in the eighteenth-century by Gregory and more fully by Skene).

The correct reading of the frequently cited Oken (1961) study and of the two crucial sources in modern Western medical ethics (Gregory 1772 [1998]; Percival 1803 [1975]) is that physicians were not directly forthcoming with bad news because giving bad news was—and remains—one of the "most disagreeable" and "alarming" obligations of physicians, one away from which they would naturally want to turn, just as Oken described. However, patients were to be informed and, as a matter of obligation, physicians should see to it that seriously ill patients were informed, but by others than the physician—by their families in the case of wealthy patients being seen in their homes, and by the house surgeon in the case of worthy-poor patients in the infirmaries. Physicians endorsed the ethical norm that patients should participate in decisions about their patient's care as early as the eighteenth century and robustly in the nineteenth century.

Physicians were not systematic paternalists before bioethics heroically exposed and corrected medical paternalism. The view that physicians for centuries had been engaging in systematic medical paternalism became not just a central claim but, perhaps, the defining claim of American bioethics in its origins and throughout its history to date. Major empirical and historical sources invoked to support this view, I have shown, do not in fact support it. The origin of this view in bioethics, and the heroic image of American bioethicists as reformers, even revolutionaries, was empirically and historically flawed. Its remarkable persistence means that American bioethics continues to be defined and deformed by an ideology of anti-paternalism that it is long past time to abandon as the defining trope of bioethics (McCullough 2011). Doing so will free the field for the important work of empirically and historically well-informed conceptual investigation of the nature and limits of physicians' intellectual and moral authority and therefore professional integrity, especially in the context of, and as shaped by, organizational culture (Chervenak and McCullough 2005).

References

Baker, R.B. 1993. The eighteenth-century philosophical background. In *The codification of medical morality: Historical and philosophical studies of the formalization of Western medical morality in the eighteenth and nineteenth centuries: Vol. 1: Medical ethics and etiquette in the eighteenth century*, ed. R.B. Baker, D. Porter, and R. Porter, 93–98. Dordrecht: Kluwer Academic Publishers.

Beauchamp, T.L. 1978. Paternalism. In *Encyclopedia of bioethics*, ed. W.T. Reich, 1194–1201. New York: Macmillan.

Beauchamp, T.L. 1995. Paternalism. In *Encyclopedia of bioethics*, 2nd ed, ed. W.T. Reich, 1914–1920. New York: Macmillan.
Beauchamp, T.L., and J.F. Childress. 1979. *Principles of biomedical ethics*. New York: Oxford University Press.
Buchanan, A. 1978. Medical paternalism. *Philosophy and Public Affairs* 7: 370–390.
Chervenak, F.A., and L.B. McCullough. 2005. The diagnosis and management of progressive dysfunction of health care organizations. *Obstetrics and Gynecology* 105: 882–887.
Dworkin, G. 2005. Paternalism [On-line]. In *Stanford encyclopedia of philosophy*, ed. E.N. Zalta. Available at http://plato.stanford.edu/entries/paternalism/. Accessed 6 July 2009.
Gorovitz, S., et al. (eds.). 1976. *Moral problems in medicine*. Englewood Cliffs: Prentice Hall.
Gregory, J. 1772 [1998]. *Lectures on the duties and qualifications of a physician*. Edinburgh: W. Strahan and T. Cadell. Reprinted in *John Gregory's writings on medical ethics and philosophy of medicine*, ed. L.B. McCullough, 161–248. Dordrecht: Kluwer Academic Publishers.
Hippocrates. 1923. The art. In *Hippocrates*. Loeb Classical Library, vol. II. Trans. W.H.S. Jones, 185–217. Cambridge, MA: Harvard University Press.
Jonsen, A.R., M. Siegler, and W.J. Winslade. 1982. *Clinical ethics*. New York: Macmillan.
Katz, J. 1984. *The silent world of doctor and patient*. New York: The Free Press. Re-issued. 2002. With an introduction by Alexander Morgan Capron. Baltimore: Johns Hopkins University Press.
McCullough, L.B. 1998. *John Gregory and the invention of professional medical ethics and the profession of medicine*. Dordrecht: Kluwer Academic Publishers.
McCullough, L.B. 2011. Was bioethics founded on historical and conceptual mistakes about medical paternalism? *Bioethics* 25: 66–74.
McCullough, L.B., J.H. Coverdale, and F.A. Chervenak. 2004. Argument-based ethics: A formal tool for critically appraising the normative medical ethics literature. *American Journal of Obstetrics and Gynecology* 191: 1097–1102.
Nuland, S.B. 2009. Autonomy amuck. [Review of] Patient, heal thyself: How the 'New Medicine' puts the patient in charge. *The New Republic* 240(4): 862. Available at http://www.tnr.com/story.html?id=3861e85c-a13a-47f3-8b34-84bb2eddd731. Accessed 6 July 2009.
Nutton, V. 2009. The discourses of European practitioners in the tradition of the Hippocratic texts. In *The Cambridge world history of medical ethics*, ed. R.B. Baker and L.B. McCullough, 359–362. New York: Cambridge University Press.
Oken, D. 1961. What to tell cancer patients. A study of medical attitudes. *Journal of the American Medical Association* 175: 1120–1128.
Percival, T. 1803 [1975]. *Medical ethics: A code of institutes and precepts, adapted to the professional conduct of physicians and surgeons*. London: J. Johnson & R. Bickerstaff. C.R. Burns (ed.). 1975. Huntington: Robert E. Krieger Publishing Company.
Post, S.G. (ed.). 2004. *Encyclopedia of bioethics*, 3rd ed. New York: Macmillan Reference USA.
Powderly, K.E. 2000. Patient consent and negotiation in the Brooklyn gynecological practice of Alexander J.C. Skene: 1863–1900. *The Journal of Medicine and Philosophy* 25: 12–27.
Reich, W.T. (ed.). 1978. *Encyclopedia of bioethics*. New York: Macmillan.
Reich, W.T. (ed.). 1995. *Encyclopedia of bioethics*, 2nd ed. New York: Macmillan.
Risse, G.B. 1986. *Hospital life in Enlightenment Scotland: Care and teaching at the royal infirmary of Edinburgh*. Cambridge: Cambridge University Press.
Veatch, R.M. 1976. *Death, dying and the biological revolution: Our last quest for responsibility*. New Haven: Yale University Press.
Veatch, R.M. 2000. *The basics of bioethics*. Upper Saddle River: Prentice Hall.
von Staden, H. 1996. 'In a pure and holy way.' Personal and professional conduct in the Hippocratic Oath. *Journal of the History of Medicine and Allied Health Sciences* 51: 404–437.

Part IV
The Future of Bioethics: Looking Ahead

Chapter 13
Themes and Schemes in the Development of Biomedical Ethics

Richard M. Zaner

13.1 Preliminary Reflections on Themes

A survey of the main issues in both public and professional settings since the early 1960s, suggests that most of them take their point from their relation to, or implications from, that special relationship between physician and patient: the clinical encounter. This is as true for solid organ transplantation as it is for in utero surgery, as true for prevention as it is for prognosis, and as true for dying persons as it is for embryos. It is also true even for the more exotic topics occasioned by the continuous outpouring of procedures, drugs and equipment from medicine's ever-expanding cornucopia, whether diagnostic or therapeutic.

This focus on the clinical event brought out a number of serious moral concerns that mainly had to do with issues at the end of life. Popular and professional attention alike became absorbed with questions raised by aid in dying, euthanasia, brain death, and, along with these, advanced directives (living wills, durable power of attorney for health care). When the Nancy Cruzan case reached the U.S. Supreme Court in 1990,[1] it not only securely established the right to sign an advance directive, but the case also motivated Congress to pass, in 1991, a law requiring all health care institutions to duly inform all incoming patients and their families of this right to refuse treatments when terminally ill.

Other difficult, even harsh, questions had, of course, already occasioned heated disputes that also led to calls for the involvement of bioethics: abortion, treatment for severely premature babies, informed consent and privacy, allocation

R.M. Zaner, Ph.D. (✉)
Center for Biomedical Ethics and Society (Emeritus), Vanderbilt University Medical Center, 2525 West End Ave., Suite 400, Nashville, TN 37203, USA
e-mail: rmorrisz@comcast.net

J.R. Garrett et al. (eds.), *The Development of Bioethics in the United States*, Philosophy and Medicine 115, DOI 10.1007/978-94-007-4011-2_13,

of scarce resources, among others. Some became part of our public iconography: providing a liver transplant for a three-year-old child living in poverty; choosing who should get renal dialysis when there are not enough machines; developing policy on novel alternative forms of pregnancy (such as surrogacy, artificial insemination, in vitro fertilization and embryo transfer); deciding whether family planning should include sex pre-determination; and many more. Some questions were raised by the development of dramatic, highly sophisticated types of diagnostic imaging technology: ultrasound (US), computer-assisted tomography (CAT), positron emission tomography (PET), or magnetic resonance imaging (MRI).

Other questions arrived in the wake of breathtaking forms of treatment: multi-drug chemotherapy, cellular transfusion, mind-altering drugs, and the like. Still others came on the heels of sensational forms of surgical intervention: intrauterine surgery, solid organ transplantation, neural cell implantation, stem cell infusion, etc. And some questions arose simply because the health care system came to the point of near-collapse from mal-distribution of resources; or from too many non- and under-insured people; or from other sources related to our society's inability to confront and resolve grievous social problems, such as the increasing numbers of children living in poverty, ghettos, and hunger. Perhaps the most compelling irony of the latter half of the twentieth century in the United States was the existence of a society with wealth and power exceeding all other nations, but which could not properly feed, clothe, house, educate, or employ many of its own citizens or properly nurture its young.

More recently, the implications of the genome project have clearly stretched our moral imagination well beyond traditional limits. If a central, governing thought in the late twentieth century was the "what-is-to-come"—that is, death, believed to be determinative for the being and life of humans—the twenty-first century is already noticeably turning the other direction in search of guidance. Birth, not death, is now definitive; genetics, not geriatrics, has rapidly become the basic medical and biomedical discipline; life-before-birth, rather than end of life, has become the focal point. This can be seen from the impact of molecular biology and genetics on medical theory and clinical practice to questions of the legitimate use of genetic information, and from cellular manipulations to embryo research, pre-natal diagnosis, and fetal interventions more broadly.

13.2 Emergence of a New Schematic

Thinking about just these matters some time ago, Hans-Jörg Rheinberger concluded that a "new medical paradigm: molecular medicine" has become evident (1995, p. 254). Already ongoing for the past century, biomedicine has blossomed more fully over the past several decades and is well on its apparently unstoppable way to taking over the entire medical landscape. As with the appearance of most new paradigms, this bodes radical changes in the way health, disease, treatment and the like will come to be understood.

Rheinberger is in any event very clear about what concerns him:

> With the acceleration of a historical, irreversible alteration of the earth's surface and atmosphere, which is taking place within the span of an individual human's lifetime; with the realization that our mankindly, science-guided actions result, on a scale of natural history, in the mass extinction of species, in a global climatic change, and in gene technology that has the potential to change our genetic constitution, a fundamental alteration in the representation of nature is taking place, which we are still barely realizing. (1995, p. 260)

To be sure, discovery and diagnosis continue to occupy the limelight of human genetics research—even with its newly acquired name, genomics—with treatments and understanding lagging far behind. Nonetheless, the regularly stated, almost mantra-like discourse about genetics projects is that clinical practice will be totally transformed as new genetic knowledge leads eventually to effective treatment modalities. This often-used justification is also the probable future reality of genomics. With that eventuality, a wholly new meaning of "health" must shortly follow: one where "health" is understood more as a matter of healthy genes (with the ability to make and keep them healthy) than as a matter of the absence of sickness, or curbing the workings of some pathological process or entity.

13.3 Issues Remain Complicated, Often Opaque

Matters are made all the more complicated by the fact, as I see it, that the discipline of ethics in medicine is still not very well understood by the general public, medical personnel, well-meaning academic colleagues, and even some of the medical humanities' more dedicated practitioners. This is even clearer when the development of that 'new paradigm' in medicine is taken into consideration—as it must be, in order for us to make sense of medicine and the ethical issues occasioned by it. Indeed, the more serious are these issues, the more we should be puzzled by many of the persistent questions, intractable conflicts, and unending disputes.

Place all of this confusion against the amazing growth and popularity of bioethics—from meager beginnings in the early 1960s to the world-wide status it enjoys only four decades later—and we face one of the most intriguing questions about this field: What can account for the striking popularity of an endeavor still not well explained, especially in what may seem to be such central and important questions about human life and death? Equally perplexing is that it remains unclear just what the, at times, raucous concern is all about.[2] And that, to be candid, is complicated by the further fact that few of its practitioners and advocates agree on what "it" really is! Bioethics, like ethics more broadly, remains seriously problematic.[3]

13.4 A Bit of History

Wonder, therefore, seems to me the proper mood for anyone who chances to look into these matters. But, first, a bit of history may be helpful. What was behind the initial idea, over 50 years ago, of asking philosophers and others in the humanities to come into medical centers (at first, only medical and, a bit later, nursing schools and, still later, into hospitals)[4] in order to participate in medical education and, in some instances, even in clinical situations?

With the astonishing new technologies and medical knowledge already at hand in the early 1960s, and even more remarkable prospects on the immediate horizon, physicians had good reason to be troubled. Furthermore, new diagnostic tools and techniques promised more accurate, and ever earlier, detection of both present and possible damage from diseases and anomalies hitherto not available. Coupled with these were emerging new surgical techniques and instrumentalities, pharmacological interventions, new types of anesthesia, and all manner of new treatments for conditions not previously treatable, as well as for those not previously treated very effectively. Resuscitative techniques along with associated technologies and new medical understanding (critical for the development of intensive care units) showed that different body-systems functioned and ceased to function in different ways and at different paces, and that some could be artificially re-started and supported, thereby allowing needed time for medications to work properly, or for healing to take place.

These raised quite awesome and, in some cases, wholly new issues, and gave new force and content to many perennial issues (Gorovitz et al. 1976). Not only was it increasingly possible to maintain patients who only a few years before would have died, often very painfully (as with end-stage renal disease), but the horizons of life's beginnings and endings were also becoming ever more understood. Though not widely realized at the time, these horizons were also being perforce re-defined—especially thanks to what were termed the "new genetics."[5]

Some perceptive physicians and researchers were already agonizing over the values and moral issues implicit in these developments.[6] Recognizing their lack of training and knowledge to grapple with, much less resolve, such issues, they quite naturally turned to others whose credentials at least seemed to bespeak competence, if not expertise.[7] Many of these physicians, too, were haunted by the horrors of the Nazi concentration camps, and seemed anxious to realize in practice what was asserted in the Medical Trials at Nuremberg[8] and by the United Nations Charter: affirming the existence of inalienable human rights, especially for those who are sick, maimed, and vulnerable.[9]

The lingo of the times is suggestive. Even while precious little energy had been expended to rectify perceived flaws, physicians and others in the so-called health care system expressed serious dismay over the bureaucratic organization of the modern health science centers and health care more generally, as well as the ways in which new technologies tended, as was often said, "to dehumanize" people. Most health care professionals exhibited genuine concern over the increasing specialization in health care after World War II, which seemed to "fragment" the "whole person" and promote greater focus on diseases and organ systems than on people. At the same time, most also recognized the remarkable advances were achieved precisely by such specialization.[10] While many tried earnestly to stay abreast of the ever-growing cornucopia of new developments, substances, sub-specialties, and the like, this often meant that physicians were obliged to be and remain technically proficient. Thus, they did not always have the time or inclination to be alert to moral issues, religious values, sensitive caring, and other such concerns (Gorovitz 1982).

The "new physician," avidly discussed as the major part of the agenda for medicine in the 1970s and beyond, it was thought, needed to be "humanized." But, as has been pointed out numerous times, it was not in the least clear what this would require, nor, in the end, why it was thought to be so important. Such "humanists," after all, were hardly the talk of the commons, nor had they made notable or recognized contributions to the common weal, much less to health care. Moreover, phrases such as "medical humanities" seemed only to confuse and bewilder.

Thus, one practicing physician, Samuel Martin, lamented publicly in 1972 about how unclear it was just who would be responsible for training that new physician. For many educators, medicine needed to call on humanists—so-called "experts in human values"—and a new name was quickly concocted for the new breed: *ethicist*, an occupation as unlikely as the name was awkward to pronounce.[11] Martin and others were dubious about the entire venture. In poignant, if inelegant, terms, Martin worried aloud:

> How can we humanize the humanist, the man who must help us all? Some are worried that our humanists are trying to get away from emotions, empathy, feeling, and other parts of our esthetic continuum, and that they are trying to outscience our scientists. At some time we must deal not only with what makes a humanist, but also with how we can facilitate the transmission of his art. (Martin 1972)

What indeed "makes a humanist," how does he transmit "his art," and what, in the end, are the "humanities" all about? To be sure, Martin perhaps should have worried not only about those trying to "outscience" the scientists, but also about those at work cultivating ever sharper and deeper divisions between what C.P. Snow termed "the two cultures" (Snow 1993). In any event, neither physicians (ever more reliant on the biomedical sciences), nor the humanist (whose craft remained opaque to most) were likely to worry much about Martin's appeal. Indeed, an appeal to supposed experts in values was not only quite implausible and unlikely at the time, but also, for the most part, highly improbable. Most so-called humanists were rarely interested in, much less competent to make recommendations about, such matters as were posed daily at the bedside. In the process of creating the new medicine and its supposed new healers, these questions included: what is death and how ought a person's last days be managed, much less how ought such persons be cared for? For that matter, is a person whose breathing occurs solely because of a ventilator's mournful chug still alive or in some halfway condition never seen before? When does a fetus become a person? Is a comatose individual still a person? The questions only proliferated and, too often, left "humanists" as bewildered as any of the rest of us. Help, in short, seemed ever further away as regards the actual, intense issues faced by health care professionals on the wards, in the nursing homes, or other places where our society tends to house the sick, maimed, and elderly.

13.5 Beneath These Themes

Few matters generate more heat and conflict in biomedical ethics than questions about what decisions are best at the "end" or "beginning" of life. Why is this? We might consider that we regularly witness remarkable passion, not only about human life, but animal life as well: for instance, the way in which most of us think about and act toward puppies and kitties as opposed to dogs and cats. What is that all about?

Of course, we might also observe that, considering human life generally, things are even more complicated. There are old people, mature adult people, young adult people, teenagers, children, youngsters, infants, and babies; further down the developmental scale, there are viable fetuses, previable fetuses, and nonviable fetuses; even further, zygotes, embryos, and blastocysts; and, still further, sperm and egg. Now, which of these various sorts of human beings or individuals should capture our concern? Which should we *really* be concerned about? And, of course, why? What reasons do we have for any concern at all? Who are 'we,' who exhibit such concern? Who are 'they,' who preoccupy that concern?

People from different societies, and different moral and religious traditions, express serious concerns about experiments on embryos, much less on fetuses: why is that? Hauerwas and Burrell put the matter somewhat differently, noting that "we assume, without good reason, that it is wrong to kill children … even more strongly, we assume that it is our duty to provide children (and others who cannot protect themselves) with care that we do not need to give the stranger." The reason we so commonly make such assumptions, they contend, is that they are "like the air we breathe—we never notice them (Hauerwas and Burrell 1977)." These assumptions are simply not subject to the sorts of decisions that the accepted view in ethics insists must be made. They are, in Alfred Schutz's well-known term, simply taken for granted without question.

This may be, but we must at the same time take careful note. Even if these are not the sorts of problems and enigmas on which the typical texts actually dwell; even if they are assumptions that are typically not subject to decisions; even if what gives current medical ethics its keen disputes and barbs; even if there is a tendency to force decisions involving matters that we otherwise take for granted as beyond decision or choice—even so, why is that the case? What, in a word, is so upsetting to us when we are asked to confront such questions?

If we consult much of the bioethics literature, we quickly learn that what is at issue, what modern medicine and biomedicine have succeeded in calling into question, is precisely that set of taken for granted assumptions about moral life. At bottom, the real issue is said to concern the *moral status* of what we have hitherto simply taken for granted: the sperm, egg, embryo, fetus, the baby, the terminally ill person, the one with persistent vegetative state. What, then, are our obligations and responsibilities concerning "life before birth" at each of these various entities or stages of life? The 'beginning' is no longer simply what we used to call 'birth,' but has now been pushed far back up the developmental scale. And while 'end' used to mean 'death,' this has become another equally shadowy point. When are these, 'beginning' and 'end,' and what is the moral status of the individual which is starting to be, or the one who is dying? If, so the story goes, we could just get clear about that *moral status*, we would then be able to figure out our responsibilities—we, that is, those of us who worry, are people whose own moral status is, of course, taken for granted as unquestionable.

13.6 Digging Further

These issues have, I think, become increasingly the central ones. Instead of merely noting what we specifically take for granted as beyond question and decision, we might instead dwell upon these, for the fact that they are taken for granted itself invites vigorous but cautious attention. Why is it, after all, that we assume only certain sorts of values and not others? What is it about children that causes us to assume that it is wrong to harm, abuse, or kill them? Simply noting that we take it for granted is surely inadequate. Of course, dwelling on these deep-lying beliefs is neither easy nor comfortable—ask any father who is asked to decide what to do now that his infant son seems destined to die.

Being with a dying person is haunting. I recall being at my mother's bedside in an intensive care unit while she was dying: who, where, what is she now? Is she somehow still there, in that failing body? Is she perhaps still locked within her non-functioning brain? If that brain collapses, where will *she* have gone? Which, it seems ineluctably, brings us witnesses to another chilling, if only momentary, realization: will I too die? If and when I die, who and where and what will I then be? I recall, too, what my mother once confided to me about my birth: I was not only 'unexpected,' but her previous pregnancy had ended in abortion. Suppose that had been *me*: what would I then have been? Nothing? A never was? Is that what must be said about the aborted fetus?

What I am suggesting is that at the very heart of these passionately contested questions is a far more intimate and deeply personal set of questions: Who, really, *are* we—we who find ourselves ill, injured, or genetically compromised? More, who am I? How did I get here, at this time, in this place, as a member of this family, and with just these specific characteristics, traits, and all of that? Who or what was that aborted fetus? If it was at all alive, where then did its life go when its development was cut short? Even more hauntingly: if, by techniques such as rDNA, gene splicing, and the like, the developing life of an embryo can be altered, *what* is it that is altered? What would it have been if it were not altered? What else is changed? Is there within that budding 'life' also a budding person, a self? If so, what happens to that self when such techniques are utilized?

We know, too, that it is currently possible to produce clones—animal or human—by blastomere separation (Hall et al. 1993), for instance, or other types of genetic engineering.[12] While John A. Robertson (1994, pp. 6–14) believes the procedure is morally acceptable (and will result in cloned human beings within a few years), others are deeply upset by the very idea—for instance, Leon Kass (1997, pp. 16–23), Richard McCormick (1994, p. 14), the Islamic scholar, Munawar Anees (1994, pp. 23–24), and of course much of the Catholic hierarchy.[13] On the one side, riveted fascination, even delight in the prospect; on the other, shocked rejection and dismay. Why?

What provokes such evident passion and concern about 'life before birth' and 'life facing death,' may be something like this: what is truly called into question, from its hitherto bedrock of assumptions, is *that, who, why* and *what* we *are*, each of us: my very own 'who I am.' What would have occurred had somebody not done what they did (give birth, abort, experiment, or whatever)? What would *you* or *I* be, if 'be' we would at all?

13.7 What Troubles the Course of Bioethics

We are deeply beset by the unique way that questions about the beginning or end of life force such searching, intense questions. However, this is not so much because of what has been taken for granted or because these questions have at times seemed merely to pose quandaries and a need for decisions. Nor are these matters merely of policy or procedure. Hauerwas and Burrell were right to insist years ago that the standard view in ethics encourages learning "to view our desires, interests, and passions as if they could belong to anyone, … [that] we view our own projects and life as if we were outsiders" (1977, p. 171). Precisely because of that, much of contemporary ethics "has the distressing effect of making alienation the central moral virtue" (1977, p. 171)—and, I would add, then having to seek ways and means to guarantee precisely what was ignored, by renewing an emphasis on autonomy. Precisely what is presupposed is at the same time ignored: the individual person who, after all, must take a stand on matters, whichever stance it may be—but what is this 'person?'

We can witness these matters especially well, for instance, in the highly animated and far-reaching dispute that has been going on over the last three decades about the prospects and implications of the new genetics. Almost from its beginning, announcements of results—some preliminary, some more firm—have been promoted by scientists and tickled into public awareness by the media. One discovery after another is reported about this or that bit of the human genome; one or another gene governing this or that human trait has been located on this or that chromosome: Huntington's, cystic fibrosis, Duchenne's muscular dystrophy, or Curler's syndrome, not to mention genes which presumably make their bearers 'susceptible' to breast cancer, Alzheimer's, schizophrenia, or alcoholism, or those that govern, say, shyness or resentment. These reports fascinate us mainly because they are supposed to concern the causes of inheritable diseases: flawed genes whose abnormal functioning (or normal malfunctioning) results or could result in the appearance of a malady at some time in a person's life. At bottom, they captivate because they bring into question who and what each of us is.

The unraveling, mapping, and sequencing of the human genome being accomplished in countless projects around the world, Gilbert avers, promises to "put together a sequence that represents … the underlying human structure … our common humanity."

Soon, he is convinced, we'll be able "to pull a CD out of one's pocket and say, 'Here is a human being; it's me'!" (Gilbert 1992, p. 95). More likely, of course, such genetic information will be encoded into some miniscule cell in the body, imprinted in the tiniest of our tiny body-parts, or tucked away like clandestine cargo in a skin, tissue, or other cell like some covert neo-modernist spy; a mole whose code is known only by those who implanted it, and is ready to re-emerge at the command of some future molecular physician, for whatever purposes may then be at hand. In pursuit of the "holy grail," presumably nothing can or should be ruled out of bounds! Indeed, quite literally everything "human" is regarded as very much *in* bounds: mind, body, self, emotion, action—the works. It is already thought by many of these advocates that the answers to the ultimate questions of our nature, duty, and destiny will be regarded, as it in many quarters is already suggested, as either to *be* the genes or to be *in* the genes.

In the late 1950s, the great British science fiction writer, James Blish (1957), wrote a charming little novel suggestively entitled, *The Seedling Stars and Galactic Cluster*. It has a simple premise, as inventive as it is remarkable for its prescience. Interstellar travel has become routine even as the population has long since burgeoned beyond Earth's and other planets' resources, and thus habitable planets have become premium. Most are fiercely uninhabitable; making them habitable requires the immensely complicated, expensive, and only rarely effective labor of what Blish calls "terra-forming." Here, not unlike the early Greek dietetic healers, attention is mainly on what is external to human beings: the environs, past or present, and using dietetics to ameliorate the effects of these.

Biology, Blish also postulates, had undergone a far-reaching revolution—the beginnings of which were already apparent when his novel appeared. We have since become acutely aware that biology is undergoing a revolution matching, if not surpassing, the earlier one in physics. In the novel, biological manipulations are routinely designed for population projects using the most elementary reproductive life-processes, including cloning and other types of genetic engineering. Now, to the contrary, the focus is on the processes of human life itself—which also bears some resemblance to the early Greek interest in dietetics for human improvement.

Blish's tale is intriguing. In his imaginative hands, the deliberate, literal re-designing of human individuals by other human individuals is an accomplished fact. Changes are brought about which need neither centuries of evolutionary change nor spontaneous mutation, but only the ingenuity and sportive inventiveness of highly powerful biomedical scientists possessing "the secret of life," avidly in pursuit of ever-new ways to design and produce people.

This is familiar territory, for several Nobel Laureates in genetics overtly advocated this very approach: Joshua Lederberg in the late 1960s,[14] and Sir John Eccles (1979) and Sir Macfarlane Burnett (1978) in the late 1970s. Cloning, of course, first became a reality for complex animal vertebrates only in the 1990s (Wilmut et al. 1997, pp. 810–813). What Blish only imagined has thus become a reality; indeed, with monkeys, sheep, dogs, and cats now successfully cloned, cloning other vertebrates—humans included—may seem inevitable.[15]

Thus, the response to the ultimate questions of human life is often said to be in the genes, and is not to be found in the quaint metaphysical quests as moved the likes of Plato or Aristotle, Aquinas or Occam, Kant, or Heidegger! Something like full circle will then be reached, for at the time of DNA's discovery—what a 1961 *Life* magazine cover coyly announced as the "secret of life," and which Kurt Vonnegut cut a literary jib about in his classic satire, *Cat's Cradle*—it was thought[16] that the new genetics was indeed "the holy grail" of science and society. The human genome, etched in electronic miniature on a CD, thus came to be regarded as the secret hiding place of self, indeed of life itself.

The motif is historically fascinating as well, for it is of a piece with one of the core convictions in medicine's long history, as articulated in two fundamental visions. Ancient physicians were struck by the ways in which the human body and soul could be changed by either medicines or, more likely, dietary regimens. Galen went so far as to assert the need to "clear the path for using bodily factors to elevate man beyond the possibilities of purely moral teaching" (Tempkin 1973). In our times, what is often found in such science fiction as that produced by Blish lends support to that view of things: Blish's biologists, after all, are far less interested in correcting the ravages of disease or genetic anomalies, and far more interested in altering the genomes of human individuals to be as immune as possible to those ravages from the start of life (as well as, of course, in enabling them to live, love and be productive in hazardous, alien places). On the other hand, reports from the Genome Project highlight the therapeutic potential of new discoveries while almost always downplaying the eugenic designs that fascinated both the ancients and much of contemporary biological science and science fiction.

The sort of power inherent in this genetic vision of human life has, in our times, become a literal reality with the success of animal, and especially vertebrate, cloning. In short, we supposedly are learning daily more about what we human beings truly *are*. For scientists and ethicists such as Robertson or Kass, much more is at stake than detecting or even treating genetic disorders. Underlying much of the professional and especially public discourse is a battle of prophetic visions: either an apocalyptic sense of impending disaster to be avoided at all costs, or a utopian dream of a promised idyllic state to be embraced and realized. The latter, however, is clearly winning the day. Within this dream, one can detect the not-so-subtle workings of the vision: a time when, not only the currently known 3–4,000 or so, but the entirety of maladies will be known down to the tiniest sub-molecular detail, along with the means to change their otherwise inexorable course: gene splicing, genetic surgery, and other promised therapies just around the corner.

13.8 Historical Precedents

It was precisely in this context that Rheinberger emphasized the fundamental change going on at the heart of medicine. Such ideas are deeply rooted in Western history (Zaner 1984, 1998), especially since the advent of science in its specifically modern character (Toulmin 1964/1990; Mumford 1964/1970). The core views of biomedical scientists like Hans Spemann,[17] Herbert Müller, Sir Julian Huxley, Lederberg,

Eccles, Watson, Gilbert, and others were already on hand by the time Blish wrote his intriguing novel.

Such views stem directly from a theme inspired by, as Jonas (1959, pp. 127–134)[18] pointed out, Bacon's sixteenth century heraldic text, *The Great Instauration*, proclaiming the need to subdue and alleviate the "necessities and miseries of humanity" by means of "a line and race of inventions" (technologies) (cited in Jonas 1959, p. 189; see also Bacon 1996). Bacon made it quite clear that nature's blind, but potent, actions on human life harbor significant practical consequences: natural disasters such as conflagrations, epidemics, floods, and famines. In light of that, he sounded an alert: nature untamed and left to its own devices most likely will devastate us; nature, itself uncaring, insensate, and indifferent to the human presence, is yet fundamentally aggressive and unpredictably destructive when not understood. It must therefore be subdued, made amenable by force to human design.

Thus, the relation of human beings to nature came to be conceived as one of power, specifically like the relation of the conqueror to the vanquished. On the basis of knowledge, human beings must act on nature for the sake of the human community. Knowledge is pre-eminently practical and must become wedded to power: accordingly, "those twin objects, *human knowledge* and *human power,*" must be combined in that specific human activity, "invention" (technology). The whole point of Bacon's "race of inventions" was to overcome the "necessities and miseries of humanity … for the benefit and use of life" (cited in Jonas 1959, p. 189; see also Bacon 1996).

Realizing that knowledge and its "inventions" are governed by the uses to which they are put, for ill or good, he insisted that "charity and benevolence" are required for that governance. Despite that caveat, with modern technology developing directly from this marriage of knowledge and power, everything radically changed. Jonas later noted the "critical *vulnerability* of nature to man's technological intervention—unsuspected before it began to show itself in damage already done" (Jonas 1974, p. 9)—which is evident in the sheer scale and speed of current technological interventions, far outstripping the wits of the humanist, the scientist, and the politician. The same urgency is ingredient to the promise of genetic research: rectifying the ravages of genetic diseases, enhancing the quality of life for future generations, controlling violent behavior, and even altering our sense of death and, therefore, life.

Yet the riveting question remains, unposed and unresolved: from whence do the charity and benevolence come in order to govern these uses? What place is there for goals in such a view—in a later vernacular (Kohler 1934), whence the place of value in a world of fact? All of this invokes fundamental social, political, and ethical issues at the heart of medicine and biomedical research. The difficult issue, H.T. Engelhardt once urged, echoing Jonas, is that we have "become more technically adept than [we are] wise, and must now look for the wisdom to use that knowledge [we possess]" (2000, pp. 451–452). As Jonas urged:

> [We are] constantly confronted with issues whose positive choice requires supreme wisdom—an impossible situation for man in general, because he does not possess that wisdom, and in particular for contemporary man, who denies the very existence of its object: viz., objective value and truth. We need wisdom most when we believe in it least. (1974, p. 18)

This is serious business. What are we supposed to understand from the decoding of the human genome? Is there something that can shed light on the naked excitement and frank passion that grips so many people, from patients to physicians, scientists to reporters, clinicians to insurers? What we are learning will presumably reveal everything essential about humanity: me, you, him, her, them, now and in the past and future. The announcements stir such dramatic responses because the Project promises finally to resolve the ancient puzzle of who and what we human beings are: the answer is in our genes. But is all this actually true?

13.9 Genes and the Personal Enigma

These disputes are hardly new in another way, as was evident at the beginning of the IVF debate decades ago. Interpreting 'infertility' as a 'disease,' thus bringing it within the prevailing medical model, the IVF industry swiftly became pervaded with this "medical treatment." (Edwards 1974; Kass 1971). To this, Paul Ramsey (1972, pp. 1480–1483) retorted that IVF was little more than "manufacture by biological technology, not medicine." But Ramsey's unease did little more than ripple the advancing wave of excitement over the burgeoning cornucopia of ever-new kinds of technology. Against the rarely mentioned backdrop of Bacon's trumpeted call of centuries ago, the recent developments in biomedicine, especially genetics, have spawned a renewal in the idea that human life is finally coming to be fundamentally conquered, and therefore understood.

In a contribution to an anthology addressing our duties to others, for instance, Albert Jonsen tackles questions he believes are at the heart of modern genetics. He points out that we are now "far beyond what we humans have always known." Genetically, while "every man and woman is part of the main," at the same time and regardless of these "genetic communalities," each of us is yet a distinctive and unique individual (Jonsen 1994, pp. 279–291).

The question posed is at once deeply familiar, yet strangely novel; it is one that genetics forces on us due to increasingly available and accessible genetic information: the question of privacy, which harbors very much the same paradox Rheinberger had emphasized. On the one hand, if human life is taken to be fundamentally precious, then it seems odd at best to be so concerned about privacy. On the other hand, each person's privacy needs serious protections just because it can, and doubtless will, be invaded and compromised.

Jonsen addresses this peculiarity, it seems to me, by re-formulating the issue as including far more than the need for protections; it is to be understood, he contends, under the rubric of obligation. Thus, he asks, "under what conditions might an individual have an obligation to others to reveal genetic information?" (Jonsen 1994, p. 282). Jeffrey Reiman's (1976) proposal is, Jonsen believes, compelling. As "a social ritual by means of which an individual's moral title to his existence is conferred," privacy constitutes a "right" that "protects the individual's interest in becoming, being, and remaining a person."[19] This acknowledgment of our moral personhood, Jonsen argues, brings the central question to the front: underlying the perennial puzzles over 'personhood,' what constitutes the "individual" as such? In terms of privacy: "What constitutes the separateness that makes it possible to designate 'this person' and distinguish between 'this' and 'that' person?" (Jonsen 1994, p. 283; Campbell and Lustig 1994).

In a sense, Jonsen is clearly correct: the core questions of ethics, and surely bioethics, are raised by the very thing that seems to have advanced our understanding of ourselves: genetics. Each of us has his or her individual history, specific situation, and social context, and each of our biographies is at once 'my own' and 'yours;' what I am comes both from me and my history, and from others. So profoundly intermeshed is each person with multiple others from the earliest stirrings of life, that even the most cautious reflection confronts an enigma: even while 'I' and 'you' are clearly different, "where the me ends and the mine begins" still seems locked in mystery (Jonsen 1994, p. 284).

Although this deeply personal enigma is, I am convinced, the real issue buried within the Human Genome Project, Jonsen apparently agrees with many geneticists that the way out of the mystery has also become evident: "at the core of my 'individual substance' is a repeated molecular structure that is mine alone," but at the same time "much of that structure is the same in other individuals who have been generated by my ancestors and by my siblings." (Jonsen 1994, p. 284). The moral question of privacy is therefore a central issue and concerns whether genetic information about me in one sense belongs exclusively to me; yet, my genetic makeup is pretty much the same as others within my kinship group, and thus seems also to belong to these others. Since what might be diagnostically determined about me also gives information about them, Jonsen thinks that I am obligated to share what I learn about myself with them. Similarly, as you belong to my kinship group, you have a similar obligation to share with me whatever you might learn, accidentally or deliberately, about your genetic makeup.

Issues of screening and testing, disclosure and ownership, and confidentiality and privacy depend strictly on making sense of me and thee, and us and them. These in turn ineluctably bring Jonsen to the truly difficult question: "What constitutes the separateness that makes it possible to designate 'this person' and distinguish between 'this' and 'that' person?" (Jonsen 1994, p. 283). Although he is right on point with this question, it is with his response that matters quickly become very murky: that "I am my genome" merely retreats to the very notions that were supposed to lead to the key to the "mystery."

That turn, however, is not a response at all, since it presupposes precisely what was supposed to be at issue: Who is this "I?" And, since the "my" (in "my genome") signifies 'belongs to me,' what does 'belonging' signify? The fundamental question is not at all what it is to be "this" person or "this" self, to use Jonsen's terms, but rather, what is it to be *me-myself*? What is it to be not simply 'that' person or 'that' self, but *you-yourself*? To wish, as Jonsen does, to distinguish between "what is mine" and "what is yours," the central questions must first be pursued: Who and what am I? Who and what are you? Simply because people in a kinship share (whatever that may mean) some genomic characteristics does not in the least imply any duty to share genetic information. Indeed, the very identification of you and I as 'belonging to the same kinship group' already presupposes the core questions have been formed and pursued: the fundamental questions about this 'I' and this 'you' (as well as 'we,' 'them,' etc.).

Nor is Jonsen on target when he notes that each of us "is constituted by a body and by certain mental phenomena associated, in a still mysterious way, with that body," (1994, p. 283) for it is not merely 'a' or 'that' body; the question is what constitutes *my own animate organism as mine*, as what *embodies me-myself*? (Zaner 1981, pp. 144–241). Each embodied person has his or her own birth, history, specific situation, and social context, and each of our biographies is both emergent from our own experiences and assimilated from others (parents, siblings, friends, teachers, as well as the extant socio-historical milieu with its nexus of folkways, mores, laws, institutions, etc.). What is this 'body' that is 'mine' and even 'me' (if something strikes my body, it hits *me*), yet is so uncanny and perplexing ('this' hair, which is 'mine,' grows all by itself), even alien ('I' want to jump 6 ft, but 'my body' just won't do it)? 'My body' is 'mine,' yet what does 'belong' really mean? And how do other persons figure in what I am? How did 'I' get here, in this world, this place, this family, this body?

So profoundly intermeshed is each person with multiple others (who are also embodied) from the earliest stirrings of life on, that even the most cautious reflection seems stymied: even while 'I' and 'you' are clearly different, "where the me ends and the mine begins" seems locked not merely in mystery, but instead in a profound enigma. Indeed, the perplexing wonder here is a labyrinth at once deeply personal, and yet also social, historical, and even conceptual. Who are we? Who am I? How is each of us connected to that body we experience as our own? *These* are the actual, pressing issues nestled deeply in and driving the development of biomedical ethics, especially since the establishment of the Human Genome Project. It is only by getting clear about them that it becomes possible to make sense of such matters as privacy, confidentiality, promise-keeping, autonomy, beneficence, or any of the other questions that are so frequently posed in biomedical ethics—much less begin to make sense of any of the other ones posed by the beginnings or ends of our lives.

13.10 A Conclusion of Sorts

Our very uncertainty, our inability to know—e.g. not only the long-term effects of our technologies on future human life, but equally how interventions into embryonic life will affect embryos, how interventions into some piece of the genome may affect other pieces of it—surely underlies the possibility of making good sense of the idea of 'moral status.' Without, first of all, addressing both questions of the ends to which our splendid means should be placed, and, beneath that, the anthropology buried within our individual and collective sense of means and ends, there is no way to make sense of that 'status.'

All I have attempted here is to gain some degree of clarity about those matters and the wholly novel kinds of ethical questions that confront us as we probe, even gently, into human life at its beginning or its ending—both of which have become all the more problematic by the developments in the new genetics. We should probably realize, too, that if we accede uncritically to the excitement and challenge of the new, we then risk cutting ourselves off from the genuine tasks of medicine and biomedical research: that both are ultimately for the benefit of our common lives and preservation of our heritage, for which we desperately need that wisdom that, despite our best efforts, may yet be, as Jonas laments, beyond our ken.

Notes

1. The 26-year-old woman who suffered a single car accident in Missouri and, resuscitated by the emergency personnel who found her partially submerged in water, was eventually diagnosed with persistent vegetative state. Her parents finally decided to ask courts for permission to have the only life-support being used, a feeding tube, removed, allowing her to die. Opposed by so-called "right to life" groups, she did manage to die. Her parents had taken their request through the Missouri court system and then to the U. S. Supreme Court.
2. I think especially the continuing disputes over abortion, persistent vegetative state and, more recently, stem cell research.
3. See, for instance, MacIntyre (1981, the first few pages). See also the lovely, sharp essay by Toulmin (1982, pp. 736–750).
4. I first became "involved," as is said, in 1971, when I was invited by Edmund Pellegrino to become Director of the newly established Division of Social Sciences and Humanities in The Health Sciences Center at The State University of New York at Stony Brook. Only much later did it become clear to me that it would be essential, to me and, I thought, to this "field," to become an actual clinical practitioner; I was finally able to take up the challenge at Vanderbilt University Medical Center in 1981.
5. Burnett (1978); Eccles (1970); Eccles (1979); Penfield (1975)—to mention only a few books replete with announcements of the new, different and dramatic discoveries being made almost daily.
6. See, for instance, Beecher, the first Dorr Professor of Research in Anesthesia, at Harvard Medical School (1959a, b); also, his widely influential article, "Experimentation in Man," (1959a, b, pp. 461–478).
7. This common attitude was displayed, for example, by Grant Liddle (1967, pp. 1028–1030) in his outgoing Presidential Address to the Society for Clinical Investigation, and the subsequent article, "The Mores of Clinical Investigation."
8. See especially Howard-Jones (1982, pp. 1429–1448); Curran (1982); and, later, Annas and Grodin (1992).
9. Although just why vulnerability should function so powerfully was not made thematic as a major phenomenon in ethics until much later.
10. See, for instance, Pellegrino (1979), which contains a collection of his articles from the 1960s to the late 1970s.
11. After all, it is no mystery that the most demandingly practical of human enterprises—clinical medicine—found itself calling on what is, along with poetry, surely among the most impractical of disciplines, philosophy. The challenge to both still sets the tone for many of their interactions.
12. See Wilmut et al. (1997, pp. 810–813) and Pence (1998).
13. Buittiglione (1994, p. 20); also Concetti (1993; 1994, p. 21).

14. Cited in Kass (1997); also (1972). Lederberg's argument appeared in an article published in the *Washington Post*.
15. Just this is the attitude of Leon Kass, the chair of the President's Commission on Bioethics under President George W. Bush; see Kass (1997, pp. 17–26). Kass' article was later published along with a supposed rebuttal article by James Q. Wilson (1998, pp. 3–59).
16. See especially Eccles (1970) and his numerous references, in particular in Chapters I and IV.
17. See The President's Council on Bioethics (2002, Chapter 2). The German embryologist and Nobel Prize winner, Hans Spemann, conducted what many consider to be the earliest "cloning" experiments on animals (1938).
18. Bacon's theme is a principal topic of this fine essay.
19. Reiman (1976, p. 44); cited in Jonsen (1994, p. 283).

References

Anees, M.A. 1994. Human clones and God's trust: An Islamic view. *NPQ: New Perspectives Quarterly* 11: 23–24.

Annas, G.J., and M.A. Grodin (eds.). 1992. *The Nazi doctors and the Nuremberg code*. New York: Oxford University Press.

Bacon, F. 1996. *Great instauration and the Novum organum*. Whitefish: Kessinger Publishing.

Beecher, H. 1959a. *Experimentation in man*. Springfield: Charles C. Thomas.

Beecher, H. 1959b. Experimentation in man. *Journal of the American Medical Association* 169: 461–478.

Blish, J. 1957. *The seedling stars and galactic cluster*. Hicksville: Gnome Press, Inc.

Buittiglione, R. 1994. Immoral clones: A Vatican view. *NPQ: New Perspectives Quarterly* 11: 20.

Burnett, M. 1978. *Endurance of life*. Cambridge: Cambridge University Press.

Campbell, C., and A. Lustig (eds.). 1994. *Duties to Others*. Dordrecht: Kluwer Academic Publishers.

Concetti, G. 1993/1994. Editorial. *L'Osservatore Romano*. Reprinted in *NPQ: New Perspectives Quarterly* 11: 21.

Curran, W. 1982. Subject consent requirements in clinical research: an international perspective for industrial and developing countries. In *Human experimentation and medical ethics*, ed. Z. Bankowski and N. Howard-Jones, 35–79. Geneva: Council for International Organizations of Medical Sciences.

Eccles, J. 1970. *Facing reality*. Berlin: Springer.

Eccles, J. 1979. *The human mystery*. Berlin: Springer.

Edwards, R.G. 1974. Fertilization of human eggs in vitro: Morals, ethics and the law. *The Quarterly Review of Biology* 40: 3–26.

Engelhardt, H.T. 2000. *The philosophy of medicine*. Dordrecht: Kluwer Academic Publishers.

Gilbert, W. 1992. A vision of the grail. In *The code of code: Scientific and social issues in the Human Genome Project*, ed. D.J. Kevles and L. Hood, 83–97. Cambridge, MA: Harvard University Press.

Gorovitz, S. 1982. *Doctors' dilemmas: Moral conflict and medical care*. New York: Macmillan.

Gorovitz, S., et al. (eds.). 1976. *Moral problems in medicine*. Englewood Cliffs: Prentice-Hall, Inc.

Hall, J.L., D. Engel., P.R. Gindoff, et al. 1993. Experimental cloning of polyploid embryos using an artificial zona pellucida. The American Fertility Society jointly with The Canadian Fertility and Andrology Society, Program Supplement, *Abstracts of the Scientific Oral and Poster Sessions*: S1.

Hauerwas, S., and D. Burrell. 1977. From system to story: An alternative pattern for rationality in ethics. In *Why narrative? Readings in narrative theology*, ed. S. Hauerwas and L.G. Jones, 158–190. Grand Rapids: William B. Eerdmans Publishing Company.

Howard-Jones, N. 1982. Human experimentation in historical and ethical perspectives. *Social Science and Medicine* 16: 1429–1448.

Jonas, H. 1959. The practical uses of theory. *Social Research* 26: 127–134.
Jonas, H. 1974. Technology and responsibility: Reflections on the new tasks of ethics. In *Philosophical essays: From ancient creed to technological man*, ed. H. Jonas, 3–30. Englewood Cliffs: Prentice-Hall, Inc.
Jonsen, A. 1994. Genetic testing, individual rights, and the common good. In *Duties to others*, ed. C. Campbell and A. Lustig, 279–291. Dordrecht: Kluwer Academic Publishers.
Kass, L. 1971. Babies by means of in vitro fertilization: Unethical experiments on the unborn? *The New England Journal of Medicine* 285: 1174–1179.
Kass, L. 1972. New beginnings in life. In *The new genetics and the future of Man*, ed. M.P. Hamilton, 13–63. Grand Rapids: William B. Eerdmans.
Kass, L. 1997. The wisdom of repugnance. *The New Republic*, June 2, 16–23.
Kass, L., and J.Q. Wilson. 1998. *The ethics of human cloning*. Washington, DC: The AEI Press (American Enterprise Institute).
Kohler, W. 1934. *'The place of value in a world of fact', in the William James lectures*. Cambridge, MA: Harvard University Press.
Liddle, G. 1967. The mores of clinical investigation. *The Journal of Clinical Investigation* 46: 1028–1030.
MacIntyre, A. 1981. *After virtue*. Notre Dame: Notre Dame University Press.
Martin, S. 1972. The new healer. In *Proceedings of the 1st session, Institute of Human Values in Medicine*, 5–27. Philadelphia: Society for Health and Human Value.
McCormick, R. 1994. Blastomere separation: Some concerns. *The Hastings Center Report* 24: 14.
Mumford, L. 1964/1970. *The myth of the machine, vol. I: Technics and human development; vol. II: The pentagon of power*. New York: Harcourt Brace Jovanovich.
Pellegrino, E. 1979. *Humanism and the physician*. Knoxville: University of Tennessee Press.
Pence, G. 1998. *Who's afraid of human cloning?* New York: Rowman & Littlefield.
Penfield, W. 1975. *The mystery of the mind*. Princeton: Princeton University Press.
President's Council on Bioethics. 2002. *Human cloning and human dignity: An ethical inquiry*. Washington, DC: President's Council on Bioethics.
Ramsey, P. 1972. 'Shall we 'reproduce'? II. Rejoinders and future forecast. *Journal of the American Medical Association* 220: 1480–1485.
Reiman, J. 1976. Privacy, intimacy and personhood. *Philosophy and Public Affairs* 6: 26–44.
Rheinberger, H. 1995. Beyond nature and culture: A note on medicine in the age of molecular biology. *Science in Context* 8: 249–263.
Robertson, J.A. 1994. The question of human cloning. *The Hastings Center Report* 24: 6–14.
Snow, C.P. 1993. *The two cultures*, Canto edition. Cambridge: Cambridge University Press.
Spemann, H. 1938. *Embryonic development and induction*. New Haven: Yale University Press.
Tempkin, O. 1973. *Galenism: Rise and decline of a medical philosophy*. Ithaca: Cornell University Press.
Toulmin, S. 1964/1990. *Cosmopolis: The hidden agenda of modernity*. New York: The Free Press.
Toulmin, S. 1982. How medicine saved the life of ethics. *Perspectives in Biology and Medicine* 25: 736–750.
Wilmut, I., A.E. Schnieke, J. McWhir, A. Kind, and K. Campbell. 1997. Viable offspring derived from fetal and adult mammalian cells. *Nature* 385: 810–813.
Zaner, R. 1981. *The context of self*. Athens: Ohio University Press.
Zaner, R. 1984. The phenomenon of medicine: Of hoaxes and humor. In *The culture of biomedicine*, ed. D.H. Brock, 55–69. Newark: University of Delaware Press.
Zaner, R. 1998. Surprise! You're just like me! Reflections on cloning, eugenics, and other utopias. In *Human cloning*, ed. J. Humber and R.F. Almeder, 103–151. Totowa: Humana Press.

Chapter 14
Medical Ethics and Moral Philosophy in an Era of Bioethics

Edmund D. Pellegrino

> *Moral philosophy, in a way and to a degree that has some but few historical parallels, has become recognized by those who are not professional philosophers as an important form of inquiry*
>
> *(Alasdair MacIntyre 1983).*

14.1 Introduction

Medical ethics, the traditional professional ethic of the physician, is some 2,500 years old. Bioethics, its contemporary rival and partner, is barely half a century old. Each functions under the rubric of "ethics"; each purports to provide guidance to society on the moral uses of the unprecedented powers of modern biology. Yet, each construes "ethics" in its own way. Neither supports a robust moral philosophy with which to justify its moral dicta or define the boundaries of its moral pretensions.

Whether medical ethics and bioethics will in the future coalesce under a common definition of ethics is problematic. Whether one will absorb the other or each be replaced by some new form of ethics is equally uncertain. The need for a robust moral philosophy to provide rational answers to these questions grows while the intellectual status of moral philosophy itself declines.

What seems clear is that bioethics seems destined to continue its prodigious growth and influence while its ancient predecessor, medical ethics, is undergoing slow but definite erosion. Both, however, are moving progressively away from classical ethics as the disciplined study of right and wrong, and good and bad human conduct. Each seeks normative guidance beyond philosophy. As a result, customs, values, affects and behaviors too easily become moral norms without principled justification.

E.D. Pellegrino, Ph.D., M.A.C.P. (✉)
Center for Clinical Bioethics, Georgetown University, Box 571409, 4000 Reservoir Road NW, Suite 238, Washington, DC 20057-1409, USA
e-mail: patchelm@georgetown.edu

J.R. Garrett et al. (eds.), *The Development of Bioethics in the United States*, Philosophy and Medicine 115, DOI 10.1007/978-94-007-4011-2_14,

This essay examines this evolving state of affairs from four points of view: (1) The sociocultural environment which begat bioethics and reshaped medical ethics; (2) The evolving character of bioethics as a new field of study; (3) The transformations of traditional medical ethics by the same sociocultural forces that begat bioethics; (4) The place, if any, of moral philosophy in preserving ethical integrity of both medical and bioethics.

14.1.1 The Sociocultural Context Within Which Bioethics Was Born

It is always hazardous to extract selectively from any sociocultural matrix those forces we take to be responsible for a putative shift in world view. In the birth of bioethics and the transformations of medical ethics in the mid-sixties of the last century two sources for change seem to have been most crucial. One is the unprecedented expansion of power of biological science, and the second was the transformation of sociocultural attitudes away from traditional sources of moral, spiritual and political authority.

14.1.2 The Unprecedented Expansion of Bioscience and Biotechnology

Scientific knowledge of human biology has been growing rapidly in the West at least from the sixteenth century. In the mid-twentieth century every aspect of the human body, from the molecular to the psychic, became subject to the probings of the scientific method. The depth and power of today's bioscience have made it a most powerful agent for social and cultural change. As the twentieth century came to a close, a cascade of discoveries enabled humans to cure disease, prevent illness, and reshape their genetic and cellular identities in the future. As humans took more control over their own destiny, the moral questions they faced grew more numerous and more complex. Should we do all that biotechnology permits us to do? Are there any moral boundaries we must never cross? Are there things we should never do? Who has authority to answer such questions? Is it science itself, popular opinion, physicians, the church, government, or is every individual his own judge? Have our moral values simply become the domain of the scientific method (Harris 2010; Hawkins and Milodanimow 2010)?

14.1.3 The Drift Away from Traditional Moral Authority

As the power of biological science to re-shape human life grew, the gravity of moral decisions in its use also grew proportionately. At the same time respect for the authority of traditional sources of morality—custom, formal religion, the prestige and authority of the medical profession, or government, became less influential.

Increasingly the Western world began to look to empirical science for its moral, as well as, its technical compass points. Many religious believers became more tolerant of secular ideas and favored more latitude in moral choice. Others found their "spirituality" outside the confines of formal church affiliation. Willingness to accept personal moral guidance declined concurrently. Looking to the future, some foresaw a "post secular" society in which religion and secular world views would live side by side (Eder 2002).

The ancient domains of religion—e.g. reproduction, death, dying and sexuality became secular matters, open to public and personal choice. The illusion of a disease-free, pain-free, even a death-free future, took hold of the public imagination. What was legal increasingly became the measure of what was morally and socially acceptable. For many, the successes of science seemed more realistic than the promises of a spiritual future. Moral relativism became more attractive. Transcultural and global issues in bioethics were more often the domain of technological resolution.

In this anti-authoritarian milieu, the traditional moral deontologic commitments of the medical profession to the Hippocratic Ethos were judged as culturally outmoded. Many urged that the physician's moral focus on the individual patient should be broadened to include cultural and social issues (Fox and Swazey 2008; Dzur 2008). Physicians came under pressure to provide the putative opportunities of biological science by their patients. The latent Promethean promise of human control of human destiny on human terms seemed a genuine, even an imminent, possibility. An inchoate illusion of a disease-free, pain-free, immortality through science encouraged many to support the new biology, unreservedly.

Physicians wedded to the precepts of traditional medical morals together with the teachings of established religions were seen as obstacles to human progress. Answers to the most profound ethical dilemmas, and definitions of the good life, were transformed from ethics to the arena of social or political ideology. The Enlightenment philosophy of the eighteenth century entranced moralists whose driving ideals were freedom and liberty. For them moral restraints were arbitrary obstacles to the imminent human progress they saw in biotechnology. The freedom of individual choice moved from private life to social and public demand for access to each new medical technology as it appeared. For many the sovereignty of personal values begin to replace more general moral precepts.

These trends were enhanced by the concomitant surge of civil and consumer rights and participatory democracy. The demand for patient autonomy in medical ethics strengthened these trends (Veatch 2009). The principle of autonomy provided re-affirmation and re-enforcement for the dominant social trends toward personal interpretation of moral values.

14.2 The Birth and Evolution of Bioethics

Bioethics, in America, was born into this sociocultural ambiance. Several academic loci, each with its special emphasis, served as the birthplace of its origin. One was Georgetown University's Kennedy Institute of Ethics, the focus of which was bioethics as an extension of traditional medical ethics on a philosophical foundation. A second locus was at the University of Wisconsin where the orientation was toward the ethical issues of ecology, environment and the global challenge of technology. The third was the Hastings Center which took a broad-based, generally communitarian approach, guided by an ideal of the Common Good. A fourth was the Center at the University of Texas at Galveston, where the humanities and interdisciplinarity were emphasized. Each founding center contributed its particular view of ethics in the emerging field of bioethics. Each center provided a model embodied or reshaped by other later bioethics centers.

From the beginning, the relationship between the new bioethics and traditional medical ethics was tentative and occasionally querulous. A wide range of opinions about what bioethics should be was evident even a few years after its appearance. A 20-year retrospective on the meaning of philosophy of medicine and bioethics was sponsored by the University of Texas Medical Branch in Galveston in 1997. It was attended by some of the early workers in the field (Carson and Burns 1997). A sampling of their recommendations reveals the wide range of shapes the contributors thought bioethics should take: Toulmin (1997) urged emphasis on clinical practice rather than theory; Wartofsky (1997) opined that the study of medicine was essential to any sound epistemology; Engelhardt (1997) urged bioethics to "…wield political influence in direct day by day decision making"; ten Have (1997) placed emphasis on national traditions and cultural values; Churchill (1997) opted for a deeper immersion in social contexts; Carson (1997) called for better communication between patients and caregivers; Jones (1997) urged a turn to narrative ethics; still others made pleas for inclusion of anthropology, ethnography, empathy, law, policy or economics.

In my contribution I called for a more rigorous adherence to classical philosophical ethics. I, like others, argued that bioethics must be an interdisciplinary enterprise to be effective. But I argued that it must also be faithful to ethics as the formal study of right and wrong, good and bad human conduct (Pellegrino 1997). I emphasized the contributory importance of the humanistic disciplines for the range of concrete, existential detail they could add. But I warned that the meaning of bioethics and medical ethics as ethical enterprises would be lost if either abandoned the centrality of philosophical ethics. In 1981 Thomasma and I argued for a moral philosophy of medical practice as the grounding for medical or bioethics. We did not define a separate moral philosophy for bioethics but implied that it was a broader field than medical ethics but not one that absorbed or replaced medical ethics (Pellegrino and Thomasma 1981).

14.3 The "New" Progressivist Bioethics

The latest development of consequence in the evolution of bioethics has been the emergence of "progressivist" bioethics. This development moves bioethics away from traditional medical ethics, toward direct participation in social and political policy formulation. Ethics involving society's many possible uses of biological knowledge and biotechnology took center stage. The breadth and commitments of "progressivist bioethics" are well summarized in a recent collection of articles edited by Jonathan Moreno and Sam Berger (2010). These articles reveal a wide panorama of ways progressivist bioethicists might combine the values of liberalism and pragmatism to advance their socio-political agendas. Harold Shapiro, former chairman of President Clinton's bioethics commission, defined this view as a "distinctive attitude":

> ...a disposition to embrace change as a vehicle to enhance the social, political, and economic conditions they believe to be appropriate for allowing individuals, and through them society, to realize, in the fullest manner, their individual and joint destinies (Shapiro 2010, p. xi).

In a variety of different ways each of the contributors to the Moreno-Berger volume grounded progressivist bioethics on the values of liberalism. Berger and Moreno give emphasis to "...social justice, protection of the least among us, and engagement with everyday problems" (2010, p. 21). Lempert sees the need to develop "...political effectiveness needed to advance the application of sound bioethical principles to public life" (2010, p. 42). Charo, on the other hand, clearly puts political philosophy before moral philosophy (2010, p. 49). Hinsch urges progressives to embrace a more compelling values-based approach not limited to bioethical values (2010, p. 86). Zoloth asks for a deeper exploration into what it is to be human, to be free and to suffer (2010, p. 103). Wolpe urges professionalization, the better to provide a forum for debate and legitimization of the biotechnology industry (2010, p. 110), while Darnovsky calls progressives to promote regulation, oversight, social justice and the common good (2010, p. 212). Caplan forthrightly urges progressives "...to learn to live with power and operate from an explicit ideological perspective" (2010, p. 223).

It would make an interesting study in its own right to compare the conceptual differences between recommendations in the earlier Galveston volume with those in the recent Moreno/Berger volume. A little over a decade separates these two studies. References in both are to a wide spectrum of sociological, cultural, or politico-economic agendas. In both volumes "ethics" extends to transnational sociocultural issues. Both reports, and especially the Moreno/Berger volume, should give some solace to those social scientists who have strongly criticized bioethics for its overly philosophical emphases (Fox and Swazey 2008).

From its origins bioethics has had a salutary effect on various aspects of health and medical care. New dimensions of expertise and opinion emerged from better public awareness of the practical problems and processes of applied biology. Democratization of the debates in public and policy venues engaged public interest in the quality and level of medical care. National councils, commissions, and cooperative policy efforts alerted the public on a global scale to the benefits and dangers of commercialization, unregulated competition and the uncritical promotion of therapeutic panaceas. The new bioethicists also alerted the public to the need to avoid the extremes both of technological utopianism and irresponsible technophobia. Policy makers were alerted to the need for public control of decisions that affect humanity on a broad scale. In 2005 the UNESCO International Bioethics Committee released its report on bioethics (2009) and human rights emphasizing the international significance of bioethics.

These positive contributions notwithstanding, improving our knowledge of the socio-political and global dimensions of either bioethics or medical ethics is insufficient to qualify as "ethics" properly speaking. Neither deontological nor praxiological guidance can be gained solely from sociocultural data. The danger of conflation of custom and politics with morals is genuine. To qualify as "ethics," which the names of both bioethics and medical ethics imply, each must attend to the moral desiderata of good and bad, right and wrong. In the end, ethical discourse must go beyond activism or political ideology if it is to give moral substance to its action or policy proposals. This caveat applies equally to "conservative" as it does to "progressivist" bioethicists. Both must derive ethical credibility from ethical deliberation—not earnest commitment however genuine.

Where, one may fairly ask, is the ethics, to say nothing of the moral philosophy in bioethics or medical ethics today? McCullough and Baker (2007) have given what I think is an accurate answer. They find it in "appropriations" from existing philosophical systems, preferentially selected to justify some change in bioethics. These appropriations, important as they are by themselves, are insufficient to sustain the depth and integrity of the analysis valid ethics or valid moral philosophy require.

Lacking these ethical constraints, public policy formulations about right and wrong policy easily become polarizing ideologically based political agendas. The more "ethical" justifications are intermingled with activist agendas, the more distant they become from a valid notion of ethics. The trend of progressivist bioethics away from deontologic, axiologic, or moral justification inevitably tends toward activist exhortation. Indeed the task of bioethics, in general, is to sort out the wide range of emphases outlined in the Galveston and Moreno-Berger volumes if their contributions are to be most helpful in public deliberations about the optimal usages of the powers of biotechnology.

14.4 The Erosion of Medical Ethics

Medical ethics, traditionally understood as the professional ethic of physicians, has not done well in the era of bioethics. This is not because of any intentional or unintentional influence of bioethics. Rather, it is the result of the same confluence of sociocultural forces which begat bioethics. At the beginning of the era of bioethics the Hippocratic ethic was widely accepted as the moral signature of the physician's profession and the measure of its commitment to the welfare of the patient. Most physicians then, took the Oath with its moral commitments in classical form as their guide to ethical practice. Even physicians who had reservations still regarded medicine as a vocation and a way of life. They still spoke of "my" responsibility to "my" patient while patients still spoke warmly of "my" physician. Seriously to violate the Oath was to become a moral pariah.

The moral status of the Oath today is drastically different. The preamble of the Oath is undeniably sexist and has the trappings of a secret society. It spoke only of loyalty between "brothers" and "sons" with no mention of "daughters" or "sisters." Some critics now argue against the taking of a solemn Oath because they say it interjects religion; others challenge beneficence as the first principle of professional ethics replacing it with non-maleficence or respect for autonomy. In 1973 the prohibition against abortion was eliminated. Later the prohibitions against assisted suicide and euthanasia were omitted; sexual relations with patients or their families were treated more tolerantly; confidentiality was subordinated to patient demand not the safety of third parties; a "life of virtue" was no longer expected. Infidelity to the Oath ceased to be a warrant for self recrimination or rejection of one's peers. Historians and ethicists questioned the authorship, dating, provenance, content and interpretation of each precept. The patient-physician relationship is as a result, being transformed from a moral covenant to a contractual or commodity transaction. The Oath continues to be taken but in a widening variety of forms (Orr et al. 1997).

Clearly, the sexist preamble and the oath to pagan deities are justifiably eliminated. However, what is being eroded but not yet lost entirely is an ancient moral edifice. For centuries this edifice characterized the physician's calling across national, cultural, geographic and religious boundaries. It provided patients with a reason to trust in the profession. It set high ideals and imposed serious moral obligations. It became, as Edelstein said, "…the nucleus of all medical ethics" (1962, p. 63, italics added).

The Hippocratic ethic was first recorded in the West in the work of Scribonius Largus, a medical attendant to the Emperor Claudius in the first century A.D. He epitomized it in these morally charged terms:

> All men and gods, in fact, should despise any physician whose heart is not full of humanity and mercy according to the purpose of his profession (*secundum ipsius professionis voluntatem*). It is because of these feelings, that a physician bound by the proper Oath of medicine will give no poison (*alum medicamentum*, literally a harmful drug) even to an enemy… Medicine in fact does not measure men's worth by their fortune or personal qualities, but offers to help all who seek it and promises never to injure anybody (Pellegrino and Pellegrino 1988; Largi 1983).

This is indeed a morally lofty commitment and over the centuries not all oaths nor all physicians aspired to this level of commitment.

Still, Scribonius' statement is for several reasons pertinent to the theses in this paper. It is the first recorded reference to the term "profession" in its original Latin meaning of a promise, declaration or commitment usually made publically. The Oath of Hippocrates in this sense is indeed an act of "profession," i.e. a public commitment (Guillen 2010; Davey 2001). Second, Scribonius correctly identifies beneficence as the primary principle of the physician. It is not "do no harm" as so many modern commentaries erroneously assert. Third, Scribonius makes humanity and humanism part of the physician's commitment. Many today erroneously claim "humanism" in medicine to be the product of the "holism" of contemporary medicine.

Many today would say Scribonius' level of commitment sets the bar too high, that it is out of harmony with today's more relaxed moral climate. Is Scribonius' commitment to beneficence, to protecting patients, and to treat patients with equal attention really too high? Or, is this not the level of commitment the existential realities of the patient physician encounter require?

To be sure, there have been recurrent periods of moral failure when Hippocrates and Scribonius' ideal of a profession was so seriously violated that morally sensitive physicians felt compelled to set themselves apart from their colleagues. One such example was the sorry state of the ancient Greek profession which prompted the Hippocratics to separate themselves from their colleagues, and to compose their communal Oath of fealty to care of the sick. Another example was the development of the AMA code of professional ethics in 1847 (Pellegrino 2000). A group of morally responsive physicians, like their ancient counterparts the Hippocratics, set themselves apart from their errant contemporaries. Their code emulated and extended the Hippocratic ethic which survived intact until its latest emendations to adapt to societal acceptance of abortion and assisted suicide.

There were other times when the moral precepts of the Hippocratic ethic were violated, rephrased, or selectively twisted to suit predetermined social, scientific or political purposes. The most egregious historical example was the complicity of the Nazi physicians in the heinous crimes of the Holocaust, their experimentation with prisoners and euthanasia of the unfit (Baumslag 2005; Pellegrino and Thomasma 2000). Equally horrendous were the contemporaneous subversions of medical ethics by the totalitarian ideologies of the Stalinist and Maoist regimes. Nor should we forget the open violations of Hippocratic ethics in our own country at Tuskegee, Willowbrook and others (Krugman 1986; Jones 1981).

The history of traditional medical ethics is remarkable for its combination of fragility and durability. Any moral system so durable in the face of repeated defections must contain a durable core of moral truth to survive. Human society itself depends for survival on some number of its members being dedicated to a life beyond raw self-interest. It is from a similar core of ethically responsible physicians that the Hippocratic ethic received its moral impetus for survival.

A significant portion of the fragility of the Hippocratic ethic lies in the method of its justification. From the beginning its precepts were sincere but free assertions of moral commitments. They were taken for granted but unjustified by robust rational demonstration. They were therefore vulnerable to attack by equally free assertions to the contrary. As the moral climate of each era shifted, the Hippocratic ethic was accordingly strengthened or weakened.

Today's societal mores are reshaping and eroding the traditional foundations of medical ethics. This is reflected in the piecemeal dissection and revision of the Hippocratic ethic. As a result, the ethical concepts of an Oath or covenant of duty and obligation have given way to changing social contracts whose legitimacy rests on popular or legal approval rather than settled moral principle. As a result some argue forcefully that the physician patient relationship should be defined not by ethical principle but by ethnography, anthropology and the social sciences.

Bioethicists are advised thus to move away from the biomedical model to an ethic defined by community, culture and societal construction (Dzur 2005).

For some today professionalism replaces the Oath or act of moral obligation by a "charter," expressed in a set of behaviors like altruism, compassion, integrity, honesty, etc. (American College of Physicians et al. 2002, p. 136). These behaviors are admirable but they lack the deontological force of a public oath publicly committing the physician to act primarily for the good of the sick. Professionalism is well intentioned educational device designed to help the student re-define himself to fill society's role of "doctor" (Smith 2005). Personally accepting this social role is presumed to convert the student from being a "layperson" to being a "physician."

Professionalism focuses strongly on the "nurture" of nascent physicians. It tends to adjust its ethics to the differences between the expectations of students and those of their forebears. The generational difference in lifestyles and expectations impels educators "...to redefine excellence and professionalism in terms that are both generationally diverse and appropriate" (Smith 2005, p. 441). Whether this shift of emphasis will be conducive to ethically sensitive patient care is problematic. The ultimate test of professionalism is whether it better prepares medical students to serve patients competently and with ethical sensitivity (Smith 2005; van Mock et al. 2009; Braddock et al. 2002; Hafferty and Castellani 2010). Professionalism is not an adequate substitute for a well grounded ethics of medicine. One hope is that professionalism with its emphasis on positive affective traits and medical ethics with its emphasis on moral duties can re-enforce each other.

14.5 Medical Ethics, Bioethics and Moral Philosophy

> If bioethics is the appropriate label for most of the studies undertaken in the philosophy of medicine, then the majority of bioethics studies can no longer be characterized as moral philosophy (ten Have 1997, p. 105).

The trajectories of both medical ethics and bioethics in the last half century have drifted away from classical ideas of medical ethics. That drift revives interest in the oldest question in ethics, medical or otherwise, i.e. is there, or can there be a rationally defensible theory of right and wrong? Such a theory could help to mitigate the erosion of medical ethics and the drift of progressive bioethics away from ethics classically considered. Both medical ethics and bioethics need to recapture the concept of ethics as a discipline—recognizing the importance of input from the other humanist and sociological disciplines but not being displaced by them.

However, since Prichard (1912) argued that moral philosophy rested on a "mistake", it has been cast into intellectual limbo. Prominent moral philosophers, like Hampshire (1949), MacIntyre (1983) and Anscombe (1958) have doubted its claim to ethical credibility. As a result, questions of moral philosophy have been reduced to debates about semantics, metaethics or the relative merits of the major philosophical systems. Lacking a credible, agreed-upon moral philosophy, medical and bioethics have turned increasingly to public opinion surveys, ethnographic and political ideologies, and political preferences for guidance in their recommendations. The tendency to convert custom, public opinion, politics and science into morals has increased proportionately.

These criticisms of moral philosophy to the contrary, moralists and bioethicists often mistake Beauchamp and Childress' system of common morality as a moral philosophy. Beauchamp and Childress (2009) themselves make it clear that they do not offer common morality as a moral philosophy. They argue clearly that: "…general theories should not be expected to yield concrete rules or judgments capable of resolving all contingent moral conflicts" (2009, p. 397). They do not rule out moral theories and have even encouraged them in defense of common morality. They put their preference this way: "We also have reason to trust norms in the common morality more than norms found in general theories" (2009, p. 397).

In some sense the concept of common morality does serve as a "moral philosophy" in Beauchamp and Childress' system. It is at least a philosophical stance within which certain prima facie principles serve as moral guideposts. It is notable here that Beauchamp and Childress share some of the difficulties Aristotle saw in deriving principles of moral rightness from other general moral principles. Aristotle (1109) insisted that particular moral decisions cannot be made outside the existential realities within which a particular moral decision must be made:

> It is not easy to determine how one ought to become angered with whom and on what sort of grounds and for how long, or up to what point one acts rightly or goes wrong … it is not easy to declare in a rule (toi logoi) by how much or what manner or variance a person becomes blameworthy. For the decision depends on the particular facts and upon perception (1109, pp. 14–23).

In these words Aristotle asserted his belief that principles of moral rightness are embedded in the concrete realities within which the agent must decide. Looked at another way, the product of practical reason (the virtue of prudence) must be reconciled with the intellectual virtues. Cooper (1986) comments *in extenso* on the nuances of Aristotle's perception as it appears in the *Nicomachean Ethics*, the *Eudemian Ethics* and the *Magna Moralia*.

Some years ago, mindful of Aristotle's dictum about practical moral decisions, Thomasma and I proposed a philosophical basis for medical practice grounded in the existential phenomenology of the patient-physician relationship (Pellegrino and Thomasma 1981; Pellegrino 2003, 2007). Over the years, together and individually, we have expanded the concept beyond medicine to include the other health professions, as well as ministry, law and education. In all of these practices prudential decision, which is the virtue of practical reason, ultimately finds its justification in the intellectual virtues.

A serious moral philosophy of medicine must encompass the ethical realities of medicine as a societal phenomenon as well. Some of the relevant questions in this domain include these: Is medicine ultimately constructed politically and economically? Are the multiple roles of physicians as businessmen, corporate employees, bureaucrats, entrepreneurs or proletarians and healers morally compatible? To whom in the end is the physician bound by a moral covenant? How do these multiple roles relate to each other and in what order of priority?

What are the ethical implications of our concepts of disease and health, normality, and causality? Medicine is often society's instrument through which its concepts of human life, death, suffering, and social worth are expressed. Medicine can provide empirical data for the elaboration of a philosophical anthropology. In what way will the religious and spiritual beliefs of Americans be given ethical voice in the post secular society now emerging?

These questions will confront all Americans in the immediate future as the reforms of the Affordable Health Care Act are implemented. The simultaneous pursuit of affordability, quality, access, accountability and commercial interest will stress the moral bonds that bind the sick to those who profess to heal them. Conflicts between the needs of the sick and the bureaucratic mechanisms devised to meet them always weigh most heavily on the most vulnerable—the very young, the very old, the very sick, the disabled and the powerless.

The recurrent and unavoidable central ethical question for any "reform" is the impact on those it presumes to help. As the United States engages the imminent moral challenge of reform it must finally ask this question. How we as a people answer it will reveal what kind of people we are and want to be. This is the question any credible medical or bioethical enterprise must struggle with. As it does, it will confront the foundational challenge Danner Clouser (1978) posed early in the era of bioethics: "We have only to scratch the surface of medical ethics and we break through to the issues of standard ethics as we have always known them" (p. 116).

To confront these issues of standard ethics, bioethics must be aware that as Loretta Kopelman (2006) said, it is a "second order discipline," lacking a method of its own. Medical ethics will have to remember its own history of success and failure. Both must remain true to the "ethics" their names connote if they are to serve society wisely and well. In that effort it seems reasonable that the possibilities of moral philosophy be revisited by each without denying the contributions of the social sciences and humanities to its maturation. To this end, both medical ethics and bioethics must each ground its social, political and economic advocacy on sound moral reason.

14.6 Conclusion

Bioethics emerged from medical ethics a half century ago as the ethical issues associated with the uses of biotechnology exceeded the ambit of medical ethics. Today medical ethics and bioethics overlap, the influence of bioethics growing and that of medical ethics contracting. To optimize their contributions to society each must define its nature and the boundaries of its pretensions. Each must be more attentive to the rubric of "ethics" it bears. Each must search for a moral philosophy on which to define its identity if they are not to convert ethical issues into problems of political or social advocacy.

References

American College of Physicians, American Society of Internal Medicine, and European Federation of Internal Medicine. 2002. Medical professionalism in the new millennium: A physician charter. *Annals of International Medicine* 136: 243–246.

Anscombe, G.E.M. 1958. Modern moral philosophy. *Philosophy* 33: 1–19.

Aristotle. 1109. NE IV 5, ii26a32-b4, 14–23. In *Nichomachean ethics, the complete works of Aristotle, the revised Oxford translation,* ed. J. Barnes, li, 26a–32b4. Princeton/Bollingen Series LXXXI–2. Princeton: Princeton University Press, 1984.

Baker, R.B., and L.B. McCullough. 2007. The relationship between moral philosophy and medical ethics. *Kennedy Institute of Ethics Journal* 17: 271–276.

Baumslag, N. 2005. *Murderous medicine*. Westport: Praeger Publishers.

Beauchamp, T.L., and J. Childress. 2009. *Principles of bioethics*, 397. New York: Oxford University Press.

Braddock, C.H., S.L. Linas, and W. Levinson. 2002. A behavioral and systems view of professionalism. *The Journal of the American Medical Association* 304: 2732–2737.

Caplan, A.L. 2010. Can bioethics transcend? (And should it?). In *Progress in bioethics, science, policy and politics*, ed. J. Moreno and S. Berger, 223. Cambridge, MA: MIT Press.

Carson, R.A. 1997. Medical ethics as reflective practice. In *Philosophy of medicine and bioethics: A twenty-year retrospective and critical appraisal*, ed. R. Carson and C. Burns, 181–189. Dordrecht: Kluwer Academic Publishers.

Carson, R.A., and C. Burns (eds.). 1997. *Philosophy of medicine and bioethics, a twenty year retrospective and critical appraisal*. Dordrecht: Kluwer Academics.

Charo, A. 2010. Politics, progressiveness and bioethics. In *Progress in bioethics, science, policy and politics*, ed. J. Moreno and S. Berger, 49. Cambridge, MA: MIT Press.

Churchill, L. 1997. Bioethics in social context. In *Philosophy of medicine and bioethics: A twenty-year retrospective and critical appraisal*, ed. R. Carson and C. Burns, 137–144. Dordrecht: Kluwer Academic Publishers.

Clouser, K.D. 1978. Bioethics. In *Encyclopedia of bioethics*, vol. I, ed. W.T. Reich, 116. New York: The Free Press.

Cooper, J.M. 1986. *Reason and human good in Aristotle*, 134. Indiana: Hackett Publishing Company.

Darnovsky, M. 2010. Biopolitics, myths in science and progressive values. In *Progress in bioethics, science, policy and politics*, ed. J. Moreno and S. Berger, 212. Cambridge, MA: MIT Press.

Davey, L.M. 2001. The oath of Hippocrates: An historical review. *Neurosurgery* 49: 1–13.

Dzur, A.W. 2005. *Democratic professionalism, citizen participation and the reconstruction of professional ethics, identity and practice*, 225–239. University Park: Penn State University Press.

Dzur, A.W. 2008. *Democrativ professionalism, citizen participation and the reconstruction of professional ethics, identity and practice*, 225–237. University Park: Pennsylvania State University.

Edelstein, L., O. Temkin, and C.L. Temkin (eds.). 1962. *Ancient medicine, selected papers*, 65. Baltimore: Johns Hopkins Press.

Eder, K. 2002. Europäische Säkularisierung-ein Sonderweg in die poatsäkular Gesellschaft. *Berliner Journal Für Soziologie* 3: 331–343.

Engelhardt Jr., H.T. 1997. Bioethics and the philosophy of medicine reconsidered. In *Philosophy of medicine and bioethics: A twenty-year retrospective and critical appraisal*, ed. R. Carson and C. Burns, 85–97. Dordrecht: Kluwer Academic Publishers.

Fox, R.C., and J.P. Swazey. 2008. *Observing bioethics*. New York: Oxford University Press.

Guillen, D.G. 2010. *The Hippocratic oath in the development of medicine*. Rome: Delentium Hominem.

Hafferty, F.W., and B. Castellani. 2010. The increasing complexities of professionalism. *Academic Medicine* 85: 288–301.

Hampshire, S. 1949. Fallacies in moral philosophy. *Mind* 58: 466–482.
Harris, S. 2010. *The moral landscape, how science can determine human values*. New York: Free Press.
Hawkins, S., and L. Milodanimow. 2010. *The grand design*. New York: Bantam Books.
Hinsch, K. 2010. Bioethics: The new conservative crusade. In *Progress in bioethics, science, policy and politics*, ed. J. Moreno and S. Berger, 86. Cambridge, MA: MIT Press.
Jones, J. 1981. *Bad blood: Tuskegee syphilis experiment*. New York: Free Press.
Jones, A.H. 1997. From principles to reflective practice. In *Philosophy of medicine and bioethics: A twenty-year retrospective and critical appraisal*, ed. R. Carson and C. Burns, 193–195. Dordrecht: Kluwer Academic Publishers.
Kopelman, L. 2006. Bioethics as a second order discipline: Who is not a bioethicist? *The Journal of Medicine and Philosophy* 31: 606–628.
Krugman, S. 1986. The Willowbrook hepatitis studies revisited. *Reviews of Infectious Diseases* 8(1): 157–162.
Largi, S. 1983. *Compositiones edidit Sergio Sconocchia, Biblioteca scriptorium gracorum et latinorum*. Leipzig: Teubner.
Lempert, R. 2010. Can there be a progressive bioethics? In *Progress in bioethics, science, policy and politics*, ed. J. Moreno and S. Berger, 42. Cambridge, MA: MIT Press.
MacIntyre, A. 1983. Moral philosophy: What next? In *Revisions: Changing perspectives in moral philosophy*, ed. S. Hauerwas and A. MacIntyre. Notre Dame: University of Notre Dame Press.
Moreno, J., and S. Berger (eds.). 2010. *Progress in bioethics, science, policy and politics*. Cambridge, MA: MIT Press.
Orr, R.D., N. Pang, E.D. Pellegrino, and M. Siegler. 1997. Use of the Hippocratic oath: A review of twentieth century practice and a content analysis of oaths administered in medical schools in the U.S. and Canada in 1993. *The Journal of Clinical Ethics* 8(Winter): 377–388.
Pellegrino, E.D. 1997. Bioethics as an inter-disciplinary enterprise: Where does ethics fit in the mosaic of disciplines? In *Philosophy of medicine and bioethics: A twenty-year retrospective and critical appraisal*, ed. R. Carson and C. Burns, 1–23. Dordrecht: Kluwer Academic Publishers.
Pellegrino, E.D. 2000. One hundred fifty years later: The moral status and relevance of the AMA code of ethics. In *The American medical ethics revolution: How the AMA's code of ethics has transformed physicians' relationships to patients, professionals, and society*, ed. R.B. Baker, A.L. Caplan, L.L. Emanuel, and S.R. Latham, 107–123. Baltimore: Johns Hopkins University Press.
Pellegrino, E.D. 2003. From medical ethics to a moral philosophy of the professions. In *The story of bioethics: From seminal works to contemporary explorations*, ed. J.K. Walter and E.P. Klein, 3–15. Washington, DC: Georgetown University Press.
Pellegrino, E.D. 2007. Professing medicine, virtue based ethics, and the retrieval of professionalism. In *Working virtue: Virtue ethics and contemporary moral problems*, ed. R.L. Walker and P.J. Ivanhoe, 61–85. New York: Oxford University Press.
Pellegrino, E.D., and A.A. Pellegrino. 1988. 'Humanism and ethics in roman medicine', translation and commentary on a text of Scribonius Largus. *Literature and Medicine* VII: 22–38.
Pellegrino, E.D., and David C. Thomasma. 1981. *A philosophical basis for medical practice*. New York: Oxford University Press.
Pellegrino, E.D., and D.C. Thomasma. 2000. Dubious premises—Evil conclusions: Moral reasoning at the Nuremberg trials. *The Cambridge Quarterly* 9(2): 261–274.
Prichard, H.A. 1912. Is moral philosophy based on a mistake? *Mind* 21: 21–37.
Shapiro, H. 2010. Bioethics as politics. In *Progress in bioethics, science, policy and politics*, ed. J. Moreno and S. Berger, xi. Cambridge, MA: MIT Press.
Smith, L.G. 2005. Medical professionalism and the generation gap. *American Journal of Medicine* 118: 339–342.
ten Have, H.A.M. 1997. From synthesis to morals and procedure, the development of philosophy of medicine. In *Philosophy and medicine and bioethics*, ed. R. Carson and C. Burns. Dordrecht: Kluwer Academic Publishers.

ten Have, H.A.M., and M.S. Jean (eds.). 2009. *The UNESCO universal declaration on bioethics and human rights: Background, principles and application*, 99–109. Paris: UNESCO Publishing.

Toulmin, S. 1997. The primacy of practice: Medicine and postmodernism. In *Philosophy of medicine and bioethics: A twenty-year retrospective and critical appraisal*, ed. R. Carson and C. Burns, 42–53. Dordrecht: Kluwer Academic Publishers.

van Mock, W.N.K.A., S.J. van Kuijk, H. O'Sullivan, V. Wass, J.H. Swaveling, L.W. Schuwirth, and C.P.M. van der Vleuten. 2009. The concepts of professionalism and professional behavior: Conflicts in both definition and learning outcomes. *European Journal of Internal Medicine* 20: 85–89.

Veatch, R. 2009. *Patient heal thyself.* Oxford/New york: Oxford University Press.

Wartofsky, M.W. 1997. What can the epistemologist learn from endocrinologists? Or is philosophy of medicine based on a mistake. In *Philosophy of medicine and bioethics: A twenty-year retrospective and critical appraisal*, ed. R. Carson and C. Burns, 55–68. Dordrecht: Kluwer Academic Publishers.

Wolpe, P.R. 2010. Professionalism and politics: Biomedicalization and the rise in bioethics. In *Progress in bioethics, science, policy and politics*, ed. J. Moreno and S. Berger, 110. Cambridge, MA: MIT Press.

Zoloth, L. 2010. Justice that you must pursue: A progressive American politics. In *Progress in bioethics, science, policy and politics*, ed. J. Moreno and S. Berger, 103. Cambridge, MA: MIT Press.

Chapter 15
Prolegomena to Any Future Bioethics

Albert R. Jonsen

A few years ago, I taught a bioethics course to Yale undergraduates. On their way to class, the students passed through the Noah Porter Gate. The plaque on that venerable gate identified Porter as President of Yale (1875–1886) and Professor of Moral Philosophy. I checked his *Elements of Moral Science* (1885) out of Sterling Library. The book was typical of the times. It was written for the students in President Porter's class. It provided long, learned reviews of the current theories of ethics, particularly intuitionism and utilitarianism. It gave detailed explanations of ideas that are essential to understanding moral philosophy, such as free will, duty, and obligation. In its 300 or so pages, it contained barely a case or an example of a moral issue (although the President-Professor could not refrain from commenting on gambling and drinking by students). Professor Porter's course was very unlike the ethics course my twenty-first century Yalies were taking: a course filled with analyses of the arguments pro and con on a variety of cases and issues. We barely touched the topics exposed so carefully by President Porter.

Yet I could not suppose that my students had ever been taught those topics. They had not come through an orderly progression of philosophy courses. Few of them had taken a general ethics course. So I was assuming that I could teach them the ethics of death and dying, of experimentation with humans, of genetic screening, and so forth, without needing any of that material, or at least, not any of it that I could not squeeze into this course packed with scientific information and perplexing cases. Well, who am I to complain? I have a reputation as a casuist. A casuist is reputedly an ethicist who can do without theory, or at least without much of it. So if I were true to my casuistical creed, I should be quite happy that my Yale students were innocent of ethical theory.

A.R. Jonsen, Ph.D. (✉)
Program in Medicine and Human Values, California Pacific Medical Center,
2395 Sacramento Street, San Francisco, CA 94115, USA
e-mail: JonsenA@sutterhealth.org; Ethics@sutterhealth.org

J.R. Garrett et al. (eds.), *The Development of Bioethics in the United States*,
Philosophy and Medicine 115, DOI 10.1007/978-94-007-4011-2_15,

In fact, I do believe that one can teach a lot of bioethics without much attention to basic ethical theory. However, I do not believe that much progress can be made in bioethics without it. The historical casuistry that Stephan Toulmin and I described in the *Abuse of Casuistry* did not float untethered above the ground of basic moral philosophy (Jonsen and Toulmin 1987). Toulmin and I had borrowed the title of our *Abuse of Casuistry* from Kenneth Kirk's *Conscience and Its Problems. An Introduction to Casuistry* (1927, p. 125). (Kirk had written, "the abuse of casuistry is properly directed, not against all casuistry, but only against its abuse.") This estimable volume reviews a multitude of cases, but it opens with chapters on the nature of conscience, moral judgment, moral acts, and the relativity of the moral law. These were topics that the venerable casuist felt he must explain and clarify before plunging into cases.

These topics were some of the "commonplaces" of moral philosophy for centuries (commonplace in the rhetorical sense of a "place" or set of definitions and arguments common to a certain proposal or problem). The first three chapters of Aristotle's (1926) *Ethics* set out the commonplace arguments about the "good for man," the norm of morality and the characteristics of voluntary action. Sidgwick's *The Methods of Ethics* (1877) begins with an examination of the motives of human action. William Frankena's *Ethics* (1963) is entirely about these commonplaces: the nature of morality, theories of moral judgment, responsibility, the good life. The introductory chapters of Beauchamp and Childress's *Principles of Biomedical Ethics* cover the meaning of morality and moral norms. So, from book to book, moral philosophers select the commonplaces that they believe will start a student on the philosophical study of morality. Their choices differ somewhat, but, in general, are very similar.

These commonplaces are the familiar arguments that are necessary to support the claims that there is such a thing as distinctive moral judgment, that moral judgments are, in some important sense, objective, and that moral judgment is properly applied only to voluntary acts. Calling these arguments "commonplaces" invokes the classic rhetorical concept that certain structures of definition and of argument surround certain sorts of claims, and that those structures will be consistent regardless of the particular circumstances about which argument is being made (McKeon 1987, p. 256). For example, an argument may be made about whether a particular dying patient has the capacity to request termination of life support. The particular argument will draw on the patient's current state of consciousness, the medical facts about her diagnosis and prognosis, and so forth. However, the structure of any argument about mental capacity must contain certain definitions and move through certain logical steps. In this sense, there is a "commonplace" about mental capacity. Every bioethics consultant should know that commonplace.

I am about to claim that bioethicists largely ignore the commonplaces of moral philosophy, yet I have just cited Beauchamp and Childress. Their book is, unquestionably, the premier text for bioethics. So my claim may seem quite untenable if the major text (and, indeed, many other texts) does attend to the commonplaces. I have to parse my claim: I have not made a meticulous examination of the indices of most bioethical books, so my claim may not be supported by documentary evidence (Perhaps a graduate student might undertake this task). I am not claiming that bioethicists do not know about or understand the commonplaces. Some bio-ethicists are quite good and well trained philosophers. Others are not philosophically trained, but are quite competent and well read in ethics. Also, I have but small evidence of whether these commonplaces are taught in graduate courses. I only know that I directed a graduate program for 15 years and did not insist that my students master them.

My claim is based on my participation in many bioethical arguments over the years. I hardly remember anyone diving deeply into the commonplaces to make a point. My claim is also based on a bias that bioethics has followed during its whole history: to deal with particulars rather than universals. From its beginnings, it has aspired to be practical philosophy. When its practitioners do turn from cases to theory, they largely focus on the implications of a single commonplace: the relation of principles to choices (which among the classical commonplaces was often designated "conscience"). Finally, my claim arises from recent reading about certain current bioethical issues. I have noticed that behind the practical arguments tossed back and forth in these issues lurk some of the neglected commonplaces. It is my impression that bioethicists glance away when they glimpse these skulking ghosts of old arguments. Yet it may be that the issue under discussion cannot be properly understood without summoning them out of the shadows.

Three issues particularly impress me: the stem-cell debate, the growing interest in neuroethics, and the appreciation of cultural bioethics. In these three debates, I notice the obscure presence of three of the classical commonplaces: finality, voluntariness and moral relativity. These three commonplaces have been around in ethical literature for a long time: the first two were born in Aristotle's pages; the third is a child of enlightenment thought. All three appear regularly in nineteenth and early twentieth century ethics texts. Yet they have become rather dusty and unused.

This is not to say that these terms are never mentioned. "Voluntary" shows up in discussions of mental capacity and of volunteering for research or transplant donation. "Relativism" may appear in a treatment of culturally diverse views about the propriety of medical interventions. However, they are absent as commonplaces. A commonplace, in its classical rhetorical use, is a full exposition of the arguments in favor of and against a claim. It would appear not simply as an attempt, for example, to prove that free will is necessary to an ascription of responsibility, but also to expose the contrary opinions and the objections to that claim as well. The author may, of course, choose to defend certain of those arguments as more plausible, but will also show the range of arguments on both sides and from different perspectives. (C.D. Broad's fine little book, *Five Types of Ethical Theory* (1930) is an excellent example of exposition of certain commonplaces of moral philosophy.)

So I propose a *Prolegomena for Any Future Bioethics*. This is a grand title for a modest proposal. Unlike Immanuel Kant and T.H. Greene, who both christened books with a similar title, I am only making the proposal for what others, the future bioethicists, might do. I propose that bioethics revisit the commonplaces of moral philosophy, review them critically, and restore them to the extent that they can support or refute many of the claims that are appearing in bioethical discourse. It may happen that the commonplaces, in their classical forms, may be thoroughly antiquated. Still, they may suggest ways in which arguments may be reborn to suit the current issues. Also, the commonplaces have themselves been subjected to serious philosophical criticism, almost to the point of demolition. Classical free will arguments, for example, are no longer what they were before David Hume battered them. Decades of deconstruction have made it almost embarrassing to propose a return to basic moral notions. Still, I believe there are lessons about forms of argument, pitfalls, and promising directions that lie beneath the old commonplaces. I offer several suggestions about how the commonplaces might contribute to current bioethical debates. The commonplace about finality and nature may have relevance for the stem cell debate. The commonplace about responsibility may help as neuroscience raises questions about human choice. The commonplace on moral relativism may be useful in understanding cultural bioethics.

15.1 Stem Cells and Finality

The stem-cell debate has raged for several decades. A huge literature exists; a multitude of arguments defend or condemn stem-cell research. Among them, the most central may be the debate over the moral status of the human embryo. That debate frequently invokes the notion of potential life or potential personhood. For example, the President's Council on Bioethics Report, *Monitoring Stem Cell Research*, contains many sentences like these, "They [the proponents of continuity between embryo and person] contend that a human embryo already has the biological potential needed to enable the exercise, at a later stage of development, of certain functions… They [those who deny continuity] suggest that genetic identity and organismic continuity are not sufficient; what matter is present form and function, more than mere potential" (2004, pp. 78–79). I cannot find the words "finality" or "end" anywhere in this document. This is hardly surprising. Those terms have been banished from philosophical discourse for a long time. Despite the venerability of Aristotelian *telos*, the idea that beings have some internal purpose, or that physical process is explicable by certain goals that must be fulfilled, has long been out of style. Galileo rendered risible the idea that heavy objects fell to the ground because their *telos* is to seek the center of the earth. Darwin thoroughly replaced organic teleology with random selection. Finality plays no place in the physical and biological sciences. Purpose resides only in the intentional projects of individuals and, even there, modern psychology, particularly neuropsychology, have made it into something other than it seems to those who form them and act upon them.

Yet, finality seems to retain some sort of moral relevance. Language like "potential for development" seems inevitable in the moral debates about abortion, cloning, and stem cell research. This language raises the specter of finality, of purpose inscribed in structure. From fertilization, an embryo has a genetic structure that will push on to fetus, to infant, to child, to man or woman. Is that not potential in some significant sense? Is that potential, whatever it be, of moral relevance? Does such an ascription of potential imply any moral imperative not to disturb or distort the process that is somehow "destined" for this being?

Similarly, the debates over "enhancement" skirt the supposition of nature or normalcy, closely related to finality. Can therapy of disease be clearly seen as correction of a deviation from normal characteristics, while enhancement of strength, intelligence, or beauty is criticized as a deliberate departure from the normal and natural? These debates have been inconclusive (Parens 1998). It would it be advisable to return to the commonplaces about finality and nature, not to resuscitate dead and dying arguments, but to seek in their remains any clue to moral relevance. A distinguished Catholic theologian, reflecting on the supremacy of these ideas in Catholic thought over many centuries, said to me that, although he was now skeptical of their classical formulation, he felt they had a message. He said, "an embryo has potential to become a baby; stem cells have the potential to become life-saving cardiomyocytes. Perhaps that is the basic moral justification for stem cell research."

15.2 Neuroscience and Free Will

The neurosciences have progressed with great rapidity. The technical ability to image with great specificity the functioning of a person's brain as they live through various experiences and emotions and perform tasks has revealed not merely the loci of cerebral reaction, but also the patterns that preexist and are activated and shaped by experience. Now that brain imaging reveals real time pictures of the brain responding to stimuli and forming electrochemical patterns that correspond to experience and emotion and intellection, much of the mystery of our lives seems reducible to that capsule of cells, neurons and chemicals called the brain. Since human acts of choice show up vividly in those patterns, it is reasonable to ask whether the moral characteristics that we associate with choice, such as autonomy, free will, and responsibility, have the ethical relevance that moral philosophy has ascribed to them. One author, describing the progress of neuroscience, writes about responsibility and self-determination,

> ...if self-determination lies in a specific bit of tissue, it follows that those who appear not to have it may simple be unlucky: victims of a sluggish brain module. So is it reasonable to blame [them] for their ways? Should we be unsympathetic to addicts who fail to conquer their habit or punish recidivist criminals? (Carter 1989, p. 1).

Of course, neuroscientists are hardly the first to question self-determination, free will, and responsibility. Long before the neuroscientists, philosophers such as the Epicureans and theologians such as John Calvin were convinced that free will was but an "illusion." But these ancients could not see the vivid images of neural activity, nor could they cite elaborate scientific data to support their speculative claims. Yet they quite sensibly proclaimed that if thought and choice are equivalent to highly complex functions of material bodies or, in more modern terms, "hard wiring in a chemical analog computer," the claims we make about responsibility, freedom, moral obligation and religious belief may not be what they seem (Wolfe 2002, p. A26). Do the theories and data of the neurosciences inevitably lead to such firm declarations of determinism? Does modern neuroscience negate common understanding of free will and erase our vaunted sense of human uniqueness? Attribution of moral responsibility for one's actions is the most urgent question for practicing bioethicists. Is this person—someone consenting to a medical procedure, or deciding to forego life support, or declining genetic information, or, in another area of ethics, accused of a crime—responsible?

Free will is not "free-wheeling," that is, choice rolling along without brake or impulse. Almost all philosophers have acknowledged that choice is caused by prior events and influences. Nevertheless, free choice, in some special way, seems to come from ourselves; we seem to be in control. Aristotle wrote, "a voluntary act would seem to be one of which the originating cause lies in the agent himself, who knows the particular circumstances of his action" (*Nicomachean Ethics*, III, iii, 111a20). What might this originating cause be? Is it some spirit floating above brain and body? Or is it somehow embedded within brain, body, and physical world, but in a way quite different than the cause and effect relationships we find in nature? Is it nothing but an illusion? Since the debut of philosophy, these questions have been asked. We now know what no ancient thinker knew, namely, that learning right from wrong—acquiring a conscience—is dependent on neural tissue in the ventromedial pre-frontal cortex and on circuitry in the hypothalamus, amygdala and cingulate cortices. Damage in those regions extinguishes or diminishes the behavior we associate with free, voluntary choice; drugs and devices, such as implanted microchips, seem to affect these behaviors. What might philosophers familiar with these facts say about free choice?

Bioethicists constantly extol the principle of autonomy. However, if they turn to issues in neuroscience, they must plunge into the *meaning* of autonomy, revisiting the time-honored philosophical debates about free will in light of the science. It is their task to re-describe these problems in ways that, as medieval philosophers used to say, "save the appearances," not denying the experience but explaining it so as to complement the scientific data. The paradoxes of free will have never abolished the insistent human need to assign responsibility for actions: praise and blame, and also reward and punishment, are universal human behaviors. These behaviors are, of course, the subject matter of ethics. So, neuroethics must accommodate these activities. What can be made of the daily problems of moral duty and dereliction in the light of the brain sciences?

The assertion that "self-determination lies in a specific bit of tissue" needs the most careful examination. Is the behavior we describe as self-determination in the tissue? Does the tissue cause self-determining acts? Is the bit of tissue a necessary and sufficient cause of behavior? Does the tissue perform certain cellular and neurochemical functions when a person performs a self-determining act? What, indeed, is a self-determining act? An act without any influences other than the mysterious will? These are questions for the complementary dialogue between philosophers and neuroscientists.

The neuroscientists will contribute new data and interpretations of neurological physiology and activity to the dialogue. They can propose, for example, that persons have dispositions to act in certain ways because neural tracks and pathways are built by every experience and every choice. The brain approaches new situations with mutiplex circuitry in place. However, the new experience also reforms that circuitry and does so through processing new information, interpretation, and imagination. The philosophers must take these explanations of the data and explain what choice, freedom, and responsibility might mean in light of them. It may be valuable to critically reexamine the commonplaces of moral philosophy.

The questions and answers of this dialogue will inform the policies and practices of major social institutions, such as the criminal justice system, the mental health system, and the educational system. Although these questions and answers will never be definitive, their quality, as properly asked, well informed, prudently skeptical, and open to revision, is essential. Although many philosophers have claimed that the problem of free will and determinism is a "pseudo-problem" because it has no solution, some moral philosophers with a practical bent have not been disconcerted. Aristotle proposes no clear thesis about free will, but uses the question of uncoerced choice to open many crucial questions about moral education. William James, a philosopher who was deeply immersed in empirical psychology, did not believe that the "free will problem" could be solved in any theoretical or metaphysical way. Rather, he insisted that an affirmation of free will is necessary to make sense of vast tracts of our moral life and to make moral education and social ethics into useful enterprises. Free will is expressed in the conviction that our decisions here and now "in soul trying moments … [give] palpitating reality to our moral life and [make] it tingle" (James 1977, p. 610). A practicing bioethicist may not know for certain whether self-determination lies in a specific bit of tissue, but must certainly know that praising and blaming, and also punishment and reward, are practices inherent to the moral life and must be done intelligently and responsibly.

15.3 Cultural Bioethics and Ethical Relativism

A third question is being asked incessantly in contemporary bioethics: are there principles of bioethics that should be honored in all nations and cultures? Early bioethics lingered somewhat chauvinistically in the English speaking world. It quickly became obvious that its subject matter could be found wherever modern medicine and science had penetrated (or, in some developing nations, where it *should* have penetrated but had not). American bioethicists early discovered that the vaunted American value of autonomy did not have universal credence. The ethical standards for research with human subjects, polished to perfection by American commissions, seemed ill-fitting in Asia and Africa. A charming book relating Hmong beliefs about illness and medicine demonstrated differences of cultural values that American bioethicists had hitherto ignored (Fadiman 1997). Those differences were appearing in American clinics and emergency rooms as patients from many other cultures sought care. One particular question, female genital mutilation, forced a serious consideration about how American caregivers should think about a practice repugnant to themselves but often respected by those who had inherited them.

Moral philosophers commonly prefaced their texts with a discussion of moral relativism. The question had been set to moral philosophy by Protagoras whose maxim, "man is measure of all things," implied that the right and the good were as individuals and societies perceived them to be. Montesquieu's observations about diversity of mores posed the question for Enlightenment moral philosophers and the data of anthropology illustrated it vividly. The Eskimos who set their elders adrift posed a perfect example. And, of course, the moral skepticism of untutored college students, seeking to liberate themselves from moral strictures, was always a challenge to the ethics professor. Moral philosophers since Socrates have devised many arguments to demonstrate or refute ethical relativism. It is my impression that many bioethicists may no longer be familiar with these commonplaces. I feel that the entire argument about relativity was banished, in the early days of bioethics, by Alasdair MacIntyre's powerful repudiation of the comparability of moral systems and concepts (1966). Recently, Ruth Macklin has resuscitated the arguments about relativity and woven them into a discussion of the many intercultural bioethical issues that she encountered in her work as a world-travelling bioethicist (1999).

The treatise on ethical relativism should shed light not only on the diversity of values in cultures other than our own, but also within our own. Policy debates, such as those over abortion and stem cell research, reveal contrasting ethical commitments. Policymakers confronted by these diverse commitments are as stymied as the college student confronted by purported Eskimo gerontology. Their reports list the opinions in language as neutral as possible, almost as if they balance out. They then politely excuse themselves from making any judgment (usually because the opinions have some religious tinge and thus are not allowed into public discourse). Protagoras has won in the public space of American bioethics. A more penetrating understanding of the philosophical problem of moral objectivity might make it possible to evaluate these opinions more precisely.

15.4 Conclusion

My *Prolegomena to Any Future Bioethics* is nothing but a plea for serious thought about pieces of moral philosophy that seem to have been forgotten as bioethics has grown into a discipline. It proposes that parts of moral theory that have faded from view be restored to study in contemporary terms and in light of critical methods. Many bioethicists today have studied ethics in a broad context of philosophy; they may know these commonplaces rather well. Some who have come through an academic course in philosophy may not have been exposed to these commonplaces. Other bioethicists have become competent in the theory and practice of bioethics without plunging into the philosophical depths. I would hope that all who work in bioethics at the academic level should have some exposure to the commonplaces as they might be reworked to conform to contemporary critical style. I cannot say what that Prolegomena would look like, but, if it were done well, it would provide a firmer foundation for bioethical discourse.

References

Aristotle. 1926. *The Nicomachean ethics*. Trans. H. Rackham. Cambridge, MA: Harvard University Press.

Broad, C.D. 1930. *Five types of ethical theory*. London: Routledge and Kegan Paul.

Carter, R. 1989. *Mapping the mind*. Berkeley: University of California Press.

Fadiman, A. 1997. *The spirit catches you and you fall down*. New York: Farrar, Straus and Giroux.

Frankena, W. 1963. *Ethics*. Englewood Cliffs: Prentice Hall.

James, W. 1977. The dilemma of determinism. In *Writings of William James*, ed. J.J. McDermott, 587–610. Chicago: University of Chicago Press.

Jonsen, A.R., and S.E. Toulmin. 1987. *The abuse of casuistry: A history of moral reasoning*. Berkeley: University of California Press.

Kirk, K. 1927. *Conscience and its problems: An introduction to casuistry*. London: Longmans, Green.

MacIntyre, A. 1966. *A short history of ethics*. New York: Macmillan.

Macklin, R. 1999. *Against relativism: Cultural diversity and the search for ethical universals in medicine*. New York: Oxford University Press.

McKeon, R. 1987. *Rhetoric: Essays in invention and discovery*, ed, M. Backman. Woodbridge: Ox Bow Press.

Parens, E. 1998. *Enhancing human traits: Ethical and social implications*. Washington, DC: Georgetown University Press.

Porter, N. 1885. *Elements of moral science, theoretical and practical*. New York: Charles Scribner's Sons.

President's Council on Bioethics. 2004. *Monitoring stem cell research*. Washington, DC: President's Council on Bioethics.

Sidgwick, H. 1877. *The methods of ethics*. London: Macmillan.

Wolfe, T. 2002. Commencement speech. *New York Times*, June 2, A26.

Author Biographies

George J. Annas, J.D., M.P.H. is the Edward Utley Professor and Chair of the Department of Health Law, Bioethics and Human Rights at the Boston University School of Public Health.

Howard Brody, M.D., Ph.D. is the John P. McGovern Centennial Chair in Family Medicine and Director of The Institute for the Medical Humanities at the University of Texas Medical Branch at Galveston.

Eric J. Cassell, M.D., M.A.C.P. is Emeritus Professor of Public Health at Weill Medical College of Cornell University and Adjunct Professor of Medicine at McGill University. His writings center on moral problems in medicine, the care of the dying and the nature of suffering.

H. Tristram Engelhardt, Jr., M.D., Ph.D. is Professor of Philosophy at Rice University and Professor Emeritus at Baylor College of Medicine. He is the Senior Editor of *The Journal of Medicine and Philosophy,* the journal *Christian Bioethics,* and the *Philosophy and Medicine* book series.

Edmund L. Erde, Ph.D. retired from his position as a Professor at the School of Osteopathic Medicine of the University of Medicine and Dentistry of New Jersey, in July of 2010. There, for 28 years, he taught courses on professionalism, and ethics and the humanities in all 4 years of the undergraduate medical curriculum. He also conducted seminars in residency and fellowship programs at that institution. Previously, he was a member of The Institute for the Medical Humanities and of the Department of Preventive Medicine and Community Health at the University of Texas Medical Branch Galveston from 1975 to 1981.

Jeremy R. Garrett, Ph.D. is Research Associate with the Children's Mercy Bioethics Center at Children's Mercy Hospital in Kansas City, Missouri. He also is Assistant Professor of Pediatrics and Adjunct Assistant Professor of Philosophy at the University of Missouri-Kansas City. He has published both invited and peer-reviewed articles and has recently edited *The Ethics of Animal Research: Exploring*

J.R. Garrett et al. (eds.), *The Development of Bioethics in the United States*, Philosophy and Medicine 115, DOI 10.1007/978-94-007-4011-2, *the Controversy*, published in 2012 by MIT Press. His areas of specialization include bioethics, ethical theory, and social, political and legal philosophy.

John Collins Harvey, Ph.D. is a Professor at the Center for Clinical Bioethics, Georgetown University.

Albert R. Jonsen, Ph.D. is Senior Ethics Scholar in Residence and Co-Director of the Program in Medicine and Human Values at the California Pacific Medical Center in San Francisco. He also presently teaches at the Fromm Institute for Lifelong Learning of the University of San Francisco. He is Emeritus Professor of Ethics in Medicine at the School of Medicine, University of Washington, where he was Chairman of the Department of Medical History and Ethics from 1987 to 1999. From 1972 to 1987, he was Chief of the Division of Medical Ethics, School of Medicine, University of California, San Francisco. Prior to that, he was President of the University of San Francisco, where he taught in the Departments of Philosophy and Theology. Professor Jonsen has written chapters in over 70 books on medicine and health care, and is a member of the Institute of Medicine, National Academy of Science.

Fabrice Jotterand, Ph.D., M.A. is Senior Researcher at the Institute for Biomedical Ethics, University of Basel, Switzerland and Lecturer in the Department of Philosophy and Humanities at the University of Texas Arlington, Arlington, Texas. He has written multiple articles and reviews, and has edited two books: *The Philosophy of Medicine Reborn: A Pellegrino Reader* and *Emerging Conceptual, Ethical and Policy Issues in Bionanotechnology*. His areas of expertise include neuroethics, nanoethics, bioethics, and the philosophy of medicine.

Loretta M. Kopelman, Ph.D. (philosophy) is a Professor Emeritus at the Brody School of Medicine, where she founded and chaired its Department of Bioethics and Interdisciplinary Studies, and faculty affiliate at Georgetown University. She was founding president of the American Society for Bioethics and Humanities, president of the Society for Health and Human Values, and the winner of the 2007 William G. Bartholome Award for Ethical Excellence from The American Academy of Pediatrics. She was a member of the Institute of Medicine's Committee on Research with Children, and of the American Philosophical Association's Committee on Philosophy and Medicine. She serves on many editorial boards and has published widely, including on topics in bioethics, the rights of disabled persons, research ethics, philosophy and medicine, the fair allocation of health care resources, and especially on children's rights and welfare. She has just been appointed to Ethics Advisory Panels at NIH and the FDA.

Laurence B. McCullough, Ph.D. is the Dalton Tomlin Chair in Medical Ethics and Health Policy, Professor of Medicine and Medical Ethics, and Associate Director for Education at the Center for Medical Ethics and Health Policy of Baylor College of Medicine.

Edmund D. Pellegrino, Ph.D., M.A.C.P. is a Professor at the Center for Clinical Bioethics, Georgetown University.

D. Christopher Ralston, Ph.D., M.A., is a former Assistant Managing Editor of the *Journal of Medicine and Philosophy*. He co-edited a Philosophy and Medicine volume entitled *Philosophical Reflections on Disability* (2009), and recently defended successfully a doctoral dissertation, entitled "The Concept of Disability: A Philosophical Analysis," at Rice University. His areas of specialization include ethics, bioethics, the philosophy of medicine, and disability studies.

Warren T. Reich, S.T.D. is Distinguished Research Professor of Religion and Ethics and Professor Emeritus of Bioethics in the Theology Department at Georgetown University. He was Founding Research Scholar in the Kennedy Institute of Ethics (beginning in 1971), Founder and Director of the bioethics and medical humanities program in the Medical Center of Georgetown University, and currently Founder and Director of Georgetown's Project for the History of Care and the Georgetown Chivalry Initiative.

Carson Strong, Ph.D., is Professor of Medical Ethics in the Department of Medicine at the University of Tennessee Health Science Center in Memphis. He is coauthor of *A Casebook of Medical Ethics* (1989) and author of *Ethics in Reproductive and Perinatal Medicine: A New Framework* (1997). Currently he is doing research for an NIH grant on human subjects protections in prenatal gene transfer. His recent publications address the nature and structure of moral justification.

Robert M. Veatch, Ph.D., is Professor of Medical Ethics and the former Director of the Kennedy Institute of Ethics at Georgetown University, where he also holds appointments as Professor of Philosophy and Adjunct Professor in the Department of Community and Family Medicine at Georgetown Medical Center. He is the Senior Editor of the *Kennedy Institute of Ethics Journal* and a former member of the Editorial Board of the *Journal of the American Medical Association*. He served as an ethics consultant in the preparation of the legal case of Karen Ann Quinlan, the woman whose parents won the right to forego life-support (1975–1976) and testified in the case of Baby K, the anencephalic infant whose mother insisted on the right of access to ventilatory support.

Richard M. Zaner, Ph.D., is A.G. Stahlman Professor Emeritus of Medical Ethics and Philosophy of Medicine in the Medical School of Vanderbilt University. His writings focus on human life, pursued through a philosophy of medicine, biomedical research, and ethics in clinical life. He has held appointments in Philosophy, Religion, and other areas. He is the author, most recently, of *Conversations on the Edge* (Georgetown, 2004), and *Voices and Visions: Clinical Listening, Narrative Writing*, published in Chinese (National Cheng-Chi University Press, Taipei, Taiwan, 2008). A *Festschrift* on his work, entitled *Clinical Ethics and the Necessity of Stories*, was published by Springer in 2011.

Index

J.R. Garrett et al. (eds.), *The Development of Bioethics in the United States*, Philosophy and Medicine 115, DOI 10.1007/978-94-007-4011-2,

CPSIA information can be obtained at www.ICGtesting.com
Printed in the USA
LVOW102202080213

319383LV00007B/154/P

9 789400 740105